FLORA ZAMBESIACA

Flora terrarum Zambesii aquis conjunctarum

VOLUME THREE: PART SIX

FLORA ZAMBESIACA

MOZAMBIQUE

MALAWI, ZAMBIA, ZIMBABWE

BOTSWANA

VOLUME THREE: PART SIX

Edited by
G.V. POPE

on behalf of the Editorial Board:

S.J. OWENS
Royal Botanic Gardens, Kew

M.A. DINIZ
Centro de Botânica, Instituto de Investigação
Científica Tropical, Lisboa

G.V. POPE
Royal Botanic Gardens, Kew

Published by the Royal Botanic Gardens, Kew,
for the Flora Zambesiaca Managing Committee
2000

© *Flora Zambesiaca Managing Committee, 2000*

Typesetting and page make-up by Media Resources, Information Services Department, Royal Botanic Gardens, Kew

Printed in Great Britain by
Whitstable Litho Printers Ltd., Whitstable, Kent

ISBN 1 84246 007 2

CONTENTS

ARRANGEMENT OF THE TRIBES IN VOLUME 3   *page* vi

LIST OF NEW NAMES PUBLISHED IN THIS PART   vi

GENERAL SYSTEMATIC TEXT   3.6.**1**

INDEX TO VOLUME 3, PART 6   3.6.**171**

ARRANGEMENT OF TRIBES IN VOLUME 3

vol. 3 part 1   *Mimosoideae* · · · · · · · · · · · · · · · published July 1970

vol. 3 part 2   **Caesalpinioideae**
           Tribe 1. *Caesalpinieae*
           Tribe 2. *Cassieae*
           Tribe 3. *Cercideae*
           Tribe 4. *Detarieae*

vol. 3 part 3   **Papilionoideae**
           Tribe 1. *Swartzieae*
           Tribe 2. *Sophoreae*
           Tribe 3. *Dalbergieae*
           Tribe 4. *Abreae*
           Tribe 5. *Millettieae*
           Tribe 6. *Robinieae*

vol. 3 part 4   **Papilionoideae**
           Tribe 7. *Indigofereae*

vol. 3 part 5   **Papilionoideae**
           Tribe 8. *Phaseoleae*

vol. 3 part 6   **Papilionoideae** · · · · · · · · · · · · · published May 2000
           Tribe 9. *Desmodieae*
           Tribe 10. *Psoraleeae*
           Tribe 11. *Aeschynomeneae*

vol. 3 part 7   **Papilionoideae**
           Tribe 12. *Loteae* incl. *Coronilleae*
           Tribe 13. *Galegeae*
           Tribe 14. *Vicieae*
           Tribe 15. *Cicereae*
           Tribe 16. *Trifolieae*
           Tribe 17. *Podalyrieae*
           Tribe 18. *Crotalarieae*
           Tribe 19. *Genisteae*

LIST OF NEW NAMES PUBLISHED IN THIS PART

	Page
Aeschynomene pawekiae Verdc. sp. nov.	3.6.**88**
Droogmansia pteropus var. **giorgii** (De Wild.) Verdc. comb. et stat. nov.	3.6.**33**
Otholobium foliosum subsp. **gazense** (Baker f.) Verdc. comb. et stat. nov.	3.6.**48**
Ormocarpum zambesianum Verdc. sp. nov.	3.6.**57**

Acknowledgements

The Flora Zambesiaca Managing Committee thanks M.A. Diniz and E. Martins of the Centro de Botânica, Lisbon, for their valuable help in reading and commenting on the text.

Tribe 9. DESMODIEAE

by B. Verdcourt

Desmodieae (Benth.) Hutch., Gen. Fl. Pl. **1**: 477 (1964).
Subtribe Desmodiinae Benth. in Bentham & Hooker f., Gen. Pl. **1**, 2: 449 (1865).

Herbs, less often shrubs. Leaves pinnately 3(9)-foliolate or 1-foliolate; stipules mostly striate; stipels present. Flowers in terminal or axillary racemes or more generally pseudoracemes (the flowers paired or clustered at the nodes). Calyx usually 5-toothed, the 2 upper lobes often joined higher up, rarely 2-lipped. Petals often disorientated or caducous after pollen release. Vexillary filament free or joined to the others, sometimes forming a closed tube; anthers uniform. Fruits transversely jointed into articles (loments), rarely of only 1 article, or 2-valved (valves not twisting). Seeds with a well-developed radicular lobe, rarely arillate, mostly with a rim membrane.

Nearly 30 genera, mainly tropical, most diverse in the Sino-Indian region. Molecular evidence suggests that the tribe is nested within Phaseoleae, but the fruits are characteristic and the genera occurring in Africa form a coherent group.

1. Pod 2-valved, linear-oblong, not breaking transversely into articles · · · · · 67. **Pseudarthria**
 – Pod breaking transversely into articles · 2
2. Pod folded like an accordion, hardly exceeding the calyx in length; leaflets (1)3–9, up to 8 cm long or more · 69. **Uraria**
 – Pod not as above; leaflets 1 or 3 · 3
3. Pod on a long plumose stipe; petiole markedly winged; leaflet 1 · · · · · · 68. **Droogmansia**
 – Pod not on a long stipe; petiole not winged · 4
4. Calyx not glumaceous; leaflets 1 or 3; pod flattened · · · · · · · · · · · · · 66. **Desmodium**
 – Calyx ± glumaceous; leaflet 1, very rarely 3; pod subcylindrical or somewhat flattened · 70. **Alysicarpus**

66. DESMODIUM Desv.

Desmodium Desv. in J. Bot. Agric. **1**: 122, t. 5, fig. 15 (1813) *nom. conserv.* — Verdcourt in Kirkia **9**: 506 (1974).

Shrubs or erect or prostrate subshrubs or herbs. Leaves 1-foliolate or pinnately 3–5-foliolate; leaflets often large; stipules free, somewhat joined, or joined and leaf-opposed, striate; stipels present. Inflorescences axillary or terminal, falsely racemose or paniculate rarely subumbellate, the flowers solitary or fasciculate on the rhachis; primary bracts striate, persistent or sometimes membranous and early deciduous; secondary bracts often present but bracteoles usually absent. Calyx 5-lobed, the lobes mostly ± 2-lipped or subequal; the upper lip entire or bidentate composed of 2 lobes joined together, the lower of 3 larger lobes, the central one the longest. Corolla mostly small, yellow or red; standard oblong to round or transverse, narrowed into a short claw, sometimes with small callous appendages; wings ± attached to the keel; keel petals clawed, partly joined. Stamens mostly diadelphous, the vexillary filament free or partly joined; anthers uniform. Ovary sessile, (1)2–many-ovuled; style inflexed or incurved, glabrous; stigma terminal, capitate or minute. Fruits usually stipitate or sometimes sessile, well-exserted from the calyx, compressed, (1)2–many-jointed into articles (loments), indehiscent or at length splitting up, the articles 1-seeded, membranous or leathery, almost flat or more rarely inflated; or in a few cases fruits dehiscing and not or scarcely breaking into articles. Seeds oblong, reniform or subquadrangular, compressed; aril not developed.

3.6.1

A large genus variously estimated to comprise 200–350 species occurring in the tropics and warmer temperate regions of both hemispheres, but not in Europe or New Zealand. Several species have been cultivated in the Flora Zambesiaca area, some having become very locally naturalised. The taxonomy of these is difficult but an attempt has been made to key out all those that I have seen. Those which have become naturalised are treated in the numbered sequence but the others are briefly treated below in alphabetical order.

Desmodium discolor Vogel, a species from South America, has been grown in Zambia at Choma (21.ii.1963, *van Rensburg* 1378), and Mt. Makulu Research Station (31.iii.1966, *van Rensburg* 3100), and in Zimbabwe at Harare (9.iv.1942, *Hopkins* in *GHS* 8945), and Mlezu Govt. School in Gweru (4.iv.1965, *Biegel* 640). It is very similar to *D. distortum* but the whole plant, particularly the undersides of the leaflets, is densely velvety; fruits with 4–7 articles, 4 × 3 mm, mostly glabrous.

Desmodium distortum (Aubl.) Macbr., a species from Mexico and northern South America, has been grown at Mt. Makulu Research Station in Zambia (13.v.1962, *van Rensburg* 3108). It is a shrubby herb 2–3 m tall with extensively branched inflorescences of pink to blue flowers and fruits of 1–6 articles; the articles 1.5–2.5 × 1.5–2 mm but occasionally, when reduced to 1, much larger, 5 × 4 mm with a false stipe up to 3.5 mm long partly made up of undeveloped articles.

Desmodium incanum DC. (*Desmodium canum* (J.F. Gmel.) Schinz & Thell., nom. illegit.), an American species, has been grown at the Matopos Research Station in Zimbabwe (10.iv.1962, *S.S.D.* 39), and occurs as an escape in other parts of Africa. This is a spreading or erect herb or shrub 0.3–3 m tall; leaflets 2–9 × 1.5–4.5 cm, ± round, elliptic or obovate, sparsely hairy or glabrescent above, grey and hairy beneath particularly on the nerves. Flowers in terminal or axillary racemes, blue, red or purple. Fruits of up to 8 articles, upper margin ± straight, lower margin indented for about two-thirds the width of the articles; articles 3.5–5 × 2–3.5 mm, oblong-elliptic, uncinulate-pubescent.

The complicated nomenclatural tangle surrounding this species has been elucidated by Nicolson in Taxon **27**: 365–370 (1978). The name *Desmodium incanum* DC. is treated as a new name dating from 1825 and the parenthetical author as in *D. incanum* (Sw.) DC. must not be used. Both the basionyms *Hedysarum canum* J.F.Gmel. and *Hedysarum incanum* Sw. are illegitimate.

Desmodium intortum var. *pilosiusculum* (DC.) Fosberg (*D. pilosiusculum* DC., 1825; *D. sandvicense* E. Mey., 1851), long known from Hawaii and widely cultivated elsewhere, is clearly of tropical American origin; it has been cultivated in Zimbabwe at the Matopos Research Station (17.xi.1967, *Mangena* in *MRSH* 4330 & 4.iii.1968, *Mangena* in *MRSH* 4372). The Zimbabwe plant is more slender, less hairy, and with smaller leaves than the type. This is a trailing plant, lightly appressed pubescent, with more constricted branches; leaflets mostly 3 × 1.5 cm, up to 4.5 × 2 cm, elliptic, very shortly obscurely hairy above, more densely so beneath, judging by some dried material sometimes with a pale median mark above; inflorescences less branched and pedicels mostly shorter. Fruits with 9–10 articles, the basal suture usually much more indented than the upper but sometimes subequally so; articles 3.5–5.5 × 2.5–3.5 mm, oblong-elliptic in outline, densely covered with hooked hairs.

Corby 1031 (Zimbabwe, Marondera, 19.ii.1962) has been annotated *Desmodium* cf. *leiocarpum* G. Don by Schubert but is only in flower and its identity is not certain. This species is vegetatively similar to *D. discolor* but the articles are 5–6 × 3.5–4 mm, elliptic, and much more distinctly reticulately veined.

Desmodium nicaraguense Oersted, a central American species, has been grown in the Harare District of Zimbabwe (2.x.1940, *Arnold* in *GHS* 7761). Shrub 2–3 m tall; stems hairy; leaflets 3–6 × 1.2–4.5 cm, elliptic, obtuse to rounded at the apex, appressed silvery-grey silky. Inflorescences densely branched, flowers lilac to purple. Fruits of 5 articles, each 3–4 × 3.4 mm, ± round or broadly elliptic, densely finely pubescent. The only other specimen I have seen from Africa is *da Silva* 2846 collected as an escape in Angola, Huambo (Nova Lisboa), Chianga, in degraded secondary savanna. The Zimbabwe specimen was presumably cultivated although this is not specifically stated.

1. Leaves all 1-foliolate, or at least some 1-foliolate* · 2
 – Leaves 3-foliolate · 9
2. Stems prostrate; leaflets ovate-oblong, rounded-oblong or almost round, shallowly cordate at the base; inflorescences lax · 18. *cordifolium*
 – Stems not prostrate or if somewhat decumbent then other characters different · · · · · · 3

---

* Some species are included twice since there is variation in this character.

3. Leaflets large, 2.5–19 × 1.1–13 cm, velvety pubescent on both surfaces but particularly beneath; fruits densely hairy; pedicels short, even when fruiting only 3 mm long · · · · · · · 4. *velutinum*
 – Leaflets usually smaller, pubescent to silky-pilose especially beneath but never velvety · · 4
4. Inflorescences short, dense and rather compact, leaflets mostly appressed silky-pilose beneath (very densely so in var. *argyreum*); pedicels mostly bent just below the calyx; stem hairs not tubercular-based nor glandular · 24. *barbatum*
 – Inflorescences longer or much laxer · 5
5. Inflorescences covered with tubercular-based glandular hairs; each primary bract subtending only one flower · 11. *helenae*
 – Inflorescence hairs not tubercular-based nor glandular; primary bracts usually subtending 2 or more flowers · 6
6. Stipules broadly amplexicaul at the base, strongly striate; leaves often 3-foliolate; both margins of the fruit approximately equally indented between the articles · · 3. *dichotomum*
 – Stipules not broadly amplexicaul at the base, much narrower; one margin of the fruit straight or at least one much more indented than the other · · · · · · · · · · · · · · · · · · 7
7. Pedicels 3–6 mm long; leaflets 1.3–17.5 cm long, rounded to acuminate; articles of fruit 2.5–3.5 × 2–2.5 mm · 5. *gangeticum*
 – Pedicels 4–13 mm long; leaflets 0.8–5 cm long, rounded to emarginate; articles of fruit 2.5–4 × 2.3–3 mm · 8
8. Hairs on stem (not inflorescences) white and appressed; leaves nearly all 1-foliolate · · · ·
 · 13. *appressipilum*
 – Hairs on stem ferruginous and spreading (not or less so in var. *delicatulum*); leaves mostly 3-foliolate · 12. *hirtum*
9. Fruits very characteristic, slightly indented on both margins, with 4–7 narrow elongate articles 4–5 × 2 mm with raised anastomosing ribs (Malawi S, naturalised) · · · 1. *scorpiurus*
 – Fruit articles not as above · 10
10. Leaves practically sessile, the rhachis very much longer than the petiole which never exceeds 5 mm in length; inflorescences long-branched terminal panicles; articles 7–8 × 6 mm · 20. *tanganyikense*
 – Leaves not practically sessile (save in *D. fulvescens*); rhachis not or scarcely exceeding the petiole in length ·11
11. Fruit somewhat thickened when mature, scarcely indented between the oblong articles*; leaflets mostly acute, less often rounded, 3–17.5 cm long; bracts 3–9 mm long, lanceolate, comose at the apices of the young inflorescences · · · · · · · · · · · · · · · 19. *salicifolium*
 – Fruit not so thickened and one margin at least strongly indented · · · · · · · · · · · · · ·12
12. Terminal leaflets broadly rhombic, acuminate, the lateral leaflets similar but asymmetric, all leaflets 2–11 cm long with margin often slightly undulate; pedicels 1–5 cm long; articles 8–11.5 × 3.5–4 mm, obliquely lunate, densely covered with uncinulate hairs · · 2. *repandum*
 – Terminal leaflets not rhombic, mostly smaller and never undulate, or if rhombic then articles not as above · 13
13. Leaflets 5–7.7 × 1.8–2.8 cm, oblong-elliptic, ± rounded at both ends, with petiole 4–10 mm long; inflorescences extensive and densely spreading brown hairy; calyx with 5 subequal lobes (Zambia W) · 23. *fulvescens*
 – Not as above; if calyx subequally 5-lobed then inflorescences short and condensed but similarly brown hairy ·14
14. Inflorescences short and condensed, often scarcely exceeding the leaves; pedicels mostly strongly bent just below the calyx; calyx subequally 5-lobed · · · · · · · · · · · · · · · · · · ·15
 – Inflorescences laxer and longer (except in *D. triflorum*); pedicels not so prominently bent; calyx ± bilabiate, the upper lobe usually 2-fid ·16
15. Leaflets sensitive to touch; inflorescence very dense; secondary bracts present · · · · · ·
 · 22. *dregeanum*
 – Leaflets not sensitive to touch; inflorescence looser; secondary bracts lacking · · · · · · · ·
 · 24. *barbatum* var. *procumbens*
16. Leaflets triangular-rhombic, pointed, broadest near the base; plants procumbent or erect; stipules linear-lanceolate, fruits rather contorted with terminal article the largest; articles ± rhomboid (except in the terminal 1–2 articles) joined near upper margin but often centrally when young · 8. *ospriostreblum*
 – Not as above ·17

---

* Note that sometimes an aborted article can resemble a narrow neck.

17. Stems densely (except in *D. intortum* var. *pilosiusculum*) covered with hooked hairs which render the plant harshly adhesive; articles of fruit joined near their upper margins but sometimes appearing centrally attached; upper inflorescence bracts obvious and comose, 10 × 3–8 mm · 18
– Stems glabrous to hairy but without hairs rendering the plant adhesive; bracts not so obvious · 19

18. Leaflets without a pale median mark on the upper surface, (2)3–12 × (1)1.5–7 cm (cultivated and naturalised) · 7. *intortum**
– Leaflets with a marked silvery median mark on the upper surface, 2.5–8.5 × 1–4.5 cm (cultivated and naturalised) · 6. *uncinatum*

19. Fruit with both margins ± equally indented, the articles joined by narrow ± centrally placed necks · 20
– Fruit with only one margin markedly indented, necks between articles not centrally placed · 25

20. Leaflets velvety pubescent, at least on lower surface · 21
– Leaflets glabrous to finely pubescent (a dubiously recorded cultivated species has ± densely appressed pilose leaflets beneath but hardly velvety) · 22

21. Articles glabrous or with very few hairs; leaflets 2.5–13 × 1–7 cm, rhombic, narrowed to a blunt apex (cultivated) · *discolor*
– Articles densely puberulous; leaflets 3–6 × 1.2–4.5 cm, elliptic, ± rounded at both base and apex (cultivated) · *nicaraguense*

22. Articles glabrous, 5–6 × 3.5–4 mm, elliptic, strongly reticulate; leaflets ± densely pilose beneath, up to 13 × 5 cm but can be small (one cultivated plant from Marondera may be this) · *leiocarpum*
– Articles not glabrous; leaflets usually small and glabrescent, finely pubescent or puberulous · 23

23. Articles 7–8.5 × 5–6 mm when mature, strongly reticulate, puberulous (naturalised) · 10. *psilocarpum*
– Articles smaller · 24

24. Stipules very obliquely ovate-lanceolate, falcate, up to 8 mm wide at base; articles 1.5–2.5 × 1.5–2 mm but occasionally, when reduced to one, 5 × 4 mm · · · · · · · · · · · · · · · *distortum*
– Stipules less oblique and more attenuate, up to 3–4 mm wide at base; articles 3–6.5 × 3–4 mm, uncinulate-pubescent · 9. *tortuosum*

25. Articles 5–9.5 × 2.5–3.5 mm, lunate; leaflets round to broadly obovate; pedicels 8–12 mm long; plant spreading, straggling or prostrate (widespread) · · · · · · · · · · · · 21. *adscendens*
– Articles smaller · 26

26. Stipules partially connate, at least when young; articles of fruit 3.5–5 × 2–3.5 mm, uncinulate-pubescent, the fruit indented on lower margin for two-thirds; upper margin of articles straight (cultivated) · *incanum*
– Stipules not connate; articles of fruit usually smaller and fruit less indented on lower margin, but if as large and indentations as deep then upper margin of article concave (wild) · 27

27. Flowers 1–8 in very short axillary inflorescences; prostrate creeping plant with small obovate retuse leaflets · 14. *triflorum*
– Flowers in more elongated inflorescences · 28

28. Articles 4.5–7 × 2–4 mm, with upper margin concave; strictly erect herb mostly with several slightly branched or unbranched stems from a woody rootstock; pedicels 1–3 cm long · 16. *stolzii*
– Articles smaller with straight upper margin · 29

29. Inflorescence rhachis and young stems with appressed hairs; articles uncinulate-puberulous; lower margin of fruit indented to about half way · · · · · · · · 17. *ramosissimum*
– Inflorescence rhachis and young stems with spreading hairs; articles very finely puberulous; lower fruit margin indented about half to one-fifth · 30

30. Fruit dehiscent along the lower margin, articles later separating from each other; lower margin of the fruit indented by only about one-fifth · · · · · · · · · · · · · · · · · · 12. *hirtum*
– Fruit not dehiscent along lower margin; lower margin of the fruit indented by about one-third to half · 15. *setigerum*

---

* Var. *pilosiusculum* has much less hairy stems and leaflets much smaller, 3–4.5 × 1.5–2 cm, possibly sometimes with a median mark.

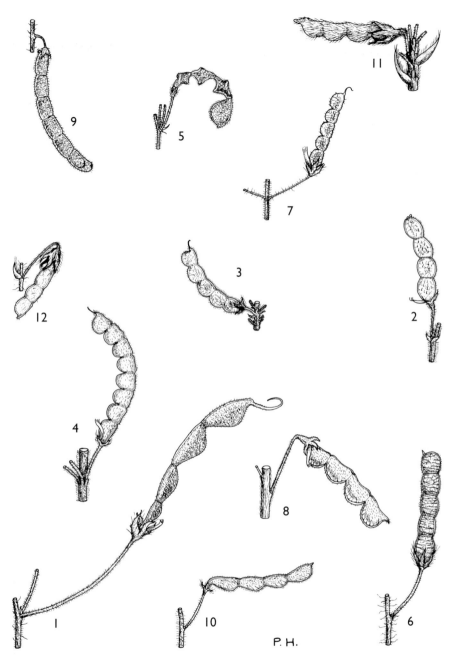

Tab. 3.6.**1**. DESMODIUM. Fruits. 1, D. REPANDUM (× 1), from *Oteke* 78; 2, D. DICHOTOMUM
(× 2), from *Pappi* 366; 3, D. VELUTINUM (× 1), from *Richards* 17723; 4, D. GANGETICUM
(× 2), from *Faulkner* 2927; 5, D. OSPRIOSTREBLUM (× 1), from *Polhill & Paulo* 1971;
6, D. HIRTUM var. HIRTUM (× 2), from *Richards* 18656; 7, D. SETIGERUM (× 2), from
Purseglove 3377; 8. D. RAMOSISSIMUM (× 2), from *Tweedie* 2425; 9, D. SALICIFOLIUM
(× 2), from *Tanner* 5011; 10, D. ADSCENDENS var. ROBUSTUM (× 1), from *Pawek* 1598;
11, D. DREGEANUM (× 2), from *Jackson* U63; 12, D. BARBATUM var. PROCUMBENS
(× 2), from *Maitland* 649. Drawn by Pat Halliday. From F.T.E.A.

1. **Desmodium scorpiurus** (Sw.) Desv. in J. Bot. Agric. **1**: 122 (1813). —Verdcourt in Man.
New Guinea Leg.: 407, fig. 94G (1979). —Lock, Leg. Afr. Check-list: 247 (1989). Type
from Jamaica.
   Hedysarum scorpiurus Sw., Nov. Gen. Sp. Pl. Prodr.: 107 (1788).

Prostrate perennial herb with slender ± spreading hairy stems up to 50 cm long.
Leaflets 3, 0.5–4.5 × 0.2–1.8 cm, elliptic or less often round or oblong, mostly
rounded at both ends or very slightly emarginate, hairy; petiole 0.3–2.7 cm long;
stipules 4 × 3 mm, leaf-like with ovate bases and long attenuate tips, often
amplexicaul at the base. Flowers c. 5 mm long in terminal and axillary
inflorescences 7–33 cm long; pedicels 5–7.5 mm long. Standard pink or pale purple.
Fruits 2–3.5 cm long, of 4–7 articles which eventually separate; each article 4–5 ×
1.5–2 mm, narrowly oblong-elliptic to linear-oblong, pubescent and with a raised
reticulation of ribs; constrictions between the articles developed or not. Seeds pale
buff, 2.3 × 1 mm, oblong-ellipsoid.

Malawi. S: Zomba, near Salisbury Bridge, 12.x.1987, *Kaunda & Nachamba* 698 (K; MAL).
A native of West Indies and Mexico to Peru now naturalised in West Africa, Taiwan and
Malesia. Short grass by roadsides; 915 m.
This is the type species of the name *Desmodium*.

2. **Desmodium repandum** (Vahl) DC., Prodr. **2**: 334 (1825). —Brenan in Mem. New York Bot.
Gard. **8**: 255 (1953). —Schubert in F.C.B. **5**: 193, pl. 14 (1954). —Hepper in F.W.T.A., ed.
2, **1**: 584 (1958). —White, F.F.N.R.: 150 (1962). —Laundon in C.F.A. **3**: 221 (1966). —
Schubert in F.T.E.A., Leguminosae, Pap.: 465, fig. 65/11 (1971). —Drummond in Kirkia
8: 219 (1972). —Verdcourt in Kirkia **9**: 517 (1974). —Lock, Leg. Afr. Check-list: 247
(1989). TAB. 3.6.**1**, fig. 1; TAB. 3.6.**2**. Type from Yemen.
   Hedysarum repandum Vahl, Symb. Bot. **2**: 82 (1791).
   Desmodium scalpe DC., Prodr. **2**: 334 (1825). —J.G. Baker in F.T.A. **2**: 164 (1871). —E.G.
Baker, Legum. Trop. Africa: 328 (1929). Type from Mauritius.

Prostrate or ascending herb, subshrubby herb or weak shrub, 0.3–4 m tall. Stems
simple to strongly branched, finely shortly pubescent and with many longer bristly
hairs when young, later glabrescent. Leaves 3-foliolate; leaflets 2–11 × 1.5–9.5 cm,
rhombic or ovate, the lateral ones oblique, shortly to distinctly and sharply
acuminate at the apex, cuneate or rounded at the base, pilose with long and short
hairs admixed on both surfaces, margin entire or slightly undulate; petiole 1–10.5
cm long; rhachis 0.3–2 cm long; petiolules 1.5–5 mm; stipules brown and somewhat
chaffy, 10–18 × 2–4 mm, oblong-lanceolate, acuminate, finely densely puberulous
outside, and also with scattered long hairs, particularly dense near the node,
eventually deciduous. Inflorescences terminal, unbranched, lax, 7–30 cm long, with
same indumentum as the stem; peduncle 7–9 cm long; pedicels 1–5 cm long, all
similarly hairy; primary bracts each subtending 2–5 pedicels, 3.5–10 × 2–5 mm,
elliptic to ovate-lanceolate; secondary bracts very small. Calyx pubescent; tube c. 1–2
mm long; lobes 1.5–3 mm long, triangular or lanceolate, the upper pair joined to
form an acute 2-fid lip. Standard scarlet, vermilion or orange, 8–12 × 6–12 mm,
broadly obovate; keel cream. Fruits 1.3–4.5 cm long, of (1)4–5 articles; each article
8–9(11.5) × 3.5–4 mm, obliquely lunate, densely covered with short uncinulate hairs;
venation obscure; necks between the articles c. 0.5 mm wide; stipe 2–4.5 mm long.
Seeds dark chestnut-brown, 7 × 3 × 1 mm, obliquely lunate.

Zambia. N: Mbala Distr., Chilongowelo, fl. 12.iv.1963, *Richards* 18071 (K). W: 80 km west
of Chingola, fl. & fr. 9.vii.1963, *E.A. Robinson* 5584 (K). C: Serenje Distr., Kundalila Falls,
south of Kanona, fl. & fr. 13.iii.1975, *Hooper & Townsend* 737 (K). E: Nyika Plateau, fr.
7.vi.1962, *Verboom* 635 (K). **Zimbabwe**. N: Goromonzi Distr., Chishawasha, fl. 20.i.1960,
Mitchell 568 (BM; SRGH). C: Marondera Distr., valley leading up to Wedza Mt., fl. iii.1955,
Davis 975 (COI; SRGH). E: Nyanga Distr., near Nyamingura River, fl. & fr. 22.iv.1958, *Phipps*

Tab. 3.6.**2**. DESMODIUM REPANDUM. 1, habit (× ¹/₂); 2, flower (× 2); 3, flower, longitudinal
section (× 3); 4, calyx (× 2); 5, fruit (× 1); 6, seed (× 2), 1–6 from *Bequaert* 3503. Drawn by
A. Cleuter. From Fl. Congo Belge. Reproduced with permission of Jardin Botanique
National de Belgique.

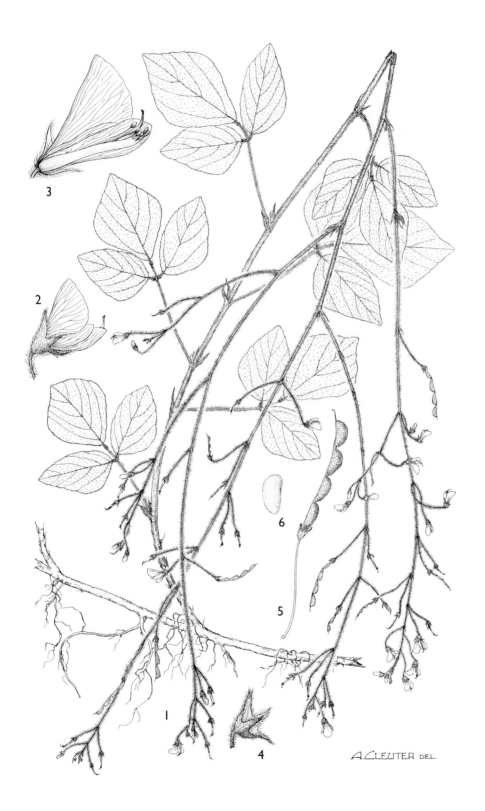

3

2

6

5

1

4

A. CLEUTER DEL.

1193 (K; PRE; SRGH). S: Masvingo (Victoria), fl. iv.1921, *Eyles* 3005 (SRGH). **Malawi**. N: Rumphi Distr., Nchenachena, fl. & fr. 28.v.1967, *Pawek* 1130 (SRGH). C: Ntchisi (Nchisi), fr. 11.vii.1960, *Chapman* 832 (BM; SRGH). S: Mulanje (Mlange) Mt., Luchenya Plateau, fl. & fr. 8.vii.1946, *Brass* 16732 (K; NY; SRGH). **Mozambique**. N: Lago Distr., Serra Jeci, fl. & fr. 29.v.1948, *Pedro & Pedrógão* 4087 (EA; K; LMA). Z: Tumbine, fl. & fr. 19.ix.1942, *A.J.W. Hornby* 1175 (EA; K; PRE). MS: Manica, base of Serra de Mavita, fl. & fr. 15.iv.1948, *Barbosa* 1466 (LISC).

Widely distributed in the tropics and subtropics of the Old World. Evergreen forest fringes and glades, and in riverine forest; 900–1850(2130) m.

3. **Desmodium dichotomum** (Willd.) DC., Prodr. **2**: 336 (1825). —Schubert in F.T.E.A., Leguminosae, Pap.: 471, fig. 65/9 (1971). —Verdcourt in Kirkia **9**: 523 (1974). —Lock, Leg. Afr. Check-list: 245 (1989). TAB. 3.6.1, fig. 2. Type from India.

   Hedysarum dichotomum Willd., Sp. Pl. **3**: 1180 (1802).
   Hedysarum diffusum Willd., Sp. Pl. **3**: 1180 (1802) non Roxb. Type from India.
   Desmodium diffusum (Willd.) DC., Prodr. **2**: 336 (Oct. 1825). —Baker in Fl. Brit. India **2**: 169 (1876). —E.G. Baker, Legum. Trop. Africa: 331 (1929) (basionym wrongly attributed to Roxb.) non *Desmodium diffusum* DC. in Ann. Sci. Nat. **4**: 100 (Jan. 1825) and Prodr. **2**: 335 (1825).
   Desmodium sennaarense Schweinf., Fl. Aethiop.: 8 (1867). —J.G. Baker in F.T.A. **2**: 164 (1871). Type from Sudan.

Annual herb, or subshrubby herb to 1 m tall. Stems ridged, uncinulate-pubescent. Leaves 1–3-foliolate; leaflets 1.5–10 × 1.3–7 cm, elliptic, oblong or obovate, obtuse or emarginate at the apex, rounded or truncate at the base, pubescent on both surfaces, the venation raised beneath; petiole 0.5–8 cm long; rhachis 0.25–3.5 cm long; petiolules 2–3 mm long; stipules 5–15 × 3.5–10 mm, ovate-falcate, striate, amplexicaul and subcordate; stipels very distinct, 5–9 × 1–2.5 mm, lanceolate. Inflorescences axillary and terminal, 4–17 cm long; peduncles 0–30 mm long; pedicels 3–5 mm long; basal bracts of inflorescence resembling the stipules; primary bracts each subtending 1–3 pedicels, 3.5 × 0.8 mm, lanceolate, puberulous and ciliate; secondary bracts similar, 3 × 0.5 mm. Calyx with long white hairs; tube 0.5 mm long; lobes 2 mm long, narrowly lanceolate, the upper pair joined for over half their length. Standard purple or light blue, 3.5 × 2.5 mm, obovate, emarginate. Fruits 7–18 mm long, of 2–6 articles; each article 2.5–3 × 2.5–3.3 mm, almost round or rounded-oblong, sparsely pubescent and reticulate, separated by rather wide necks 2 mm wide. Seeds yellow or chestnut-brown, 2.5 × 1.8–2.1 × 1–1.2 mm, rounded or oblong-ellipsoid; the hilum small eccentric or central.

Zambia. C: South Luangwa National Park, Mfuwe, fl. & fr. 9.iv.1969, *Astle* 5686 (K).

Also in Ethiopia, Sudan, Uganda, India; undoubtedly introduced into Africa from Asia. In *Echinochloa* grassland in mopane woodland on black cracking clay; 600 m.

4. **Desmodium velutinum** (Willd.) DC., Prodr. **2**: 328 (1825). —Schubert in F.C.B. **5**: 194 (1954). —Hepper in F.W.T.A., ed. 2, **1**: 584 (1958). —White, F.F.N.R.: 150 (1962). —Laundon in C.F.A. **3**: 219 (1966). —Schubert in F.T.E.A., Leguminosae, Pap.: 466, fig. 65/7 (1971). —Drummond in Kirkia **8**: 219 (1972). —Verdcourt in Kirkia **9**: 518 (1974). —Gonçalves in Garcia de Orta, Sér. Bot. **5**: 78 (1982). —Lock, Leg. Afr. Check-list: 248 (1989). —Pedley in Rev. Fl. Ceylon **10**: 183 (1996). TAB. 3.6.1, fig. 3. Type from "America calidiore".

   Hedysarum velutinum Willd., Sp. Pl. **3**: 1174 (1802).
   Hedysarum lasiocarpum P. Beauv., Fl. Oware **1**: 32, t. 18 (1805). Type from Nigeria.
   Desmodium lasiocarpum (P. Beauv.) DC., Prodr. **2**: 328 (1825). —J.G. Baker in F.T.A. **2**: 162 (1871). —E.G. Baker, Legum. Trop. Africa: 326 (1929).
   Anarthrosyne cordata Klotzsch in Peters, Naturw. Reise Mossambique **6**, 1: 39, t. 7 (1861). Syntypes: Zanzibar; Mozambique, Quirimba (Querimba) Island, *Peters* (B, syntypes; K).
   Pseudarthria cordata (Klotzsch) Walp., Ann. Bot. Syst. **7**: 765 (1868).

Subshrub or woody herb, 0.5–3 m tall. Stems densely covered with short spreading pale ferruginous hairs when young, purplish-brown and glabrescent beneath. Leaves 1-foliolate; leaflet 2.5–19 × 1.1–13 cm, ovate, oblong-ovate or almost round, bluntly acute to very rounded at the apex, broadly cuneate, truncate or subcordate at the base, densely rather roughly appressed pubescent above, velvety beneath; petiole 2–22 mm long; petiolule 1.5–3.5 mm long; stipules linear from a broad base, 5–8 × 0.5–3 mm above, 2–3.5 mm wide at the base. Inflorescences pseudoracemose, terminal and axillary, 2–30 cm long, densely covered with uncinulate hairs; peduncles short, 5–25 mm long; pedicels 1.5–3 mm long; primary bracts 1.5–4 × 0.5–1 mm, lanceolate, each supporting 3–5 flowers; secondary bracts 1–1.5 mm long, linear, persistent. Calyx spreading-hairy; tube 1 mm long; lobes 0.7–1.5 mm long. Standard violet, lilac, red or blue, 4–5 × 4–5 mm, obovate; wings and keel darker mauve. Fruits 12–24 mm long, of 2–7 articles; each article 2.5–4 × 2.5–3.5 mm, densely covered with hooked hairs, joined by necks 1.5–2 mm wide. Seeds brown, 2.5 × 1.8 × 0.9 mm, ellipsoid-reniform; hilum small, central.

Zambia. B: Kaoma Distr., Kafue road, km 8, fr. 2.iv.1964, *Verboom* 1028 (K; SRGH). N: Kasama Distr., Chibutubutu, fl. & fr. 25.ii.1960, *Richards* 12576 (K; SRGH). W: Nkana/Kitwe Sewage Works, fl. & fr. 28.iii.1959, *Shepherd* 47 (K). C: between Kafue Bridge and Gorge, fl. & fr. 2.iii.1963, *van Rensburg* 1564 (K; SRGH). S: Mapanza E, fl. & fr. 21.iii.1954, *E.A. Robinson* 629 (K; SRGH). **Zimbabwe**. N: Gokwe, near source of River Guye, fr. 28.iii.1962, *Bingham* 196 (SRGH). E: Mutare (Umtali), west base of "Cross Hill", fl. 26.ii.1956, *Chase* 5980 (BM; COI; K; LISC; SRGH). S: Chivi Distr., 6.4 km north of Runde (Lundi) Bridge, Madzivire Dip, fr. 3.v.1962, *Drummond* 7873 (K; PRE; SRGH). **Malawi**. N: Karonga Distr., fl. 25.iii.1954, *G. Jackson* 1276 (BM; K). C: Dedza Distr., Mua-Livulezi Forest Reserve, fl. 9.iii.1966, *Banda* 805 (K; PRE; SRGH). S: Chimwalira River, near Balaka (Bolaka), fr. 16.iii.1955, *Exell, Mendonça & Wild* 918 (BM; LISC; SRGH). **Mozambique**. N: Mutuáli near Nalume River bridge, fl. 12.ii.1954, *Gomes e Sousa* 4186 (COI; K; PRE). Z: Lugela Distr., Namagoa, fl. & fr. 22.iii.1948, *Faulkner* Kew 224 (K). T: Zóbuè, fr. 18.vi.1941, *Torre* 2895 (LISC). MS: Manica Distr., between Garuso (Garuzo) and Bandula, fl. 23.ii.1948, *Garcia* 325 (LISC).
Widespread in the tropics of the Old World. Grassland, wooded grassland and woodland (e.g. *Isoberlinia, Brachystegia, Combretum, Burkea* or *Pericopsis*), also riverine bushland; 100–1320 m.

5. **Desmodium gangeticum** (L.) DC., Prodr. **2**: 327 (1825). —J.G. Baker in F.T.A. **2**: 161 (1871). —E.G. Baker, Legum. Trop. Africa: 327 (1929). —Schubert in F.C.B. **5**: 196 (1954). — Hepper in F.W.T.A., ed. 2, **1**: 584 (1958). —White, F.F.N.R.: 150 (1962). —Laundon in C.F.A. **3**: 219 (1966). —Schubert in F.T.E.A., Leguminosae, Pap.: 467, fig. 65/10 (1971). — Verdcourt in Kirkia **9**: 520 (1974). —Lock, Leg. Afr. Check-list: 246 (1989). —Pedley in Rev. Fl. Ceylon **10**: 182 (1996). TAB. 3.6.1, fig. 4. Type from India.
Hedysarum gangeticum L., Sp. Pl.: 746 (1753).
Desmodium polygonoides Baker in F.T.A. **2**: 161 (1871). Type from Angola.

Subshrub or woody herb 0.2–2 m tall, mostly erect, rarely prostrate or straggling. Stems striate, angular, uncinulate-pubescent and with longer spreading hairs on the angles. Leaves 1-foliolate; leaflet 0.4–17.5 × 0.4–6 cm, round, elliptic, oblong to ovate-lanceolate, rounded to acuminate at the apex, rounded, truncate, or even slightly cordate at the base, puberulous with very short hairs above, almost glabrous, pilose with much longer hairs beneath; petiole 5–20 mm long; petiolule 2–3 mm long; stipules 7–13 × 1–2 mm, lanceolate, venose, finely puberulous and with longer hairs. Inflorescences pseudoracemose or paniculate, terminal and axillary, 7–40 cm long, hairy like the stems; peduncle 3–4 cm long, similarly hairy; pedicels 3–6 mm long, uncinulate-pubescent; primary bracts each subtending 3 or more flowers, 2–7.5 × 0.3–1 mm, linear or lanceolate, pilose; secondary bracts very small. Calyx finely uncinulate-puberulous and with long hairs on the lobes; tube 1 mm long; lobes 1–2 mm long. Standard* pink, blue, white, mauve or greenish and dark blue, 4–5 × 3–4 mm, obovate, emarginate. Fruits 8–25 mm long, of 3–7(8) articles; each article 2.5–3.5 × 2–2.5 mm, uncinulate-puberulous, sometimes with longer hairs on the margins, joined by narrow necks 0.5 mm wide. Seeds yellow or orange-brown, 1.6–2 × 1.2–1.4 × 0.5 mm, compressed reniform.

Zambia. W: Kitwe, fl. & fr. 23.i.1964, *Mutimushi* 550 (K). C: 100–129 km east of Lusaka, Chakwenga Headwaters, fr. 16.xi.1963, *E.A. Robinson* 5852 (K; SRGH). E: Chadiza Distr., Mwangazi R. valley, fl. & fr. 26.xi.1958, *Robson* 718 (BM; LISC; PRE; SRGH). S: Sinazongwe, fr. 15.iii.1960, *Fanshawe* 5485 (K). **Zimbabwe**. N: Gokwe, near source of Guye River, fr. 28.iii.1962, *Bingham* 194 (K; PRE; SRGH). C: Shurugwi (Selukwe), Ferny Creek, fl. & fr. 8.xii.1953, *Wild* 4301 (PRE; SRGH). S: Chivi Distr., Madzivire Dip, c. 6 km north of Runde (Lundi) R., fr. 16.iii.1967, *Corby* 1826 (K; SRGH). **Malawi**. N: Nyika Mts., *Whyte* (K). C: Dedza Distr., Mua-Livulezi Forest Reserve, fr. 16.v.1960, *Adlard* 359 (K; SRGH). S: Shire Highlands,

---

* Robinson (5852) reports the flowers to be pale yellow but no other collector confirms this.

Buchanan 30 (K). **Mozambique**. N: Nampula, fl. & fr. 3.ii.1936, *Torre* 672 (COI; LISC). Z: Lugela Distr., Namagoa Estate, fl. & fr. x.1943 and iii.1946, *Faulkner* PRE 278 (COI; K; LISC; PRE). MS: Sussundenga Distr., road from Mabongo to Chicuizo, fl. & fr. 22.i.1948, *Barbosa* 871 (BM; LISC).
   Widespread in tropical Africa. *Brachystegia, Burkea–Pericopsis, Bauhinia–Acacia*, etc. woodland, bushland, also in thickets of *Craibia, Mimusops*, etc. on dunes near sea, dry sandy ground near water and on kopjes; 10–1800 m.

6. **Desmodium uncinatum** (Jacq.) DC., Prodr. **2**: 331 (1825). —Verdcourt in Man. New Guinea Leg.: 411 (1979). —Lock, Leg. Afr. Check-list: 248 (1989). Type is a specimen grown at Vienna.
   Hedysarum uncinatum Jacq., Pl. Hort. Schoenbr. **3**: 27, pl. 298 (1798).

Prostrate or sprawling subshrubby herb 0.6–2 m long with ascending flowering stems from a stout rootstock. Stems covered with hooked hairs rendering plant harshly adhesive. Leaves 3-foliolate; leaflets green, the upper surface mostly with a lanceolate or somewhat pear-shaped silver marking along the midrib, 2.5–8.5 × 1–4 cm, ovate to ovate-lanceolate, acute at the apex, rounded at the base, fairly densely appressed pilose on both sides or only sparsely so above; petiole 1.8–6 cm long; stipules 5–8 × 2–4 mm, ovate, long-acuminate at apex, slightly auriculate at the base. Inflorescences terminal and axillary, up to 30 cm long. Calyx tube 3 mm long; lobes 2.5 mm long. Corolla pale mauve or pink turning blue on fading, or white. Fruit of 4–7 articles, each article 5.5–7 × 2.5–4 mm, ± semicircular or subtriangular, narrowly joined near upper margin, densely covered with hooked hairs rendering them very adhesive.

   Zimbabwe. C: Harare, outside National Herbarium mounting room, fl. & fr. 26.iv.1976, *Biegel* 5286 (K; SRGH). **Malawi**. S: Mt. Mulanje, Lichenya Path, Likhubula Valley, fl. & fr. 3.v.1989, *Rutherford-Smith et al.* 5842 (K).
   South American species widely introduced into the tropics. Pathsides in open *Brachystegia* woodland, riverine transitional woodland, also as a casual in built-up areas; 860–1500 m.
   Widely cultivated under name "*Desmodium silverleaf*" as a pasture legume but is frequently found as an escape.

7. **Desmodium intortum** (Mill.) Urb., Symb. Antill. **8**: 292 (1 Feb. 1920). —Fawcett & Rendle, Fl. Jamaica **4**: 34 (25 Mar. 1920). —Verdcourt in Man. New Guinea Leg.: 400 (1979). — Schubert in Ann. Missouri Bot. Gard. **67**: 646 (1980). —Lock, Leg. Afr. Check-list: 246 (1989). Type cultivated in England from seed sent from Jamaica.
   Hedysarum intortum Mill., Gard. Dict. ed. 8, Hedysarum (1768).

Prostrate subshrubby herb to 30 cm long, suberect subshrub or ± climbing, the flowering branches usually erect; lower stems often rooting at the nodes. Stems so densely covered with hooked hairs that the whole plant adheres on touch. Leaves 3-foliolate; leaflets (2)3–12 × (1)1.5–7 cm, ovate to ovate-lanceolate, acute or acuminate at the apex, rounded to truncate or cuneate at the base, hairy; petiole 2–5.5 cm long; stipules 4–6.5 × 1.5–5 mm, triangular or ovate-acuminate, soon falling. Inflorescences terminal and axillary, up to 30 cm long, covered with hooked hairs; bracts 5.5–10 × 3–4 mm, ovate. Calyx with the central tooth of lower lobe 4–5.5 mm long, lateral teeth 2.5–4 mm long and upper 2-fid lobe 2.5–3.5 mm long. Corolla pink or violet becoming bluish or all white; standard 6–7.5 × 4–4.5 mm. Fruits with (3)6–11 articles, each article 3–5.5 × 3–4 mm, elliptic or rounded-rhomboid, densely covered with hooked hairs.

   Malawi. N: Mzimba Distr., 16 km SW of Mzuzu, Mbowe Dam, fl. & fr. 2.viii.1970, *Pawek* 3654 (K).
   Pond and dam edges and plantation edges; 1200–1350 m.
   Apart from being an escape in Malawi it has also been grown at Marondera in Zimbabwe, (2.vii.1960, *Corby* 924 & 17.vi.1958, *West* 3606) and at Mt. Makulu Research Station in Zambia (13.v.1966, *van Rensburg* 3110).

8. **Desmodium ospriostreblum** Chiov. in Ann. Ist. Bot. Roma **8**: 428 (1908). —Schubert in F.T.E.A., Leguminosae, Pap.: 475, fig. 65/2 (1971). —Verdcourt in Kirkia **9**: 527 (1974). — Lock, Leg. Afr. Check-list: 246 (1989). TAB. **3.6.1**, fig. 5. Type from Ethiopia.

Desmodium terminale sensu Guill. & Perr. in Guillemin, Perrottet & Richard, Fl. Seneg.
Tent.: 207 (1832) non (L.C. Rich.) DC.
Anarthrosyne abyssinica A. Rich., Tent. Fl. Abyss. **1**: 204 (1847). Type from Ethiopia.
Desmodium spirale sensu Baker in F.T.A. **2**: 160 (1871) pro parte non DC.
Desmodium abyssinicum (A. Rich.) Hutch. & Dalziel, F.W.T.A. **1**: 418 (1928); in Bull. Misc.
Inform., Kew **1929**: 18 (1929) non DC. (1825) nom. illegit.
Desmodium tortuosum sensu Hepper in F.W.T.A., ed. 2, **1**: 585 (1958) pro parte non (Sw.) DC.

Procumbent or erect herb, 0.3–1.5 m long, sometimes with stems rooting at the
nodes. Stems somewhat angled, uncinulate-pubescent. Leaves 3-foliolate; leaflets
1.5–7 × 0.7–3 cm, elliptic-lanceolate, rhombic-lanceolate, mostly distinctly acute and
mucronulate at the apex, cuneate to rounded at the base, with short bristly hairs on
margins and main venation below and also more scattered on upper surface together
with some very minute hairs; petiole 0.7–4 cm long; rhachis 4–11 mm long; petiolules
2 mm long; stipules subulate or narrowly lanceolate from a broadened base, 4–6 × 1
mm at the base, shortly pilose, persistent. Inflorescences terminal and axillary, 6–14
cm long; uncinulate-pubescent and finely puberulous; peduncle 1.5–3.5 cm long;
pedicels 8–12 mm long; primary bracts each supporting 1–3 pedicels, often
including one undeveloped flower, setaceous, 3–4 mm long, persistent; secondary
bracts about half the size. Calyx puberulous and with some longer bristly hairs; tube
0.5–1 mm long; lobes 0.5–1 mm long, triangular, the upper pair joined to form a 2-
fid lip. Standard blue, white or pink, 4 × 3.5 mm, obovate, emarginate. Fruits very
similar to *Desmodium intortum*, articles 5–9 × 3.5–5.5 mm, elliptic or subtriangular, the
apical one longest, puberulous and uncinulate-pubescent particularly on the
margins, reticulate; fruiting pedicels sometimes persistent after fruits have fallen.
Seeds yellow-brown, compressed, 3–3.5 × 2.5 × 1 mm, ellipsoid.

Zambia. C: Luangwa Bridge, fl. & fr. 10.iv.1963, *Verboom* 818 (K; SRGH). S: Nega Nega Hills,
Munali Pass, fr. 4.iv.1957, *Angus* 1553 (K; PRE; SRGH). **Zimbabwe**. N: Gokwe, near source of
the River Guye, fr. 28.iii.1962, *Bingham* 199 (K; LISC; PRE; SRGH). W: Hwange Distr.,
Gwayi/Imakulwe R. confluence (Lutope River junction), fl. & fr. 26.ii.1963, *Wild* 6007 (K; LISC;
SRGH). E: Mutare Distr., Burma Valley West, escarpment east of Chishakwe Hill, fl. 19.iii.1982,
Pope & Müller 2062 (K; SRGH). S: Chivi Distr., Runde (Lundi), near Madzivire Dip, fl. & fr.
16.iii.1967, *Corby* 1824 (SRGH). **Malawi**. N: Karonga, fl. 23.iv.1975, *Pawek* 9530 (K; MAL; MO;
SRGH; UC). S: Zomba Distr., Chilwa Island, fr. 2.iii.1955, *G. Jackson* 1658 (K). **Mozambique**.
N: Montepuez, encosta do Monte Matuta, c. 5 km a sul do Rio Messalo (M'salo), próximo de
Nairoto (Nantulo), fl. & fr. 9.iv.1964, *Torre & Paiva* 11801 (LISC).
    Also in Cape Verde Islands, Senegal, Sierra Leone, Ghana, Nigeria, Annobon Island and
Ethiopia. Riverine woodland and thicket, mopane and *Brachystegia* woodland, *Kirkia–Adansonia–
Combretum–Sterculia* woodland, thicket and dry bushland on granite soils; 300–1100 m.
    Probably of tropical American origin and scarcely differing from *Desmodium procumbens*
(Mill.) Hitchc. (*D. spirale* DC.) a native of the New World, with which it is mostly confused. *D.
ospriostreblum* differs only in the enlarged or flattened terminal article of the fruit; it differs from
D. tortuosum in its more rhombic pointed leaflets, narrower stipules and bracts and with the fruit
necks often not so centrally placed.

9. **Desmodium tortuosum** (Sw.) DC., Prodr. **2**: 332 (1825). —Schubert in F.C.B. **5**: 202 (1954).
    —Hepper in F.W.T.A., ed. 2, **1**: 585 (1958). —Laundon in C.F.A. **3**: 223 (1966). —Schubert
    in F.T.E.A., Leguminosae, Pap.: 474 (1971). —Drummond in Kirkia **8**: 219 (1972). —
    Verdcourt in Kirkia **9**: 526 (1974). —Lock, Leg. Afr. Check-list: 247 (1989). —Pedley in
    Rev. Fl. Ceylon **10**: 191 (1996). Type from Jamaica.
    Hedysarum purpureum Mill., Gard. Dict. ed. 8, Hedysarum (1768). Type from Mexico.
    Hedysarum tortuosum Sw., Nov. Gen. Sp. Pl. Prodr.: 107 (1788); Fl. Ind. Occid. **3**: 1271
    (1806).
    Meibomia tortuosa (Sw.) Kuntze, Revis. Gen. Pl. **1**: 198 (1891).
    Meibomia purpurea (Mill.) Vail in Small, Fl. S.E. U.S.: 639 (1903).
    Desmodium purpureum (Mill.) Fawc. & Rendle, Fl. Jamaica **4**: 36 (1920) non Hook. & Arn.
    (1832) nom. illegit.
    Desmodium spirale sensu E.G. Baker, Legum. Trop. Africa: 331 (1929) non (Sw.) DC.

Erect perennial herb or undershrub 0.4–1.5 m tall. Stems often several from the
base, branched or unbranched, puberulous and with tubercular-based glandular
hairs. Leaves 3-foliolate; leaflets 1.2–9.5 × 0.6–5.5 cm, elliptic, rhombic or elliptic-
oblong to ovate-lanceolate, rounded and usually mucronulate at the apex, cuneate
to rounded at the base, puberulous on both surfaces and with some longer

pubescence on the nerves beneath; petiole 0.7–7 cm long; rhachis 4–22 mm long; petiolules 1.5–3 mm long; stipules 6–13 × 1.5–4 mm, obliquely ovate, broad at the base and with attenuate tips, puberulous and ciliate, persistent. Inflorescences terminal and axillary, 11–30 cm long; peduncle 5–10 cm long; pedicels 8.5–17 mm long, puberulous and with longer hairs; primary bracts each subtending 1–4 pedicels, 3–6 × 0.8–1 mm, lanceolate, puberulous, soon deciduous; secondary bracts also present. Calyx pubescent and puberulous and with some tubercular-based glandular hairs; tube 1–1.5 mm long; lobes 1.5–2 mm long, narrowly triangular, the upper pair joined to form a slightly 2-fid triangular lip. Standard pink, blue-mauve or blue-green, 5–6 × 3–3.5 mm, narrowly obovate. Fruits 10–32 mm long, of (1)3–7 articles; each article 3–6.5 × 3–4 mm, elliptic or almost round, uncinulate-pubescent and minutely glandular, reticulate, strongly constricted between articles, the necks being centrally placed, c. 0.8 mm wide; stipe 2.5 mm long. Seeds chestnut-brown or greenish or yellow-brown, 3 × 2 × 1 mm, compressed-ellipsoid, slightly wider at one end; hilum minute, central.

Zambia. N: Mbala Distr., shore of Lake Tanganyika, fl. & fr. 18.iv.1963, *Verboom* 827 (LISC; SRGH). W: Kitwe, fr. 10.iv.1968, *Fanshawe* 10343 (K). C: Lusaka, fl. & fr. 5.ii.1965, *Fanshawe* 9157 (K). **Zimbabwe**. N: Shamva, fl. & fr. 17.ii.1961, *Rutherford-Smith* 544 (LISC; PRE; SRGH). C: Arcturus, fr. 22.iii.1945, *Christian* (K; PRE). E: Chiping Distr., Tanganda Halt on Chipinga–Birchenough Bridge Road, fl. & fr. 5.viii.1973, *Mavi* 1474 (K; SRGH). **Malawi**. S: Zomba Distr., Mlunguzi (Mulunguzi) River, 15.vi.1954, *Banda* 1 (BM). **Mozambique**. Z: Namacurra, ao km 11 de Furguia (Furguilha) para Macuze (Macuzi) na estrada para o Rio Licungo (via Maganja da Costa), fl. & fr. 27.i.1966, *Torre & Correia* 14215 (LISC). M: Marracuene, Umbelúzi Experimental Station, fr. 23.xi.1946, *Pedro & Valente* in *Pedro* 3101 (LMA).

Native of tropical America now widespread in tropical Africa and Asia. As a weed on railways, in storm drains, on cultivated ground, etc., also in *Brachystegia* woodland and wooded grassland. Cultivated (Florida beggar weed) as a nitrogen-enriching plant; 30–1230 m.

10. **Desmodium psilocarpum** Gray, Pl. Wright **2**: 48 (1853). —Lock, Leg. Afr. Check-list: 247 (1989). Type from Mexico (Sonora).

Thinly woody shrub to 90 cm tall; young stems pubescent with short hooked hairs, becoming glabrous. Leaves 3-foliolate; leaflets 0.7–5.5 × 0.4–3.5 cm (up to 3 × 1.5 cm in African material seen), elliptic to ovate, rounded to subacute and mucronulate at the apex, rounded at the base, finely puberulous and with some longer hairs on the venation, distinctly ciliate around the margin; petiole 0.6–3 cm long with rhachis almost equal in length; stipules 5–12 mm long, obliquely ovate-attenuate. Inflorescences axillary and terminal 10–23 cm long; bracts and bracteoles conspicuous in young stage but soon deciduous. Calyx tube 1 mm long; lobes 2 mm long, narrowly triangular. Corolla c. 5 mm long. Fruits with 1–5 articles joined centrally, each article 7–8.5 × 5–6 mm when mature, elliptic, strongly reticulate, puberulous. Seeds pale brown, 3.5 × 2.5 × 0.8 mm, ellipsoid, very compressed; hilum minute, not quite central.

Zambia. E: Chipata (Fort Jameson), fl. & fr. 23.ix.1968, *Mutimushi* 1424 (K; NDO). **Zimbabwe**. C: Kwekwe (Que Que), fl. & fr. 24.i.1966, *Biegel* 829 (K; SRGH).

Native of Central America. Weedy vegetation on cinders at railway station with *Tribulus, Oxygonum*, etc., also beside seasonal rivers; 1230 m.

11. **Desmodium helenae** Buscal. & Muschl. in Bot. Jahrb. Syst. **49**: 472 (1913). —E.G. Baker, Legum. Trop. Africa: 326 (1929). —Schubert in F.C.B. **5**: 183 (1954). —Laundon in C.F.A. **3**: 218 (1966). —Verdcourt in Kirkia **9**: 509 (1974). —Lock, Leg. Afr. Check-list: 246 (1989). Type: Zambia, Lake Bangweulu, *von Aosta* 961 (B†, holotype; BR, photo of drawing).

Perennial usually subshrubby herb, 0.45–1.2 m tall. Stems shortly spreading pubescent, mostly reddish- or purplish-brown, distinctly woody and glabrescent below. Leaves 1-foliolate; leaflet 0.5–3.4 × 0.25–2 cm, elliptic or slightly obovate-oblong, obtuse to subacute and mucronulate at the apex, narrowly rounded at the base, mostly drying a pale rather yellowish-green, glabrous above, shortly appressed pubescent beneath; petiole (1)6–12 mm long, sometimes persistent; petiolule (0.5)1.5–2 mm long; stipules 3–12 × 1–2.5 mm, narrowly triangular-lanceolate, persistent. Inflorescences mostly terminal, much-branched, 7–20 cm long, densely pubescent with yellowish somewhat glandular hairs; peduncle 2 cm long; pedicels (1.5)3–6.5 mm long; primary bracts 5.5–8 × 3–4 mm, broadly ovate, apiculate, each supporting one flower, shortly glandular hairy outside and with ciliate margins; secondary bracts absent. Calyx glandular hairy; tube 2.5 mm long; lobes 2–2.5 mm long. Standard purplish, red or pale rose-coloured, later turning blue, 8–8.5 × 9.5–10 mm, broadly oblate-obovate; wings similarly coloured to the standard; keel cream or green, often purple at the tip (dry). Fruits 15–27 mm long, of 1–5 articles, each article 4–6.5 × 4–4.5 mm, joined by narrow necks (1.5)2–2.5(3) mm wide, venose, pubescent with dense hooked hairs and some straight hairs; stipe 1–3 mm long. Seeds dark blackish-purple, 2.5 × 2.2 × 0.7 mm, oblong-reniform; the hilum slightly eccentric.

Zambia. B: 64 km from Zambezi (Balovale) on Kabompo road, fl. & fr. 26.iii.1961, *Drummond & Rutherford-Smith* 7366 (K; PRE; SRGH). N: Kasama Distr., Mungwi, fl. 14.iii.1962, *E.A. Robinson* 5009 (K; SRGH). W: Mufulira, fl. & fr. 28.iv.1934, *Eyles* 8150 (BM; K; SRGH).

Also in the Dem. Rep. Congo and Angola. *Brachystegia* and other mixed deciduous woodlands, also in somewhat swampy places; 1200–1560 m.

12. **Desmodium hirtum** Guill. & Perr. in Guillemin, Perrottet & Richard, Fl. Seneg. Tent.: 209 (1832). —J.G. Baker in F.T.A. 2: 163 (1871) pro parte. —E.G. Baker, Legum. Trop. Africa: 329 (1929) pro parte. —Schubert in F.C.B. 5: 186 (1954). —Hepper in F.W.T.A., ed. 2, 1: 585 (1958). —Schubert in F.T.E.A., Leguminosae, Pap.: 457 (1971). —Verdcourt in Kirkia 9: 510 (1974). —Lock, Leg. Afr. Check-list: 246 (1989). Type from Senegal.

Erect or prostrate, annual or perennial herb or subshrubby herb, 0.1–2 m tall or long. Stems striate, distinctly or sparsely covered with long spreading bristly mostly pale ferruginous hairs 1.5–2 mm long. Leaves 1–3-foliolate; leaflets 0.8–4 × 0.5–2.2 cm, elliptic-oblong or the terminal ones often distinctly obovate or round, truncate or slightly emarginate at the apex, rounded to cuneate at the base, appressed pilose on both surfaces or practically glabrous; petiole 1–2 cm long; rhachis 3–7 mm long; petiolules 0.5–1.5 mm long, densely bristly; stipules 5–7 × 1.5–2 mm, ovate-lanceolate, striate, glabrous except for long spreading marginal cilia. Inflorescences terminal and axillary, lax, 2–8 cm long, hairy like the stems; peduncle 0.5–5 cm long; pedicels 9–13 mm long, finely puberulous; primary bracts each subtending 2 pedicels, 3.5–4.5 × 1.5 mm, ovate-lanceolate, pilose; secondary bracts absent. Calyx bristly pilose; tube 0.5 mm long; lobes 2 mm long, narrowly triangular. Standard pink or orange or yellow and crimson, 3.5 × 3 mm, obovate, emarginate. Fruits 10–18 mm long, of 3–6 articles but actually dehiscing as a whole and the articles not or scarcely separating although they are clearly evident, each article 2.5–3 mm long and wide, oblong with a straight edge and a very rounded edge, glabrous or uncinulate-puberulous, joined by necks 2 mm wide, reticulately veined. Seeds reddish-brown or yellowish mostly speckled with black, 1.5–2 × 1.2–1.5 × 0.5–0.8 mm, reniform; the small hilum almost central.

Var. **hirtum** —Schubert in F.C.B. 5: 186, fig. 11B (1954); in F.T.E.A., Leguminosae, Pap.: 458, fig. 65/14 (1971). —Verdcourt in Kirkia 9: 511 (1974). TAB. 3.6.1, fig. 6.

Plant sometimes perennial. Stems distinctly spreading hairy. Fruit uncinulate-puberulous.

Zambia. B: Mongu floodplain, fr. 10.iv.1964, *Verboom* 1233 (K). N: Mbala Distr., Chilongowelo, Plain of Death, fr. 5.v.1955, *Richards* 5539 (K). E: Chipata Distr., Ngoni Area, fl. & fr. v.1962, *Verboom* 600 (K; SRGH). S: Namwala Distr., Bambwe (Baambwe), fr. 17.iv.1963, *van Rensburg* 2011 (K; SRGH). **Zimbabwe**. W: Hwange (Wankie), fr. v.1915, *Rogers* 13301 (BM; PRE). **Mozambique**. N: Lalaua, fr. 22.vi.1948, *Pedro & Pedrógão* (EA; LMA).

Widespread in tropical Africa. Grazed grassland, in dambos and mixed deciduous woodland; 400–1158 m.

Var. **delicatulum** (A. Rich.) Harms ex Baker f., Legum. Trop. Africa: 329 (1929). —Schubert in F.C.B. 5: 186, fig. 11C (1954). —Verdcourt in Kirkia 9: 511 (1974). Type from Ethiopia.

Desmodium delicatulum A. Rich., Tent. Fl. Abyss. 1: 205 (1847). —J.G. Baker in F.T.A. 2: 163 (1871).

Plant mostly annual. Stems less distinctly hairy. Fruit glabrous.

Zambia. N: 10 km east of Kasama, fl. & fr. 12.iv.1961, *E.A. Robinson* 4589 (K; SRGH). W: c. 7 km east of Chizela (Chizera), fl. & fr. 27.iii.1961, *Drummond & Rutherford-Smith* 7417 (K; PRE; LISC; SRGH).
Widespread in tropical Africa. Swampy grassland, dambos, laterite pans, etc., sometimes in bare mud at pond edges; 900–1520 m.

13. **Desmodium appressipilum** B.G. Schub. in Kew Bull. **26**: 61, fig. 1 (1970); in F.T.E.A., Leguminosae, Pap.: 458 (1971). —Verdcourt in Kirkia **9**: 511 (1974). —Lock, Leg. Afr. Check-list: 245 (1989). Type from Tanzania.

Low spreading or trailing perennial herb 20–40 cm long, branched at the base. Stems slender, appressed pubescent with upwardly directed silvery-white hairs and also uncinulate-puberulous. Leaves 1-foliolate (or rarely 3-foliolate); leaflet 1–4.5 × 0.6–1.7 cm, oblong-elliptic or elliptic, rounded or even slightly emarginate and shortly mucronulate, rounded to subcordate at the base, glabrous or very minutely puberulous above, densely appressed silvery pilose beneath; venation raised beneath; petioles 5–13 mm long; rhachis when present 2.5 mm long; petiolules 1–1.5 mm long; stipules 4–8 × 1–2 mm, obliquely narrowly triangular-lanceolate, persistent. Inflorescences terminal, 4–12 cm long, uncinulate-pubescent; peduncles 2–4 cm long; pedicels 4–11 mm long; primary bracts 3.5–5 × 1.5–2.5 mm, ovate-elliptic, each supporting 2 pedicels, dense and overlapping in young inflorescences, acuminate, ciliate with silvery or ferruginous hairs, soon deciduous; secondary bracts absent. Calyx tube 1 mm long, ± glabrous; lobes 2.5–5 mm long, lanceolate, pilose with silvery or ferruginous hairs, the upper pair joined for less than one-quarter of their length. Standard pale blue, pale pink or white, 3–6 × 2.5–4.5 mm, obovate. Fruits 8–20 mm long, of (1)2–6 articles, each article 3–4 × 3 mm, oblong with a ± straight upper margin and a rounded lower margin, puberulous with short and long hairs, reticulate, separated by rather wide necks, 2–2.5 mm wide. Seeds yellow-brown, 2–2.5 × 1.5–2 × 0.5 mm, oblong-reniform; hilum small, somewhat eccentric.

Zambia. N: Mbala Distr., Kalambo Falls, fl. & fr. 12.iv.1968, *Richards* 23216 (K).
Also in southern Tanzania. Grassland and woodland; 900–1100 m.

14. **Desmodium triflorum** (L.) DC., Prodr. **2**: 334 (1825). —E.G. Baker, Legum. Trop. Africa: 327 (1929). —Schubert in F.C.B. **5**: 187 (1954). —Hepper in F.W.T.A., ed. 2, **1**: 584 (1958). —Laundon in C.F.A. **3**: 221 (1966). —Schubert in F.T.E.A., Leguminosae, Pap.: 459 (1971). —Verdcourt in Kirkia **9**: 512 (1974). —Lock, Leg. Afr. Check-list: 248 (1989). — Pedley in Rev. Fl. Ceylon **10**: 178 (1996). Type from Sri Lanka.
  Hedysarum triflorum L., Sp. Pl.: 749 (1753).

Annual (or ?perennial) herb, often forming dense mats. Stems prostrate, sometimes rooting at the nodes, much branched, 8–19 cm long, covered with spreading bristly yellow-brown hairs. Leaves 3-foliolate; leaflets 4–14(20) × 4–12(15) mm, obovate-oblong, obovate or obcordate, rounded to emarginate at the apex, rounded to cuneate at the base, glabrous above, pilose beneath with scattered white hairs, with a raised reticulation of veins above; petiole 4–11(15) mm long; rhachis 1–5 mm long; petiolules 0.5–1 mm long; stipules 3.5–6 × 1–1.5 mm, obliquely lanceolate, ± glabrous. Flowers 1–3 in leaf-axils (or, fide Schubert, inflorescences short, mostly under 6–8-flowered); pedicels 5–10 mm long, pilose or glabrous; primary bracts 2–3.5 × 1–1.5 mm; secondary bracts mostly absent. Calyx with bristly hairs, or pilose with long white hairs; tube 0.8 mm long; lobes 1–1.5 mm long, narrowly triangular-lanceolate. Standard blue, purple, reddish or pink, up to 4.5 × 2–3 mm, ± round, but long-clawed. Fruit 1.2–1.8 cm long, of 2–5 articles, held eccentrically on the pedicel which is continuous with the straight suture; other margin constricted between the articles; articles 2–3.5 mm long and wide, glabrous or uncinulate-pubescent and with a strong reticulation of raised nerves. Seeds brown, finely speckled with black, 2.1 × 1.5 × 0.8 mm, reniform; hilum small eccentric.

Zambia. N: Kawambwa Distr., Mushota (Mushoto), fl. & fr. 16.iv.1961, *Verboom LK* 238 (PRE). W: Luanshya, fl. & fr. 24.iii.1955, *Fanshawe* 2210 (K). E: Chipata (Fort Jameson) area, fl. & fr.

i.1963, *Verboom* 685 (K; SRGH). **Malawi**. N: Karonga, fl. & fr. 15.vii.1970, *Brummitt* 12131 (K; LISC). S: Blantyre, fr. 11.i.1966, *Agnew* 154 (SRGH).
Widespread in the tropics. Dambos, weed in lawns and other grassy places; 480–1350 m.
Also cultivated in Zimbabwe at Grasslands Research Station, Marondera, 21.ii.1967 *Corby* (1770).

15. **Desmodium setigerum** (E. Mey.) Benth. ex Harv. in F.C. **2**: 229 (1861–2). —Milne-Redhead in Bull. Misc. Inform., Kew **1937**: 417 (1937). —Schubert in F.C.B. **5**: 187, fig. 11E (1954). —Hepper in F.W.T.A., ed. 2, **1**: 585 (1958). —Laundon in C.F.A. **3**: 222 (1966). —Schubert in F.T.E.A., Leguminosae, Pap.: 460, fig. 65/5 (1971). —Drummond in Kirkia **8**: 219 (1972). —Verdcourt in Kirkia **9**: 513 (1974). —Lock, Leg. Afr. Check-list: 247 (1989). TAB. 3.6.1, fig. 7. Type from South Africa.
Nicolsonia setigera E. Mey., Comment. Pl. Afr. Austr. **1**, 1: 124 (1836).
Desmodium hirtum sensu J.G. Baker in F.T.A. **2**: 163 (1871) pro parte and sensu E.G. Baker, Legum. Trop. Africa: 329 (1929) pro parte.

Prostrate or trailing perennial herb 0.5–1 m long. Stems mostly several from a somewhat woody base, sometimes rooting at the nodes, densely covered with whitish, ferruginous or yellowish spreading or more rarely appressed hairs. Leaves 3-foliolate, leaflets 5–35 × 4–30 mm, the laterals usually distinctly smaller, mostly obovate or rounded to elliptic, truncate, rounded or slightly emarginate at the apex, cuneate or rounded at the base, glabrescent or sparsely pilose above, densely appressed pilose beneath; petiole 0.5–3 cm long, densely pilose; rhachis 2.5–7 mm long; petiolules 1–2.5 mm long; stipules 5–10 × 1.5–2(2.5) mm, lanceolate, brown, striate, hairy, persistent. Inflorescences terminal and axillary, 2–7(25) cm long, uncinulate-pubescent and with longer appressed hairs; peduncles 1–3 cm long; pedicels 5–14 mm long, puberulous; primary bracts each supporting 2 pedicels, 2.5–6 × 1.5–2 mm, ovate-lanceolate, hairy; secondary bracts usually absent. Calyx with spreading bristly hairs; tube 1 mm long; lobes 1–1.5 mm long, narrowly triangular. Standard white, yellow, pale mauve, pink or purple turning blue on drying, c. 4 mm long and wide, broadly obovate, emarginate; wings and keel purple at apex. Fruit 5–15 mm long, of 1–6 articles, separating and not dehiscing as in previous species; articles with the upper margin slightly curved and lower margin strongly curved, 2–3 × 1.6–2.2 mm, glabrescent or uncinulate-pubescent, joined by necks 0.5–1.5 mm wide, reticulately veined. Seeds yellow-brown to dark chestnut-brown, 1.8 × 1.1 × 0.5 mm, compressed reniform; the small hilum slightly eccentric.

Zambia. N: Kawambwa, fl. & fr. iv.1960, *Verboom* LK 207 (K; SRGH). W: Solwezi, fl. 9.iv.1960, *E.A. Robinson* 3468 (K). S: Choma Distr., Mapanza, fl. 8.iii.1958, *E.A. Robinson* 2789 (K; SRGH). **Zimbabwe**. E: Nyanga (Inyanga), fl. & fr. 20.i.1931, *Norlindh & Weimarck* 4486 (EA; K; LD; M; PRE; SRGH). **Malawi**. N: Rumphi Distr., above Nchenachena, fl. & fr. 29.iv.1971, *Pawek* 4769 (K). S: Blantyre, *Buchanan* (K). **Mozambique**. Z: Gurué, R. Licungo, fl. & fr. 7.v.1943, *Torre* 5110 (BM; LISC). MS: between Skeleton Pass and Martin's Falls, fl. & fr. 8.iv.1967, *Drummond* 9121 (K; LISC; SRGH).
Widespread in tropical Africa and South Africa (mostly KwaZulu-Natal and Transvaal). Dambo grassland, swamp forest, riverine grassland, roadsides and also in shallow soil on granite outcrops, etc.; 680–1700 m.
West 2359 from Zimbabwe N: is a dubious record.
This or something very similar has been cultivated in Lusaka, (fl. & fr. 30.iii.1994, *Bingham* 10028 (K)).

16. **Desmodium stolzii** Schindl. in Repert. Spec. Nov. Regni Veg. **21**: 9 (1925). —E.G. Baker, Legum. Trop. Africa: 328 (1929). —Schubert in F.C.B. **5**: 191 (1954); in F.T.E.A., Leguminosae, Pap.: 463 (1971). —Verdcourt in Kirkia **9**: 515 (1974). —Lock, Leg. Afr. Check-list: 247 (1989). Syntypes: Malawi, Mt. Mulanje (Mlanje), *Whyte* (B†, syntype; G; K), and Tanzania.

Perennial herb 20–75 cm tall, with several stems from a woody rootstock. Stems angular, uncinulate-pubescent and sparsely hairy with longer hairs. Leaves 3-foliolate; leaflets 10–46 × 4–17 mm, elliptic-oblong or oblong, rounded to cuneate at the base, rounded and mucronulate at the apex, glabrous on upper surface, sparsely appressed pilose beneath together with a very fine puberulence; petiole 12–25 mm long; rhachis 3–8 mm long; petiolules 0.5–2 mm long, hairy; stipules 5–11 × 1–2.5

mm, lanceolate, brown, chaffy, striate, finely puberulous outside, persistent. Inflorescences terminal and occasionally also from the uppermost axil, 8–30 cm long, with similar indumentum to the stem; peduncle 4–26 cm long; pedicels 1–3 cm long, puberulous; primary bracts each subtending 2 pedicels, 4–6 × 1.5–2 mm, ovate-lanceolate, attenuate, striate, puberulous and with longer hairs outside, ciliate, very soon deciduous. Calyx puberulous; tube 1.2 mm long; lobes 1–2 mm long, triangular to lanceolate, the upper pair joined to form a 2-fid lobe. Standard pink or rose-coloured, 5 × 5.5 mm, oblate-obovate, emarginate; keel pale. Fruits 10–23 mm long, of 1–4 articles, each article 4.5–7 × 2–4 mm, semicircular, densely uncinulate-pubescent, reticulate, necks between the articles c. 1.5 mm wide. Seeds (immature) dark reddish-brown, 3.7 × 2 mm, compressed ellipsoid, narrower at one end; hilum minute, central.

Zambia. N: Kasama Distr., Chibutubutu, fl. & fr. 26.ii.1960, *Richards* 12599 (K). W: Mwinilunga Distr., just east and south of Dobeka Bridge, fl. 5.xi.1937, fr. 17.xii.1937, *Milne-Redhead* 3109 and 3109A (BM; K; PRE). **Zimbabwe**. E: Nyamkwarara (i Nyumquarara) Valley, fr. ii.1935, *Gilliland* 1578 (BM). **Malawi**. N: Livingstonia, Kaziwiziwi (Kaziweziwe) River, fl. & fr. 8.i.1959, *E.A. Robinson* 3147 (K). S: Zomba Mt., ix.1891, *Whyte* (BM).
Also in Dem. Rep. Congo and Tanzania. *Brachystegia* woodland, evergreen thickets, boggy grassland on *Cryptosepalum* woodland edges; 700–1400 m.

17. **Desmodium ramosissimum** G. Don, Gen. Syst. **2**: 294 (1832). —Schubert in F.C.B. **5**: 191 (1954). —Hepper in F.W.T.A., ed. 2, **1**: 584 (1958). —White, F.F.N.R.: 150 (1962). — Laundon in C.F.A. **3**: 223 (1966). —Schubert in F.T.E.A., Leguminosae, Pap.: 464, fig. 65/12 (1971). —Verdcourt in Kirkia **9**: 516 (1974). —Lock, Leg. Afr. Check-list: 247 (1989). TAB. 3.6.**1**, fig. 8. Type from S. Tomé.
        Desmodium mauritianum sensu Baker in F.T.A. **2**: 164 (1871). —E.G. Baker, Legum. Trop. Africa: 330 (1929). —Exell, Cat. Vasc. Pl. S. Tomé: 158 (1944) non (Willd.) DC.

Erect slender perennial herb or subshrub 0.3–2 m tall. Stems several from a woody base, with a short pubescence and also longer white hairs, or densely appressed strigose. Leaves 3-foliolate; leaflets 0.7–4.8 × 0.6–2 cm, narrowly obovate, oblong-elliptic or oblong, rounded at the apex, rounded to cuneate at the base, glabrous above, appressed pilose beneath, reticulate on both surfaces; petiole 8–15 mm long; rhachis 2–7 mm long; petiolules 0.5–1 mm long; stipules 4–15 × 0.5–1.5 mm, obliquely lanceolate, pilose, persistent. Inflorescences terminal and axillary, 4.5–20 cm long, with indumentum similar to the stems; peduncle 0–8 cm long; pedicels 2–7 mm long, puberulous; primary bracts each supporting 2(3) pedicels, 4–6 × 1–2.5 mm, ovate-lanceolate, long attenuate, pilose; secondary bracts absent. Calyx puberulous and with some longer hairs; tube 1 mm long; lobes 1.5 mm long, triangular. Standard mauve, red or rose-coloured, 4.2 × 3 mm, obovate. Fruits 7–25 mm long, of 3–5 articles, each article 3.5–5 × 2–3.5 mm, uncinulate-pubescent, reticulately veined, joined by narrow necks 1–1.5 mm wide. Seeds chestnut-brown, 2.5–3 × 1.5–2 × 0.7–1 mm, reniform; hilum small, practically central.

Zambia. N: Luwingu, fl. 24.x.1911, *Fries* 1095 (UPS). **Mozambique**. Z: Alto Molócuè, at km 10 on road to Ribáuè, fl. & fr. 28.xi.1967, *Torre & Correia* 16256 (LISC).
Widespread in tropical Africa, also in Mascarene Islands and Madagascar. *Brachystegia* and mixed deciduous woodland; 570–1420 m.

18. **Desmodium cordifolium** (Harms) Schindl. in Repert. Spec. Nov. Regni Veg. **22**: 257 (1926). —E.G. Baker, Legum. Trop. Africa: 327 (1929). —Schubert in F.C.B. **5**: 197 (1954). — Laundon in C.F.A. **3**: 220 (1966). —Schubert in F.T.E.A., Leguminosae, Pap.: 468 (1971). —Verdcourt in Kirkia **9**: 521 (1974). —Lock, Leg. Afr. Check-list: 245 (1989). Type from Dem. Rep. Congo (Katanga Province).
        Glycine cordifolium Harms in Bot. Jahrb. Syst. **49**: 441 (1913).
        Desmodium homblei De Wild. in Repert. Spec. Nov. Regni Veg. **13**: 114 (1914). Type from Dem. Rep. Congo (Katanga Province).

Perennial herb with a tuberous rootstock. Stems prostrate or ± ascending, 0.4–1 m long, somewhat angular, covered with spreading bristly hairs. Leaves 1-foliolate; leaflet 0.6–6 × 0.6–5.2 cm, ovate-oblong or rounded-oblong to almost round, subacute, broadly rounded or even emarginate at the apex, shallowly to distinctly cordate at the base, with extremely short hairs on upper surface but much longer ones beneath; petiole 2–37 mm long; petiolule 1–2.5 mm long; stipules 7–12 × 1–2 mm, obliquely lanceolate, attenuate, strongly veined, pilose. Inflorescences very lax, with fascicles of flowers 2–5 cm apart, 2–18 cm long, finely puberulous and with long spreading ferruginous hairs; peduncle 1–4 cm long, similarly hairy; pedicels 5–13 mm long, puberulous and with a few long hairs; primary bracts each subtending 2–3 pedicels, 2.8–5 × 0.8–1.3 mm, nervose and sparsely pilose; secondary bracts lanceolate, very small, all persistent. Calyx pubescent with bristly hairs; tube 1 mm long; lobes 2 mm long, narrowly triangular-lanceolate. Standard purple or blue, 5.5 × 6.5 mm, very broadly obovate, emarginate. Fruit 8–14 mm long, of (3)5–6 articles, each article 2–2.3 mm long and wide, round, uncinulate-pubescent or glabrescent, joined by narrow necks 0.5 mm wide. Seeds reddish-brown, 2 × 1.5 mm, ± oblong.

Zambia. N: 48 km south of Kasama on Mpika road, Lukulu R., Chibutubutu, fl. 23.ii.1960, *Richards* 12536 (K; SRGH). **Malawi**. N: Chitipa Distr., 12.4 km east of crossroads towards Karonga, Songa Stream, fl. & fr. 19.iv.1969, *Pawek* 2265 (K). C: Nkhotakota (Kota Kota) road, between Kongwe Hill and Mwera Hill Agric. Station, fl. & fr. 21.ii.1959, *Robson & Steele* 1697 (BM; K; LISC). S: Namwera Escarpment, Jalasi, fl. & fr. 15.iii.1955, *Exell, Mendonça & Wild* 893 (BM; LISC; SRGH).

Also in Central African Republic and Dem. Rep. Congo, Tanzania and Angola. *Brachystegia* woodland, scrubland, etc.; 1000–1520 m.

19. **Desmodium salicifolium** (Poir.) DC., Prodr. **2**: 337 (1825). —E.G. Baker, Legum. Trop. Africa: 330 (1929). —Brenan in Mem. New York Bot. Gard. **8**: 255 (1953). —Schubert in F.C.B. **5**: 198, pl. 15 (1954). —Hepper in F.W.T.A., ed. 2, **1**: 584 (1958). —White, F.F.N.R.: 150 (1962). —Laundon in C.F.A. **3**: 224, t. 20/A–C (1966). —Schubert in F.T.E.A., Leguminosae, Pap.: 469, fig. 65/1 (1971). —Verdcourt in Kirkia **9**: 522 (1974). —Lock, Leg. Afr. Check-list: 247 (1989). TAB. 3.6.**1**, fig. 9; TAB. 3.6.**3**. Type from "l'Inde".
    Hedysarum salicifolium Poir., Encycl. Méth. Bot. **6**: 422 (1805).
    Desmodium paleaceum Guill. & Perr. in Guillemin, Perrottet & Richard, Fl. Seneg. Tent.: 209 (1832). —J.G. Baker in F.T.A. **2**: 166 (1871). Type from Gambia.

An erect somewhat woody subshrub 0.9–3.6 m tall. Stems striate or wrinkled, pilose with slightly spreading or appressed hairs together with some very short pubescence. Leaves 3-foliolate; leaflets 3–17.5 × 1–6.5 cm, oblong-elliptic to ovate-lanceolate or lanceolate, rounded to acute and mucronulate at the apex, rounded or slightly subcordate at the base, glabrous on upper surface except for the midrib, appressed-hairy on the main venation beneath and with sparse to dense shorter hairs between; petiole 0.7–6 cm long, sometimes rather thick and channelled; rhachis 0.6–2 cm long; petiolules 2.5–5 mm long; stipules 7–18 × 2.5–5 mm, ovate-lanceolate, striate, ciliate, persistent. Inflorescences terminal, mostly dense, 9–30 cm long, finely puberulous and with uncinulate pubescence; peduncle 10–25 mm long; pedicels 2.5–4.5 mm long, puberulous; primary bracts each subtending 2 pedicels, 3–9 × 0.7–2.5 mm, lanceolate, striate, pubescent and ciliate, deciduous; secondary bracts very small. Calyx puberulous and with some longer hairs; tube 1 mm long; lobes 1.5–2 mm long, triangular, the upper pair joined to form an entire ovate lip. Standard pinkish-purple becoming blue, or white, yellow or greenish-yellow, 6.5 × 5 mm, obovate; wings pink, purple or yellow-green, mauve at apex, keel pink or yellowish-green. Fruits 3–4.5 cm long, of 3–7 articles, each article 4.5–8.5 × 2.5–3.5 mm, oblong, distinctly thickened particularly at the margins, finely puberulous and with longer pubescence, with raised reticulation of veins when mature, only slightly constricted, the necks being only slightly narrower than the articles. Seeds chestnut-brown, yellow-brown or purple-brown, 2.3–4 × 1.5–2 × 0.7 mm, ellipsoid-oblong; the hilum ± central.

Var. **salicifolium** —Drummond in Kirkia **8**: 219 (1972). —Verdcourt in Kirkia **9**: 523 (1974). —Gonçalves in Garcia de Orta **5**: 78 (1982).

Inflorescence lax.

Botswana. N: Okavango River, Sepupa (Sepopa), fl. & fr. 17.iii.1965, *Wild & Drummond* 7110 (K; LISC; SRGH). **Zambia**. B: near Senanga, fl. & fr. vii.1952, *Codd* 7225 (BM; COI; K; PRE; SRGH). N: Kawimbe, fl. & fr. 25.iv.1959, *McCallum-Webster* 936 (K). W: Solwezi Distr., near River

M.BOUTIQUE DEL.

Meheba, fl. & fr. 23.vii.1930, *Milne-Redhead* 761 (K). C: Kafue River Gorge, fl. & fr. 6.x.1957, *Angus* 1752 (K; SRGH). E: 6.4 km south of Lundazi Boma, fl. & fr. 27.iv.1952, *White* 2488 (BM; FHO; K). S: Victoria Falls, fl. 25.iv.1932, *St. Clair-Thompson* 1347 (K). **Zimbabwe**. N: shores of Mazowe (Mazoe) Dam, fl. 12.iii.1958, *Corby* 853 (SRGH). W: Victoria Falls, fl. & fr. vii.1906, *Kolbe* 3152 (K). C: Wedza Mts., fl. 27.iv.1961, *Corby* 999 (SRGH). E: Mutare Distr., Vumba Mt., Nyachohwa (Nyachowa) Falls, fl. 19.iii.1950, *Chase* 1994 (BM; COI; K; LISC; SRGH). S: Bikita, upper regions of Dafana River, fr. 7.v.1969, *Biegel* 3044 (K; LISC; SRGH). **Malawi**. N: Nkhata Bay Distr., Limphasa (Limpassa) Dambo, fl. 4.vii.1952, *G. Jackson* 911 (K). C: Nkhotakota Distr., Kaombe River fr. 7.xi.1943, *Benson* 627 (K; PRE). S: Blantyre Distr., Maone, 2 km north of Limbe, fl. & fr. 10.iii.1970, *Brummitt* 9000 (K; LISC). **Mozambique**. N: Marrupa, Régulo Tlena, fl. & fr. 13.vi.1948, *Pedro & Pedrógão* 4309 (EA; LMA). Z: Mocuba, R. Licungo, fl. & fr. 19.x.1942, *Torre* 4619 (BM; LISC). T: Zóbuè, fl. & fr. 17.vi.1941, *Torre* 2865 (LISC). M: Marracuene (Vila Luiza), entre a machamba da Administração e o batelão de Marracuene, fl. & fr. 2.x.1957, *Barbosa & Lemos* in *Barbosa* 7910 (COI; K; LISC; LMA).

Widespread in tropical Africa, Mascarene Islands and Madagascar. *Brachystegia* woodland, mixed deciduous woodland and thicket, particularly by rivers and lakes, and in floodplain grassland, papyrus swamps, sometimes on mud by open water, also in swamp forest and evergreen forest; 480–1680 m.

20. **Desmodium tanganyikense** Baker in Bull. Misc. Inform., Kew **1895**: 65 (1895). —E.G. Baker, Legum. Trop. Africa: 328 (1929). —Schubert in F.C.B. **5**: 201 (1954). —Laundon in C.F.A. **3**: 224, t. 20/A3–C3 (1966). —Schubert in F.T.E.A., Leguminosae, Pap.: 472, fig. 66 (1971). —Drummond in Kirkia **8**: 219 (1972). —Verdcourt in Kirkia **9**: 524 (1974). —Gonçalves in Garcia de Orta **5**: 78 (1982). —Lock, Leg. Afr. Check-list: 247 (1989). TAB. 3.6.4. Type: Zambia, Fwambo, *Carson* 27 (K, holotype).

Erect or suberect subshrubby perennial herb 0.6–2 m tall. Stems angular, densely shortly bristly hairy on the angles and uncinulate-pubescent between them. Leaves 3-foliolate; leaflets 2–13 × 0.6–7 cm, elliptic or elliptic-oblong to lanceolate, acute to rounded at the apex, rounded to truncate or slightly subcordate at the base, margins entire or somewhat undulate, with sparse to dense longer hairs on the nerves and a mixture of very short and rather longer hairs between on upper surface, densely pilose beneath, the hairs on the nerves being mostly more ferruginous; venation strongly raised and reticulate beneath; petiole absent or not exceeding 5 mm; rhachis 1–4 cm long; petiolules 1 mm long; stipules 5–10 × 1.2 mm, lanceolate, densely pubescent outside, eventually deciduous. Inflorescences terminal, branched, 25–65 cm long, with same indumentum as the stem; peduncle 20–46 cm long; pedicels 2.5–4 mm long, uncinulate-pubescent; primary bracts 3 × 0.8 mm, lanceolate, soon deciduous, each subtending 3 or more pedicels; secondary bracts about half the size, 1–2 × 0.5 mm, deciduous. Calyx with longer bristly hairs and short uncinulate pubescence; tube 1.5 mm long; lobes 2–3 mm long, ovate-triangular to lanceolate, the upper two joined to form a very rounded lobe. Standard pink or rose-pink, rarely white or purplish, 8–9 × 6–7 mm, almost round, wings brighter rose-pink. Fruits 9–22 mm long, of 1–3 articles, each article 7–8 × 6 mm, broadly elliptic, densely uncinulate-pubescent, with rather obscurely raised venation, necks between articles 1.5 mm long. Seeds orange-brown, strongly compressed so that edge is almost keeled, 4.2–5 × 2.5–3 × 1.3 mm, reniform; hilum small, almost central, the seed margin raised to one side of it.

Zambia. N: c. 90 km south of Mpika, fl. 2.iii.1962, *E.A. Robinson* 4963 (K; SRGH). W: Solwezi, fl. 23.ii.1964, *Fanshawe* 8330 (K). C: Munshiwemba, fl. ii.1941, *Stohr* 519 (BOL; K; PRE). E: Chipata Distr., fr. 24.vi.1962, *Verboom* 649 (K). **Zimbabwe**. N: Mazowe Distr., Mtorashangu (Mtoroshanga)–Concession road, fl. 2.iii.1965, *Corby* 1251 (K; SRGH). C: Harare Distr., near Mount Hampden (Hampton), fr. 9.v.1934, *Gilliland* 120 (K; SRGH). **Malawi**. C: Dedza, Chongoni, fl. 11.iii.1961, *Chapman* 1167 (K; SRGH). S: Ntcheu Distr., Kirk Range, Chipusiri, fr. 17.iii.1955, *Exell, Mendonça & Wild* 972 (BM; LISC; SRGH). **Mozambique**. N: Lichinga (Vila

---

Tab. 3.6.**3**. DESMODIUM SALICIFOLIUM. 1, flowering and fruiting branch (× ¹/₂); 2, primary and secondary bracts and 2 buds (× 5), 1 & 2 from *Quarré* 222; 3, open flower (× 4), from *Quarré* 1166; 4, standard, internal face (× 4); 5, wing (× 4); 6, keel, spread out (× 4); 7, androecium and ovary, longitudinal section (× 5), 4–7 from *Quarré* 222; 8, fruit (× 2); 9, seed (× 4), 8 & 9 from *de Graer*. Drawn by M. Boutique. From Fl. Congo Belge. Reproduced with permission of Jardin Botanique National de Belgique.

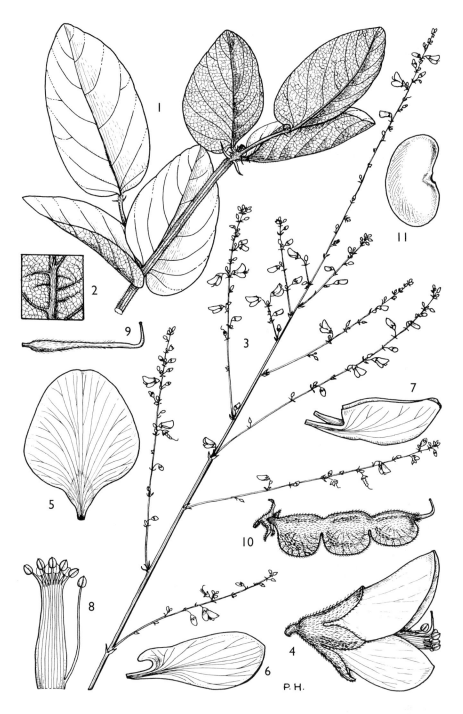

Tab. 3.6.4. DESMODIUM TANGANYIKENSE. 1, leaves (× ²⁄₃); 2, detail from underside of
leaflet (× 2); 3, inflorescence (× ²⁄₃); 4, flower (× 4); 5, standard (× 4); 6, wing (× 4); 7, keel
(× 4); 8, androecium (× 4); 9, gynoecium (× 4), 1–9 from *Richards* 12688; 10, fruit (× 2);
11, seed (× 4), 10 & 11 from *Fanshawe* 8544. Drawn by Pat Halliday. From F.T.E.A.

Cabral), Serra de Massangulo, fl. 25.ii.1964, *Torre & Paiva* 10755 (LISC). T: Macanga Distr., encosta oriental do Monte Furancungo, entre a sua base e um curso de água, fl. & fr. 17.iii.1966, *Pereira, Sarmento & Marques* 1836 (LMU).
Also in Dem. Rep. Congo, Tanzania and Angola. *Brachystegia* woodland (including mixed *Uapaca–Parinari–Julbernardia*, etc.) and grassland on sandy soil; 1140–1800 m.

21. **Desmodium adscendens** (Sw.) DC., Prodr. **2**: 332 (1825). —J.G. Baker in F.T.A. **2**: 162 (1871). —E.G. Baker, Legum. Trop. Africa: 330 (1929). —Schubert in F.C.B. **5**: 189 (1954). —Hepper in F.W.T.A., ed. 2, **1**: 585 (1958). —Laundon in C.F.A. **3**: 222, t. 20A. (1966). — Schubert in F.T.E.A., Leguminosae, Pap.: 461 (1971). —Verdcourt in Kirkia **9**: 514 (1974). —Lock, Leg. Afr. Check-list: 245 (1989). —Pedley in Rev. Fl. Ceylon **10**: 180 (1996). Type from Jamaica.
Hedysarum adscendens Sw., Nov. Gen. Sp. Pl. Prodr.: 106 (1788).

Straggling perennial herb or undershrub, or often prostrate and rooting at the nodes, 0.2–1 m long or tall. Stems densely covered with spreading white hairs when young, eventually glabrescent, red-brown and striate. Leaves 3-foliolate; leaflets 1.5–5.6 × 1.3–3.5 cm, elliptic to round or obovate to subrhombic, broadly rounded at both ends or somewhat acute, glabrous or with fine very short hairs above, densely appressed pilose beneath; petiole 8–30 mm long; rhachis 2–6 mm long; petiolules 1.5–2 mm long; stipules 7–10 × 1.5–2.5 mm, lanceolate, striate, hairy, persistent. Inflorescences mostly terminal, lax 2–11 cm long, finely uncinulate-puberulous and with longer hairs; peduncles 4–10 cm long; pedicels 7–16 mm long, uncinulate-pubescent; primary bracts each subtending 2 pedicels, 4–11 × 1.5–2.5 mm, ovate-lanceolate, long attenuate, pubescent, soon deciduous; secondary bracts mostly absent. Calyx finely puberulous and with sparse long hairs; tube 1.2 mm long; lobes 1–2 mm long, triangular. Standard whitish, greenish-blue, pink or mauve, 3.5–4 mm × 2–2.5 cm, obovate; wings bright pinkish-mauve; keel greenish-white. Fruit 2–2.5 cm long, of 2–5 articles, each article 5–9.5 × 2.6–3.5 mm, semi-elliptic with straight upper margin and strongly rounded lower margin, densely uncinulate-pubescent, joined by necks 1 mm wide, reticulately veined. Seeds chestnut-brown, 5 × 2.5 mm, elongate-reniform, narrowed towards one end.

Var. **robustum** B.G. Schub. in Bull. Jard. Bot. État **22**: 290 (1952); in F.C.B. **5**: 190 (1954). — Hepper in F.W.T.A., ed. 2, **1**: 585 (1958). —Schubert in F.T.E.A., Leguminosae, Pap.: 462 (1971). —Verdcourt in Kirkia **9**: 515 (1974). TAB. 3.6.**1**, fig. 10. Type from Dem. Rep. Congo.

Leaflets mostly larger than in typical variety, practically glabrous and usually reticulate above.

Zambia. N: Kasama Distr., Nkole Mfumu Plantation, fl. 28.i.1981, *Lawton* 2309 (K). W: Mwinilunga Distr., at junction of Matonchi and Luao rivers, fl. & fr. 18.xii.1938, *Milne-Redhead* 3720 (BM; K; PRE). **Zimbabwe**. E: Chimanimani Distr., 8 km from Haroni–Rusitu confluence, upstream of Rusitu (Lusitu) River, fl. 22.xi.1967, *Ngoni* 28 (K; SRGH). **Malawi**. N: Livingstonia, Kaziwiziwi (Kaziweziwe) River valley, fl. 9.i.1959, *Richards* 10556 (SRGH). S: Mt. Mulanje (Mlanje), Swazi, fr. 29.iii.1949, *Faulkner* Kew 390 (K). **Mozambique**. N: Serra de Ribáuè, Mepáluè, fr. 23.i.1964, *Torre & Paiva* 10179 (LISC). Z: Alto Molócuè towards Mocuba at km 15, Mt. Rurupi, fl. & fr. 2.xii.1967, *Torre & Correia* 16338 (LISC). MS: Dondo, fl. & fr. 31.xii.1943, *Torre* 6321 (LISC).
Widespread throughout tropical Africa. *Brachystegia* and other deciduous woodlands, also in riverine and coastal forest; 40–1350 m.
Material of two gatherings from Malawi, *Verboom* 671 (Nkhata Bay, 28.viii.1962 (K)) and *Pawek* 9641 (Rumphi Distr., Uzumara Forest, 1980 m fl. 25.v.1975), has some hairs on the leaflet upper surface and approaches var. *adscendens*.
The typical variety is widespread in the tropics of both hemispheres.

22. **Desmodium dregeanum** Benth. in Miquel, Pl. Jungh.: 222 in adnot. (1852). —Schubert in F.T.E.A., Leguminosae, Pap.: 476, fig. 65/3 (1971). —Verdcourt in Kirkia **9**: 528 (1974). —Lock, Leg. Afr. Check-list: 245 (1989). TAB. 3.6.**1**, fig. 11. Based on *Nicolsonia caffra* E. Mey.
Nicolsonia caffra E. Mey., Comment. Pl. Afr. Austr. **1**, 1: 123 (1836). Type from South Africa.
Desmodium caffrum (E. Mey.) Druce in Bot. Soc. Exch. Club Brit. Isles **4**: 619 (1917). —

Schindler in Repert. Spec. Nov. Regni Veg. **23**: 360 (1927). —E.G. Baker, Legum. Trop. Africa: 331 (1929). —Laundon in C.F.A. **3**: 225 (1966) non Eckl. & Zeyh.
  Meibomia caffa (E. Mey.) Kuntze, Revis. Gen. Pl. **1**: 197 (1891).
  Desmodium barbatum sensu Taub. in Engler, Pflanzenw. Ost-Afrikas **C**: 216 (1895) non (L.) Benth.
  Desmodium caffrum var. *schlechteri* Schindl. in Repert. Spec. Nov. Regni Veg. **23**: 360 (1927). Syntypes: Mozambique, Maputo (Lourenço Marques), *Schlechter* 11705 (K); Delagoa Bay, *Schlechter* 11992 (COI; K) and *Junod* 271 (Z), and also from Tanzania.

A subshrubby prostrate, decumbent, or rarely erect perennial herb, 0.2–1 m long or tall. Stems at first appressed-silky-pilose with upwardly directed hairs, later purplish-brown and glabrescent. Leaves 3-foliolate; leaflets 0.5–3.5 × 0.3–1.4 cm, obovate-elliptic or elliptic, rounded and usually slightly mucronulate at the apex, cuneate to rounded at the base, glabrous to slightly appressed pilose on upper surface, densely appressed pilose beneath; venation reticulate ± raised on both surfaces; petiole 3–15 mm long, hairy; rhachis 0–1.5 mm long; petiolules 1–2 mm long, hairy; stipules chestnut-brown, 5–10 mm long, lanceolate, hairy at the base and ciliate but otherwise glabrous, striate. Inflorescences terminal and often also axillary, 1.5–9 cm long, dense; peduncle 0–1.2 cm long, together with the inflorescence hairy like the stems; pedicels 3.5 mm long; primary bracts each subtending 2–5 pedicels, 3–5 × 1.5 mm, narrowly ovate, acuminate, striate, ciliate, comose at apex of inflorescence; secondary bracts 2.5 × 0.5 mm, lanceolate. Calyx with both white and ferruginous hairs; tube 1.5 mm long; lobes 2 mm long, lanceolate, the upper pair joined for about one-third of their length. Standard blue, purple, red or yellow, 5–6 × 4–5 mm, obovate, emarginate. Fruits 0.5–1.2 cm long, of 3–5 articles, each article c. 2.5 × 2 mm, oblong, one margin straight, the other curved, pubescent, reticulate, not strongly constricted, the necks 1.5 mm wide. Seeds yellow-brown or chestnut-brown, 1.8–2 × 1.2 × 0.6 mm, ellipsoid-reniform; hilum minute ± central.

Zambia. B: Kalabo, fl. 13.x.1963, *Fanshawe* 8056 (K; LISC). N: Mbala Distr., Lunzua Hydro Electric Power Station, fl. & fr. 19.iii.1966, *Richards* 21357 (K). **Mozambique**. MS: Cheringoma Coastal area, Zuni Sector, fl. v.1973, *Tinley* 2869 (K; LISC; SRGH). GI: R. Lumane, near the R. Limpopo, fl. & fr. 2.ii.1948, *Torre* 7280 (LISC). M: Matutuíne (Bela Vista), between Zitundo and Ponta do Ouro, fl. 10.xii.1961, *Lemos & Balsinhas* in Lemos 280 (K; LISC; PRE; SRGH).
  Also in East and South Africa and Angola. Grassland, woodland edges, dambo margins, beaches and marshes; 0–1500 m.

23. **Desmodium fulvescens** B.G. Schub. in Bull. Jard. Bot. État **22**: 296 (1952); in F.C.B. **5**: 203 (1954). —Lock, Leg. Afr. Check-list: 246 (1989). Type from Dem. Rep. Congo (Katanga Province).

Strictly erect herb up to c. 90 cm tall with angular appressed-hairy or ± woolly stems. Leaves 3-foliolate; leaflets 5–7.7 × 1.8–2.8 cm, oblong-elliptic, ± rounded at base and apex or subcuneate at base, glabrous or minutely puberulous on upper surface, densely pilose beneath; petiole 4–10 mm long; stipules 12–14 × 3.5 mm, ovate-attenuate, pubescent. Inflorescences terminal and axillary, 28 cm long; rhachis puberulous and with longer hairs (fide Schubert) but densely brown-hairy in Zambian specimen; primary bracts 5 × 3 mm, ovate, acute, densely appressed pubescent, each supporting 1–2 pedicels; secondary bracts generally not present; pedicels up to 7 mm long, densely spreading pubescent. Calyx tube 1.5 mm long, minutely puberulous; lobes ± equal, 4–6 mm long, lanceolate, pilose or long brown-hairy in Zambian material. Corolla dark crimson; petals 4–6.5 mm long. Fruits with one margin slightly incised, the other deeply so, of 1–3 articles, each article 4.5 × 3.5 mm, almost round, reticulate, uncinulate-puberulous and pubescent; neck between articles c. 2 mm wide. Seeds 2 × 1.7 mm, subrectangular.

Zambia. W: Chingola, fl. & fr. 9.vii.1957, *Fanshawe* 3341 (K).
  Also in Dem. Rep. Congo. Moist dambos; 1200 m.
  Although collected in 1957 this was not returned from loan until 1978 and not seen by me when I prepared the account for Kirkia – Schubert had annotated it in 1969 as the only collection known beside the type. Only one further sheet has been seen from Dem. Rep. Congo.

24. **Desmodium barbatum** (L.) Benth. in Miquel, Pl. Jungh.: 224 (1852). —Schubert in
F.T.E.A., Leguminosae, Pap.: 477 (1971). —Verdcourt in Kirkia **9**: 530 (1974). —Lock, Leg.
Afr. Check-list: 245 (1989). Type from Jamaica?
    Hedysarum barbatum L., Syst. Nat., ed. 10, **2**: 1170 (1759).

Erect, spreading or procumbent woody herb or subshrub, 0.07–0.9 m long or tall.
Stems with appressed upwardly directed or spreading hairs, later glabrescent.
Leaves 1–3-foliolate; leaflets 1–6.5 × 0.6–3.5 cm, elliptic, elliptic-oblong or obovate,
rarely ovate-lanceolate, obtuse to slightly emarginate and sometimes shortly
mucronulate at the apex, rounded to cuneate or even slightly emarginate at the
base, glabrous to sparsely pilose above, sparsely to very densely pilose beneath;
petiole 5–20 mm long; rhachis 2.5–5 mm long; petiolules 1–2 mm long; stipules
3.5–4.5 × 1–2 mm, ovate-lanceolate, striate, hairy outside and ciliate, persistent.
Inflorescences dense, terminal and axillary, 1–8 cm long, densely ferruginous-
pilose; peduncles 0–25 mm long; pedicels 5–10 mm long, uncinulate-ferruginous-
pubescent; primary bracts each supporting 2 pedicels, 2.5–6 × 1.5–3.5 mm, ovate,
acuminate, pilose and ciliate, comose in young inflorescences; secondary bracts
absent. Flowers deflexed at the ends of the pedicels. Calyx mostly densely pilose
with white and ferruginous hairs; tube 1–1.5 mm long; lobes 3–5 mm long, linear-
lanceolate. Standard pink, blue or mauve, sometimes with 2 purple spots inside,
5–7 × 3–5.5 mm, narrowly obtriangular-obovate; wings and keel more deeply
coloured, or all petals white tinged pink or blue. Fruit 6–15 mm long, of 3–6
articles, each article 2–3 × 1.5–2.5 mm, oblong, with upper margin straight and
lower curved, uncinulate-puberulous or pubescent and often with longer hairs,
separated by necks 1.5 mm wide. Seeds yellow-brown, 1.7–2 × 1.5 × 0.7 mm,
squarish-reniform or ellipsoid-reniform; hilum minute, ± central.

1. Leaves all 3-foliolate; inflorescences terminating the branches, of dense capitate racemes
    ·2
– Leaves 1-foliolate, or both 1- and 3-foliolate on the same plant · · · · · · · · · · · · · · · · ·3
2. Plants erect or procumbent, often with slender stems and rather long internodes; leaflets
slenderly elliptic to slightly obovate (introduced) · · · · · · · · · · · · · · · · · · var. *barbatum**
– Plants procumbent and forming mats, with much branched stems and short internodes;
leaflets usually all obovate · i) var. *procumbens*
3. Plants rarely with any 3-foliolate leaves; leaflets thick, usually densely silvery-silky on the
lower surface; inflorescences mostly spicate, sometimes the spikes occurring in panicles
    · ii) var. *argyreum*
– Plants often with some 3-foliolate leaves; leaflets thin, not usually densely silvery-silky
beneath; inflorescences capitate or spicate, usually solitary and terminal or axillary along a
whole branch · iii) var. *dimorphum*

i) Var. **procumbens** B.G. Schub. in Bull. Jard. Bot. État **22**: 297 (1952); in F.C.B. **5**: 204 (1954);
in F.T.E.A., Leguminosae, Pap.: 478, fig. 65/6 (1971). —Verdcourt in Kirkia **9**: 530 (1974).
TAB. 3.6.1, fig. 12. Type from Dem. Rep. Congo.

Plant procumbent, forming a carpet. Stems with spreading hairs. Leaves 3-
foliolate; leaflets small, 7–20 × 6–17 mm, obovate, distinctly pilose but not very
densely hairy nor silvery silky. Inflorescences short and dense.

   Zambia. N: Kaputa Distr., near Muzombwe, fl. & fr. 15.iv.1961, *Phipps & Vesey-FitzGerald* 3188
(K; SRGH). W: Ndola, fl. iii.1954, *Fanshawe* 981 (K).
   Also in Dem. Rep. Congo, Rwanda, Uganda and Tanzania. Grassland in sandy soil by lake
and marsh edges; 920–1250 m.

ii) Var. **argyreum** (Welw. ex Baker) B.G. Schub. in Bull. Jard. Bot. État **22**: 298 (1952); in F.C.B.
5: 205 (1954); in F.T.E.A., Leguminosae, Pap.: 479 (1971). —Drummond in Kirkia **8**: 219
(1972). —Verdcourt in Kirkia **9**: 531 (1974). Type from Angola.
   Desmodium dimorphum var. *argyreum* Welw. ex Baker in F.T.A. **2**: 161 (1871).

———————————

* See note under var. *dimorphum*.

Plant very similar to var. *dimorphum* but stems and undersurfaces of leaflets very densely silvery-silky.

Zambia. N: Mbala Distr., Sunzu Mt., Saisi, fl. 20.iv.1961, *Richards* 15070 (K; SRGH) (intermediate with var. *dimorphum*). W: Ndola Distr., Misaka, fr. 8.v.1969, *Mutimushi* 3095 (K). C: between Lusaka and Kapiri Mposhi, fl. 4.iv.1961, *Richards* 14918 (K; SRGH). S: Livingstone, fl. & fr. vi.1909, *F.A. Rogers* 7152 (SRGH). **Zimbabwe**. N: Trelawney Tobacco Research Station, fl. 4.v.1943, *Jack* 187* (K; PRE; SRGH). C: Harare Distr., Beatrice, fl. 8.iv.1961, *Drewe* 81 (SRGH). **Malawi**. C: Kasungu Distr., Lisasadzi Experimental Station, fl. & fr. 9.iv.1955, *G. Jackson* 1624 (K; SRGH).

From tropical Africa and Madagascar. *Brachystegia* woodland, grassland, rocky ground, also recorded as a weed; 900–1800 m.

Laundon did not recognize var. *argyreum* and it is true that it merges completely with var. *dimorphum*, but extremes are distinctive and easily recognizable. The type of var. *argyreum* is, however, not so densely silvery-silky as some of the sheets cited above.

iii) Var. **dimorphum** (Welw. ex Baker) B.G. Schub. in Bull. Jard. Bot. État **22**: 298 (1952); in F.C.B. **5**: 205 (1954). —Hepper in F.W.T.A., ed. 2, **1**: 564 (1958). —Schubert in F.T.E.A., Leguminosae, Pap.: 478 (1971). —Drummond in Kirkia **8**: 219 (1972). —Verdcourt in Kirkia **9**: 531 (1974). —Gonçalves in Garcia de Orta **5**: 77 (1982). Type from Angola (lectotype, selected by Schubert).

　　Desmodium dimorphum Welw. ex Baker in F.T.A. **2**: 161 (1871). —E.G. Baker, Legum. Trop. Africa: 332 (1929). —Brenan in Mem. New York Bot. Gard. **8**: 256 (1953).

　　Desmodium barbatum (L.) Benth. subsp. *dimorphum* (Welw. ex Baker) Laundon in C.F.A. **3**: 225, t. 20/A1 (1966).

Plant erect or spreading. Stems with appressed hairs. Leaves 1–3-foliolate, often predominantly 1-foliolate; leaflets often densely hairy but not very densely silvery-silky.

Botswana. N: near Pandamatenga, fr. 28.iii.1961, *Richards* 14893 (K). **Zambia**. B: Sesheke, fl. i.1925, *Borle* (PRE). N: Mpika Distr., 148 km south of Mpika, fl. 2.iii.1962, *E.A. Robinson* 4973 (K; SRGH). W: Mwinilunga Distr., just north of Kanyamwana, fl. 8.ii.1938, *Milne-Redhead* 4504 (BM; K; PRE). C: 28.8 km north of Kabwe (Broken Hill), fl. 13.iv.1961, *Angus* 2852 (FHO; K; SRGH). E: 4 km west of Kachalola on Great East Road, fl. 17.iii.1959, *Robson* 1743 (BM; K; LISC; PRE; SRGH). S: Namwala Distr., Kafue National Park, Ngoma, fl. & fr. 20.iii.1963, *B.L. Mitchell* 19/24 (LISC; PRE; SRGH). **Zimbabwe**. N: Gokwe, fl. 5.iii.1962, *Bingham* 161 (K; SRGH). W: Hwange National Park, Nkwasha (Ngwashla) Road, fl. & fr. 17.ii.1956, *Wild* 4757 (K; LISC; PRE; SRGH). C: SE side of Gweru (Gwelo) Kopje, fl. 5.iii.1967, *Biegel* 1985 (SRGH). E: Nyanga (Inyanga), fl. & fr. 6.ii.1931, *Norlindh & Weimarck* 4892a (EA; K; LD; SRGH). S: Mberengwa Distr., Mnene, fl. & fr. 26.ii.1931, *Norlindh & Weimarck* 5188 (EA; K; LD; SRGH). **Malawi**. N: Nkhata Bay Distr., border of Chombe 1 Estate, near Mweza Village, fl. 13.iv.1960, *Adlard* 339 (K; PRE; SRGH). C: Lilongwe, Chankhandwe Dambo, fl. & fr. 11.iv.1956, *Banda* 242 (BM; LISC; SRGH). S: Lower Kasupe, fl. 13.iii.1955, *Exell, Mendonça & Wild* 821 (BM; LISC; SRGH). **Mozambique**. N: Nampula, fl. 3.iv.1937, *Torre* 1311 (COI; LISC). Z: Mocuba Distr., Namagoa Estate, fl. 3.iv.1949, *Faulkner Kew* 424 (COI; K; PRE; SRGH). T: Macanga Distr., encosta oriental do Monte Furancungo, entre a sua base e um curso de água, fl. 17.iii.1966, *Pereira, Sarmento & Marques* 1821 (LMU). MS: Báruè Distr., Catandica (Vila Gouveia), encosta da Serra de Chôa, fl. & fr. 30.iii.1966, *Torre & Correia* 15527 (LISC). GI: between Mandlakazi (Manjacaze) and Chongoéne, Missão de S. Benedito dos Muchopes, Mangunze, fl. & fr. 2.iv.1959, *Barbosa & Lemos* in *Barbosa* 8468 (COI; K; LISC; LMA; PRE).

Also in Nigeria, Cameroon, Dem. Rep. Congo, Burundi, Tanzania, Angola, South Africa and Madagascar. Coarse grassland, bushland, *Brachystegia, Baikaea, Marquesia* and other woodland, sandy roadsides, old cultivations, etc.; 15–1890 m.

Richards 11355 (Mbala Distr., road to Itembwe Gorge, fl. & fr. 24.iv.1959 (K)) and some others have the habit of var. *procumbens*. Schubert has annotated *Richards* 8999 (Kaputa Distr., Nsama, fl. & fr. 3.iv.1957) as *D. barbatum* var. *barbatum*, and although all but the apical leaves are 3-foliolate it is identical in facies to the hundreds of other *D. barbatum* collected in northern Zambia, and I am considering it a form of *D. dimorphum*. *Richards* 11237 (Mbala Distr., Saisi R., 15.iv.1959), a curious prostrate plant with small leaves, is perhaps a teratological form as suggested by Schubert on the label.

I sympathise with Laundon's use of the term subspecies for the taxon *dimorphum* but since the only character separating the American and African races is said to be a tendency to 1- or 3-foliolate leaves and bearing in mind the presence of a distinct 3-foliolate variant in Africa, I have retained Schubert's classification.

---

* 189 on PRE sheet, no number on Kew sheet.

67. PSEUDARTHRIA Wight & Arn.

Pseudarthria Wight & Arn., Prodr. Fl. Ind. Orient. 1: 209 (1834). —Schindler in Repert. Spec. Nov. Regni Veg. Beih. 2: (1914). —Verdcourt in Kew Bull. 24: 64 (1970); in Kirkia 9: 534 (1974).

Erect perennial herbs or subshrubs. Leaves pinnately 3-foliolate (abnormally subpalmately 5-foliolate); stipules free, lanceolate, striate; stipels present. Inflorescences terminal or axillary, falsely racemose or paniculate, the flowers paired or fasciculate on the rhachis; bracts narrow; bracteoles absent or minute and soon deciduous. Calyx 5-lobed; lobes subequal, the upper pair almost entirely joined to form a lip. Corolla small, purple or white, glabrous; standard rounded or obovate, narrowed into a claw; wings free from the keel, auriculate. Vexillary stamen entirely free in fully opened flowers; anthers uniform. Ovary sessile or stipitate, many-ovuled; style filiform, recurved at the apex, glabrous; stigma terminal, capitate. Fruit narrowly linear-oblong, much flattened, the sutures often sinuate between the seeds, not articulate but splitting into 2 thin reticulate valves. Seeds subreniform, compressed, almost without an aril; funicle elongated.

A small Old World genus of about 4–6 species.

Pseudarthria hookeri Wight & Arn., Prodr. Fl. Ind. Orient. 1: 209 (1834). —J.G. Baker in F.T.A. 2: 168 (1871). —E.G. Baker, Legum. Trop. Africa: 339 (1929). —M.A. Exell in Bol. Soc. Brot., sér. 2, 12: 10 (1937). —Robyns, Fl. Sperm. Parc Nat. Alb. 1: 313 (1948). —Brenan, Check-list For. Trees Shrubs Tang. Terr.: 436 (1949). —Léonard in F.C.B. 5: 235, fig. 15A–B (1954). —Hepper in F.W.T.A., ed. 2, 1: 586 (1958). —Torre in C.F.A. 3: 232 (1966). —Verdcourt in Kew Bull. 24: 65 (1970); in F.T.E.A., Leguminosae, Pap.: 484, fig. 69/1–9 (1971). —Drummond in Kirkia 8: 225 (1972). —Verdcourt in Kirkia 9: 534 (1974). —Lock, Leg. Afr. Check-list: 250 (1989). Type from Mauritius (cultivated material originally from Zanzibar).

Erect woody herb or subshrub 0.3–3(4) m tall, from a thick rhizome. Stems strongly ribbed, pubescent to velvety. Leaflets 2.5–16.5 × 1.3–10 cm, elliptic or obovate-elliptic, acute to rounded and mucronulate at the apex, cuneate or rounded at the base, subentire or more usually broadly crenulate, somewhat rough on the upper surface, pubescent to velvety beneath; petiole 1–7 cm long, channelled; rhachis 2–30 mm long; petiolules 2–10 mm long; stipules 5–12 × (1)2–5 mm, eventually falling. Inflorescences usually much branched and very lax with numerous flowers; peduncle 0–10 cm long; rhachis 6–35 cm long; pedicels 3–8 mm long, at first erect, later reflexed; primary bracts 3–10 × 1–2 mm, ovate-lanceolate, acuminate, persistent; secondary bracts 2–3 mm long, linear. Corolla reddish-purple, blue, rose or white; standard 6–8 × 4–6.5 mm, rounded. Fruit straight, 12–38 × 3–4(5) mm, 3–12-seeded, puberulous or subtomentellous. Seeds chestnut-brown, compressed, smooth, c. 2 × 1.2 × 0.8 mm, reniform or oblong-ovoid.

Var. **hookeri** —Verdcourt in Kirkia 9: 535 (1974). —Gonçalves in Garcia de Orta, Sér. Bot. 5: 102 (1982). TAB. 3.6.5.

Stems and undersurfaces of leaves pubescent to densely velvety. Inflorescence somewhat congested to very lax with the pedicels mostly easily visible; flowers 3–4-fasciculate, the middle ones often reduced, subtended by 3–4 secondary bracts. Ovary stipitate; fruit mostly straight, 1.2–3 cm long, pubescent, puberulous or subtomentellous, stipitate, 3–12-seeded.

Zambia. N: Mbala Distr., Nkali (Kali) Dambo, fl. 26.i.1952, *Richards* 523 (K). W: Kitwe, fl. & fr. 18.iii.1955, *Fanshawe* 2143 (K; SRGH). C: 100–129 km east of Lusaka, Chakwenga Headwaters, fl. & fr. 10.i.1964, *E.A. Robinson* 6169 (K; PRE). S: Mazabuka, fl. & fr. 7.iii.1952, *White* 2219 (K). **Zimbabwe**. N: Concession, fl. 23.ii.1938, *Hopkins* in GHS 6826 (K; SRGH). W: Matopos Hills, fl. 22.ii.1903, *Eyles* 1167 (SRGH). C: Goromonzi Distr., Chinhamhora (Chindamora), fl. & fr. iv.1953, *Davies* 509 (SRGH). E: 4.8 km from Chimanimani (Melsetter) on road to Chimanimani National Park, fl. 3.iii.1963, *Loveless* in GHS 190337 (SRGH). S: 1.2 km SE of Masvingo (Fort Victoria), fl. 16.iii.1958, *Leach* 8237 (K; SRGH).

Tab. 3.6.5. PSEUDARTHRIA HOOKERI var. HOOKERI. 1, flowering branch (× ²/₃); 2, flower (× 4); 3, calyx, opened out (× 4); 4, standard (× 4); 5, wing (× 4); 6, keel (× 4); 7, androecium (× 4); 8, gynoecium (× 4), 1–8 from *Tanner* 4187; 9, fruit (× 3), from *Tanner* 4810. Drawn by Derek Erasmus. From F.T.E.A.

Malawi. N: Viphya (Vipya), Luwawa, fl. & fr. 9.iii.1962, *Chapman* 1616 (SRGH). C: Dedza Distr., Chongoni, fl. 4.iii.1967, *Salubeni* 567 (K; SRGH). S: Nsanje Distr., Malawi (Malawe) Hill, fl. & fr. 23.iii.1960, *Phipps* 2621 (K; PRE; SRGH). **Mozambique**. N: Malema Distr., Serra de Merripa, fl. 5.ii.1964, *Torre & Paiva* 10485 (LISC). Z: Lugela Distr., Namagoa, fr. 2.x.1946, *Faulkner* PRE 279 (COI; K; LMA; PRE; SRGH). T: Macanga Distr., Monte Furancungo, outskirts of Marco Geodésico, fl. & fr. 15.iii.1966, *Pereira, Sarmento & Marques* 1678 (LMU). MS: prox. serração de Chibata, fl. & fr. 21.iii. 1968, *Barbosa* 1224 (LISC). GI: Inhambane, Massinga, fl. iv.1936, *Gomes e Sousa* 1713 (COI; K; LISC). M: montes da Namaacha, fl. & fr. 2.iii.1948, *Torre* 7451 (BM; K; LISC).

Widespread in tropical Africa from Cameroon and Ethiopia to Angola and NE South Africa. Coarse lowland to upland grassland, scrub, savanna, *Brachystegia, Acacia, Uapaca* and other open woodland; sometimes also by rivers in rocky places and old cultivations and plantations; 100–1740 m.

In East Africa *P. hookeri* and *P. confertiflora* (A. Rich.) Baker intergrade but in central Africa the latter appears to be absent although certain specimens (e.g. Zambia N: Fwambo, *Nutt)* show some approach to it.

Var. *argyrophylla* Verdc. from Uganda, West Africa, Cameroon and Sudan is distinguished by the thick silvery-grey velvety indumentum of the stems and leaf undersurfaces.

68. DROOGMANSIA De Wild.

Droogmansia De Wild. in Ann. Mus. Congo, Sér. IV, Bot. [Études Fl. Katanga] **1**: 53, t. 23 (1902). —Verdcourt in Kirkia **9**: 536 (1974).

Subshrubs or small shrubs with well-developed rootstocks. Leaves 1-foliolate, frequently developing after the flowers have appeared; petioles nearly always conspicuously winged, often so much so as to appear like a second lower leaf-blade, rarely not winged; stipules striate, ciliate; stipels present. Inflorescences on leafy or leafless shoots, terminal or axillary, falsely racemose or paniculate; primary and secondary bracts present; bracteoles absent; flowers readily disarticulating from their pedicels. Calyx 5-lobed, 2-lipped; upper lip ± 2-fid, composed of 2 teeth connate for most of their length; lower lip prominently 3-fid, the central lobe the longest. Corolla mostly medium-sized, white to purple; standard rounded, produced into a claw at the base, sometimes puberulous outside when young; wings clawed, transversely rugose inside, usually shorter than the other petals; keel usually as long as the standard. Vexillary filament free at base and apex but connate with the main tube for one-third to half of its length; free parts of the filaments dissimilar, 4 short and filiform, 5 much dilated, longer and pincer-shaped at insertion of the anther; anthers uniform. Intrastaminal disk short. Ovary stipitate, 2–many-ovuled; style filiform but slightly stiffened, curved, glabrous above the narrowed hairy apex of the ovary; stigma terminal, capitate. Fruit usually markedly stipitate (the stipe often plumose), 1–several-jointed, silky pubescent. Seeds compressed-reniform; hilum minute, without appendages.

A genus usually estimated at about 28 species, one Indo-Chinese, the rest tropical African*. The "species" are, however, very poorly defined and based on sparse material; I suspect there may be only a very few true species recognizable when abundant material is available. Since it has been possible to account for the east and central African material using the oldest available epithet in the genus a revision of all the other species was not undertaken. It appears to me that the genus may contain two distinct elements since the non-stipitate fruits and much less dilated filaments of the Indo-Chinese and several of the West African species show they are much nearer true *Desmodium* than they are to those species of *Droogmansia* with long stipitate densely hairy fruits**. Since *D. pteropus* (Baker) De Wild. is the type of the generic name this problem does not concern the Flora Zambesiaca area.

---

* Despite the similar appearance of the foliage, *Vaughania* S. Moore has quite different inflorescences and flowers, and I do not agree with Hutchinson's inclusion of this genus in *Droogmansia* (Gen. Fl. Pl. **1**: 484 (1964)); I have therefore omitted Madagascar from the distribution.

** Torre (in Bol. Soc. Brot., ser. 2, **39**: 211, t. 8 (1965)) has, however, described a true *Droogmansia* with a sessile fruit, and the stipes are short in *D. lancifolia* Schindl.

Petiole much narrower than the leaflet, about one-third its width, 20 × 5 mm in Flora
Zambesiaca material · 1. *megalantha*
Petiole broader, over one-third the width of the leaflet, but in small leaves often smaller than
the above dimensions (and in some rare cases very small or lacking) · · · · · · · 2. *pteropus*

If flowering before the leaves appear see general note below, following *D. pteropus* var *axillaris*.

1. **Droogmansia megalantha** (Taub.) De Wild. in Ann. Mus. Congo, Sér. IV, Bot. [Études Fl.
 Katanga] **1**: 56 (1902). —E.G. Baker, Legum. Trop. Africa: 336 (1929). —Torre in C.F.A.
 3: 228 (1966). —Verdcourt in Kirkia **9**: 538 (1974). —Lock, Leg. Afr. Check-list: 249
 (1989). Type from Angola.
     Desmodium megalantha Taub. in Bot. Jahrb. Syst. **23**: 192 (1896).

Rhizomatous subshrub with several stems from a woody base, or a more robust
shrub with fewer stems, 0.15–1 m tall. Stems pubescent to densely appressed-silvery-
pilose or almost glabrous. Leaflets 3–12 × 0.6–3 cm, lanceolate, narrowly elliptic or
oblong, acute to rounded at the apex but mucronate, cuneate to rounded at the
base, glabrescent to densely pilose; petiole winged, or sometimes in upper leaves
unwinged, 3–20 × 2–7 mm, linear, linear-oblong or oblanceolate, mostly shallowly to
deeply emarginate at the apex; stipules 5–8 × 0.7–1.5 mm, linear to lanceolate.
Inflorescences terminal, simple, falsely racemose, 10–35 cm long, sometimes with
axillary ones as well or a few flowers in the axils, pubescent to appressed pilose.
Flowers as in *D. pteropus*. Calyx glabrescent to pilose with spreading white hairs. Fruit
densely yellowish-hairy, stipitate, the stipe rapidly elongating and attaining 2.2–3.5
cm in length; articles 3–5, strongly compressed, 6 × 5 mm (?immature), broadly
elliptic or rounded oblong. Seeds not seen.

Var. **megalantha** —Verdcourt in Kirkia **9**: 538 (1974).

Rhizomatous subshrub with many stems 15–50 cm tall from a woody stock.

Zambia. W: Mwinilunga Distr., east of Mwinilunga and 25.6 km west of River Kabompo, fl.
11.ix.1930, *Milne-Redhead* 1107 (K; PRE).
 Also in Angola. Probably in *Brachystegia* woodland or associated grassland.
 The placing of the above specimen is very unsatisfactory. It is admittedly difficult to
distinguish it from the type of *D. megalantha*, although that has the petioles shorter, wider at the
apex and more emarginate. It might be better to treat *Milne-Redhead* 1107 as a very narrow-
petioled variant of *D. pteropus* resembling var. *angustipetiolata* Verdc. from western Tanzania. The
Angolan material is not at all uniform, and until better understood I am content to follow Milne-
Redhead and Torre in considering the single Zambian gathering to be a variant of *D. megalantha*
since it differs from the type of the latter only in having longer petioles of more uniform width.

2. **Droogmansia pteropus** (Baker) De Wild. in Ann. Mus. Congo, Sér. IV, Bot. [Études Fl.
 Katanga] **1**: 54 (1902). —Schindler in Repert. Spec. Nov. Regni Veg. **22**: 271 (1926). —E.G.
 Baker, Legum. Trop. Africa: 335 (1929). —Schubert in F.C.B. **5**: 216 (1954). —Verdcourt
 in Kew Bull. **24**: 62 (1970); in F.T.E.A., Leguminosae, Pap.: 488 (1971); in Kirkia **9**: 539
 (1974). —Lock, Leg. Afr. Check-list: 249 (1989). Types: Zambia, south of Lake
 Tanganyika, Fwambo, *Carson* 94 and 117 (K, syntypes).
     Dolichos pteropus Baker in Bull. Misc. Inform., Kew **1895**: 66 (1895).

Perennial herb, shrub, or small subshrub (0.1)0.6–2.5(3) m tall, with several shoots
from a woody rootstock, very rarely a small tree to 3(?4.5) m. Stems glaucous and
glabrous to appressed pubescent or velvety. Leaflets 2–9 × 0.9–3.5(4.5) cm, elliptic,
oblong-elliptic or oblong-lanceolate, acute and mucronate or rounded at the apex,
cuneate, rounded or slightly cordate at the base, slightly to fairly densely pubescent
beneath, sometimes glaucous; petiole winged, 0.8–6 × 0.4–3.5 cm, oblong, obovate,
oblanceolate, obcordate or almost round, rounded or narrowly to widely emarginate at
the apex, obtuse to slightly cordate at the base; stipules 4–10 × 0.6–1.5 mm, linear to
lanceolate. Inflorescences usually elongate and narrow, falsely racemose, the flowers
arranged in fascicles along the rhachis, 15–50 cm long, or flowers all in axillary fascicles;
peduncle 0–11 cm long; pedicels 2–22 mm long; bracts 1.5–4 × 0.7–1.5 mm, ovate to
lanceolate, soon falling. Calyx pubescent or hairy; tube 2.5–6 mm long; lobes 3–8.5 mm
long, the upper pair joined for most of their length. Standard all white or yellow or

yellow outside and mauve inside, the veins and outer flush dark purple or crimson, (10)15–23 × (10)12–20 mm, obovate, glabrous to finely or densely puberulous outside; wings and keel bluish-purple or keel yellow-green tipped purple; rarely flowers all white. Fruit densely yellowish-brown hairy, stipitate, the stipe rapidly enlarging and attaining 3–7 cm; articles 1–5, strongly compressed, 5.5–9(13) × 5–6(7) mm, broadly elliptic or almost round. Seeds dark crimson-brown, 4 × 3 × 1 mm, compressed-reniform; hilum minute, almost round, the seed somewhat beaked on one side of it.

1. Flowers axillary · vi) var. *axillaris*
 – Flowers borne in terminal or axillary inflorescences · 2
2. Stems thickly velvety with pale brown hairs · v) var. *giorgii*
 – Stems glabrescent to ± densely pubescent but not velvety · 3
3. Inflorescences branched, consisting of a terminal and several axillary branches; pedicels mostly short, under 8 mm; both rhachis and pedicels densely pubescent · · · iv) var. *whytei*
 – Inflorescences mostly simple, lacking axillary components; pedicels mostly longer and together with the rhachis pubescent or almost glabrous · 4
4. Inflorescence rhachis, pedicels and standard pubescent; pedicels mostly shorter, c. 1 cm long; leaves mostly small · i) var. *pteropus*
 – Inflorescence rhachis, pedicels and standard glabrescent; pedicels usually longer, often 2 cm long · 5
5. Leaves smaller; leaflets c. 2.5–9 × 0.8–3.6 cm; petiole c. 1–6 × 0.5–3 cm, frequently narrowed at the apex and broadest near the base · ii) var. *platypus*
 – Leaves larger and often glaucous; leaflets up to 14 × 7 cm; petiole up to 9 × 4 cm · · · · · ·
· iii) var. *quarrei*

i) Var. **pteropus** —Drummond in Kirkia **8**: 220 (1972). —Verdcourt in Kirkia **9**: 540 (1974).

Flowers in terminal, usually unbranched, mostly rather hairy inflorescences; pedicels usually pubescent, well-developed, 0.7–1(2) cm long. Winged petiole oblong, elliptic or obcordate, half as broad to broader than the leaflet. Leaves mostly small, the leaflets c. 2–6.5 × 0.5–3.2 cm and the petiole 0.7–3 × 0.3–2.4 cm. Standard distinctly hairy outside. Stipe 3–5(7) cm long.

Zambia. N: Mbala Distr., Lumi River flats, fl. 17.viii.1956, *Richards* 5860 (K). W: Chingola, fl. & fr. 30.ix.1955, *Fanshawe* 2466 (K) (variant with transversely oblate petioles 1.5 × 2 cm). C: Fiwila, fl. & fr. 29.ix.1957, *Fanshawe* 3765 (K). **Zimbabwe**. N: Makonde Distr., Chinhoyi (Sinoia), Kapeta Road, fl. & fr. 27.ix.1965, *Corby* 1383 (K; SRGH). **Malawi**. N: Nyika Plateau, 3.6 km from Zambia Rest House, fl. & fr. 14.xi.1967, *Richards* 22530 (K) (intermediate with var. *whytei*). C: 11.2 km on road to Dowa from Nkhotakota (Kota Kota), fl. & fr. 3.x.1943, *Benson* 383 (PRE) (so far as poor specimen allows of determination). **Mozambique**. N: Sanga Distr., between Maniamba and Macaloge (Mecaloja), fl. & fr. 3.ix.1934, *Torre* 267 (BM; COI).
Also in Dem. Rep. Congo, Tanzania and ?Angola. *Brachystegia* woodland and derived areas, dambo margins, etc.; 1050–2250 m.
Merges with var. *whytei* on the Nyika Plateau.
D. hockii De Wild. does not seem distinguishable from this. *Richards* 13251 (Zambia, Mbala Distr., fl. 14.ix.1960) is interesting in showing some subsessile leaves devoid of a winged petiole and some duplicates of this gathering are specimens flowering when entirely leafless and these are ± indistinguishable from *Droogmansia tenuis* Schubert and *D. longirhachis* Schubert, both belonging to the group having leafy and flowering shoots separate. Studies of the habit in the field and its variation with climatic conditions are necessary. *Fanshawe* 1406 (Zambia W: Luanshya, fl. 29.vii.1954) and *Fanshawe* 1530 (Zambia W: Mufulira, fl. 7.ix.1954) are of small stature, 9–24 cm and *Fanshawe* 1530 at least seems to be identical with *D. munamensis* De Wild. One syntype of *D. pteropus* is scarcely 30 cm and I doubt if this small variant should be distinguished.

ii) Var. **platypus** (Baker) Verdc. in Kew Bull. **24**: 62 (1970); in F.T.E.A., Leguminosae, Pap.: 490 (1971); in Kirkia **9**: 541 (1974). Type: Zambia, L. Mweru (Moero), Kalungwishi R. (Kalungwizi), *Carson* 11 (K, holotype).
Dolichos platypus Baker in Bull. Misc. Inform., Kew **1895**: 289 (1895).
Droogmansia longipes R.E. Fr., Wiss. Ergebn. Schwed. Rhod.-Kongo-Exped.: 90, t. 9, fig. 2 (1914). Type: Zambia, River Luapula, *Fries* 547 (UPS, holotype).
Droogmansia stuhlmannii sensu R.E. Fr., Wiss. Ergebn. Schwed. Rhod.-Kongo-Exped.: 90 (1914) non Taub.
Droogmansia platypus (Baker) Schindl. in Repert. Spec. Nov. Regni Veg. **22**: 272 (1926). —E.G. Baker, Legum. Trop. Africa: 337 (1929). —Schubert in F.C.B. **5**: 217 (1954).

Flowers in terminal, usually laxer, unbranched, almost glabrous or glabrescent, inflorescences; pedicels mostly longer, 7–22 mm long. Winged petiole typically oblong-oblanceolate or at least much narrower above than at base, but sometimes oblong or elliptic in variants intermediate with the variety which follows, often broader than the leaflet. Leaflets up to 9 × 3.6 cm. Standard pubescent to glabrescent. Stipe c. 5 cm long.

Zambia. N: Kasama Distr., Mungwi, fl. 26.ix.1960, *E.A. Robinson* 3871 (K; PRE; SRGH). W: Kitwe Distr., Ichimpi, fl. 4.ix.1964, *Mutimushi* 992 (K; SRGH). **Malawi**. N: Chitipa Distr., Mafinga Hills, fl. 11.xi.1958, *Robson & Fanshawe* 567 (BM; K; LISC; PRE; SRGH).

Also in Dem. Rep. Congo and Tanzania. *Brachystegia–Uapaca, Cryptosepalum, Monotes*, etc. woodland and derived grassland, playing fields, etc.; 1200–1900 m.

iii) Var. **quarrei** (De Wild.) Verdc. in Kew Bull. **27**: 443 (1972); in Kirkia **9**: 542 (1974). TAB. 3.6.**6**. Type from Dem. Rep. Congo (Katanga Province).
    Droogmansia quarrei De Wild. in De Wildeman & Staner, Contrib. Fl. Katanga, Suppl. V: 30 (1933). —Schubert in F.C.B. **5**: 212, pl. 16 (1954). —Torre in C.F.A. **3**: 230 (1966).

Flowers in terminal lax almost glabrous or glabrescent inflorescences; pedicels (8)15–17 mm long. Winged petiole typically oblong or elliptic-oblong, often very glaucous. Leaflet typically larger than in other variants, 6–14 × 2–7 cm, oblong or elliptic-oblong, often very glaucous. Standard often almost glabrous outside. Stipe of fruit often longer than in other variants, 6–8 cm long.

Zambia. B: Kaoma (Mankoya), fl. & fr. 17.x.1964, *Fanshawe* 8965 (K). N: Kawambwa, fr. 15.xi.1957, *Fanshawe* 3981 (K). W: Solwezi Distr., west of Solwezi and 1.6 km east of River Mumbeji (Mumbezhi), fl. & fr. 15.ix.1930, *Milne-Redhead* 1129 (K). C: Serenje Distr., Kundalila Falls, fl. & fr. 14.x.1963, *E.A. Robinson* 5728 (K). S: Mumbwa Distr., Chanobi to Nambala, fl. & fr. 16.ix.1947, *Greenway & Brenan* 8096 (EA; K; PRE).

Also in Dem. Rep. Congo and Angola. *Brachystegia* woodland, floodplain grassland and dambo edges; 990–1500 m.

D. longistipitata De Wild., from Cameroon, is a variant linking var. *quarrei* with var. *angustipetiolata* Verdc.

iv) Var. **whytei** (Schindl.) Verdc. in Kew Bull. **24**: 62 (1970); in F.T.E.A., Leguminosae, Pap.: 490, fig. 70 (1971); in Kirkia **9**: 542 (1974). TAB. 3.6.**7**. Type: Malawi, Nyika Plateau, *Whyte* 194 (K, lectotype).
    "Genus novum"; Baker in F.T.A. **2**: 144 (1871) (based on a scrap collected by Kirk).
    Droogmansia whytei Schindl. in Repert. Spec. Nov. Regni Veg. **22**: 271 (1926). —E.G. Baker, Legum. Trop. Africa: 334 (1929). —M.A. Exell in Bol. Soc. Brot., sér. 2, **12**: 10 (1937). —Brenan, Check-list For. Trees Shrubs Tang. Terr.: 421 (1949). —Brenan in Mem. New York Bot. Gard. **8**: 256 (1953). —Schubert in F.C.B. **5**: 210 (1954).
    Droogmansia friesii Schindl. in Repert. Spec. Nov. Regni Veg. **22**: 272 (1926). —E.G. Baker, Legum. Trop. Africa: 335 (1929). Type: Zambia, between Mansa (Fort Rosebery) and Lake Bangweulu, *Fries* 619 (UPS, holotype) (less hairy variant).
    Droogmansia pteropus sensu De Wild. in Ann. Mus. Congo, Sér. IV, Bot. [Études Fl. Katanga] **1**: 54, t. 23 (1902) pro parte non (Baker) De Wild. sensu stricto.

Flowers in terminal hairy inflorescences and often with additional inflorescences from the upper axils giving a divaricately branched effect; pedicels usually rather short, 2–8 mm long, hairy. Winged petiole rounded, oblong or elliptic-oblong, slightly narrower to slightly broader than the leaflet, sometimes narrowest at the apex. Leaflets mostly small, up to 6 × 3 cm. Standard distinctly hairy outside. Stipe of fruit usually short, 3–4.5 cm long.

Tab. 3.6.**6**. DROOGMANSIA PTEROPUS var. QUARREI. 1, flowering branch (× ¹/₂); 2, flower, longitudinal section (× 3); 3, standard, internal face (× 2); 4, wing (× 3); 5, keel (× 3); 6, top of androecium (× 10); 7, base of staminal tube (× 5); 8, staminal tube in longitudinal section, base of ovary and stamen (× 5); 9, ovary, longitudinal section (× 3), 1–9 from *Quarré* 3003; 10, fruit (× 1), from *Becquet* 6. Drawn by J.M. Lerinckx. From Fl. Congo Belge. Reproduced with permission of Jardin Botanique National de Belgique.

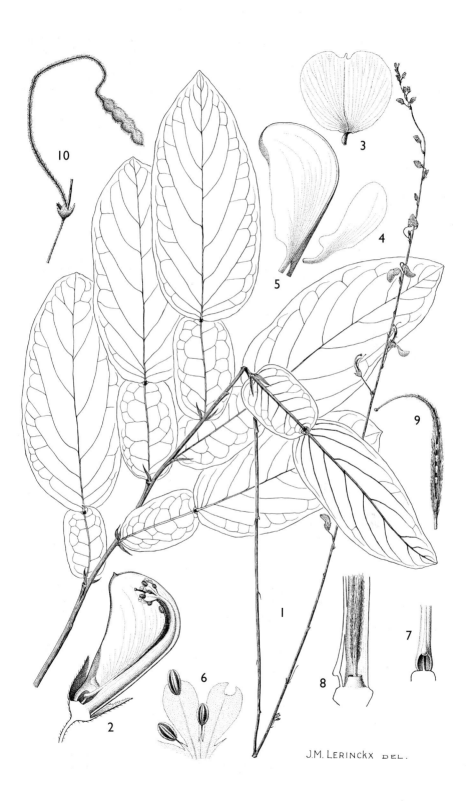

J.M. LERINCKX DEL.

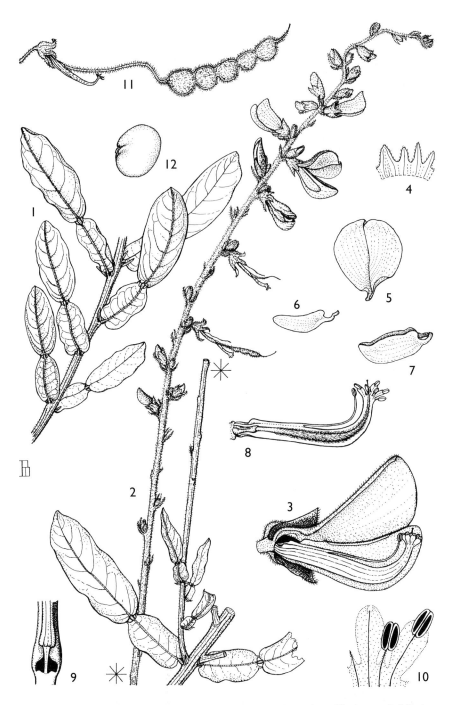

Tab. 3.6.7. DROOGMANSIA PTEROPUS var. WHYTEI. 1, portion of leafy stem (× ²/₃), from *Milne-Redhead & Taylor* 11128A; 2, flowering branch (× ²/₃); 3, half-flower (× 2); 4, calyx, opened out (× 1); 5, standard (× 1); 6, wing (× 1); 7, keel (× 1); 8, androecium with half stamens cut away to show gynoecium (× 2); 9, base of staminal tube, viewed from above (× 8); 10, detail of top of androecium (× 8), 2–10 from *Semsei* 2445; 11, fruit (× 1), from *Greenway* 6402; 12, seed (× 3), from *Napper* 866. Drawn by Diane Bridson. From F.T.E.A.

Zambia. N: Kasama, fl. l.ix.1960, *E.A. Robinson* 3793 (K; PRE; SRGH). C: Mkushi, fl. 5.viii.1959, *West* 4007 (K; LISC; SRGH). E: Nyika Plateau, fl. 24.ix.1956, *Benson* 169 (BM). S: 24–32 km north of Zimba, 18.ix.1960, *B.J. Coxe* s.n. (SRGH). **Malawi**. N: Rumphi Distr., Nyika Plateau, fl. & fr. 19.ix.1967, *Pawek* 1367 (SRGH). C: Ntchisi (Nchisi) Mt., fl. 24.vii.1946, *Brass* 16890 (BM; K; NY; PRE; SRGH). **Mozambique**. N: Lichinga Distr., Metónia, fl. & fr. ix.1933, *Gomes e Sousa* 1547 (COI).

Also in Dem. Rep. Congo, Burundi and Tanzania. Montane and dambo grassland, and *Brachystegia*, *Julbernardia*, *Uapaca* or *Marquesia* woodland; 710–2160 m.

This is possibly more distinct than I have indicated and extremes are easily named, but a number of intermediate forms are difficult to place and most routine namers of this genus have long suggested on covers, labels, etc. that *D. whytei* is not specifically distinct from *D. pteropus*.

v) Var. **giorgii** (De Wild.) Verdc. comb. et stat. nov.
  Droogmansia giorgii De Wild. in Rev. Zool. Bot. Africaines **13**: B14 (1925); in Contrib. Fl. Katanga, Suppl. 2: 68 (1929). —E.G. Baker, Legum. Trop. Africa: 334 (1929). Type from Dem. Rep. Congo.

Stems velvety. Flowers in distinctly branched inflorescences, 30–50 cm long, the rhachis ± velvety; pedicels 3–9 mm long becoming 1.5 cm long in fruit, densely pubescent. Winged petiole 8–26 × 7–23 mm, oblong or oblong-elliptic, broadly rounded to cordate at the apex, rounded at the base. Leaflet 3–5.5 × 1.4–2 cm, elliptic, finely pubescent above, pubescent to velvety beneath. Standard densely puberulous outside. Stipe attaining 2.8–5 cm in length.

Zambia. W: 65.6 km north of Kapiri Mposhi on road to Bwana Mkubwa, fl. & fr. 25.ix.1947, *Brenan & Greenway* 7983 (EA; K).

Also in Dem. Rep. Congo. Margin of *Brachystegia* and *Julbernardia* woodland; c. 1300 m.

The sheet cited does not agree too well with the type, differing in its smaller standard, longer fruit stipe, larger articles and having a leafy branch arising from the middle of the inflorescence. I originally kept it as a separate species, but it is too close to var. *whytei* to be anything but another variant of *D. pteropus*.

vi) Var. **axillaris** Verdc. in Kew Bull. **24**: 63 (1970); in Kirkia **9**: 543 (1974). Type: Zambia, Mbala (Abercorn), *Bullock* 1019 (K, holotype; SRGH).

Flowers 1–5 in axillary fascicles; pedicel mostly longer, up to 2.2 cm long, pubescent. Winged petiole 14 × 0.4–4 cm, oblong or rounded. Leaflets up to 8 × 4.5 cm. Standard hairy. Stipe of fruit 3.5–4.5 cm long.

Zambia. N: Mbala (Abercorn) to Isoko, 9.6 km from Mbala, fl. & fr. 28.ix.1967, *Richards* 22326 (K).

Not known elsewhere. *Brachystegia* woodland or bushland and marginal grassland subject to burning, also open dambos; 1410–1650 m.

Richards 1741 (Zambia, Lake Chila, fl. 23.vi.1952) shows an intermediate stage with a terminal inflorescence. The apparently terminal raceme in the typical variety is (as in the case of so many *Phaseoleae* and related tribes) in reality a series of axillary clusters in which the subtending leaves have been suppressed. The present variety may be connected in some way with different burning times; the leaves are not suppressed or only partly so and the whole of the difference in appearance can be explained by this single factor — *Pole Evans* 2958 (Zambia N: Kasama to Mbala (Abercorn), fl. 18.vii.1930) is leafless but may be a form of this variety flowering before the leaves appear or perhaps a similar variant of var. *whytei*.

General note: The arrangement used here is not very satisfactory. It might be better to associate the first three varieties and var. *giorgii* together as one species or subspecies and *whytei* as another and treating *axillaris* as a third using the characters of inflorescence indumentum, pedicel and stipe lengths as the main distinguishing features so far as the first two are concerned. A further matter which is probably best discussed in a general way is the identity of specimens which flower completely before the leaves develop. Dr. Schubert has (in F.C.B. **5**: 206–208 (1954)) divided the genus into those species that flower thus and those which flower when in full foliage. Anyone who has lived in areas subject to almost annual burning know that some species are very variable in this habit. Probably the leafless specimens are just forms of the varieties described here — e.g. *Richards* 31461 (Zambia N: Mbala Distr., Nkali Dambo, fl. & fr. 22.ix.1966), *Pole Evans* 2958 (Zambia N: Kasama to Mbala, fl. 18.vii.1930) are probably var. *axillaris* and *Fanshawe* 2466 (Zambia W: Chingola, fl. 30.ix.1955) a form of var. *pteropus*. *Richards* 11461 (Zambia N: 8 km from Luwingu, fl. & fr. 22.ix.1959 (K; LISC)) from very burnt woodland at 1500 m consists of very slender leafless stems bearing both flowers and fruit, the rhachis is distinctly pubescent, the pedicels attain 2.6 cm long and are also pubescent; the fruit stipes are 6.5 cm long; the standard varies from glabrous to pubescent. It is, I feel, no more than an extreme form of *D. pteropus* although it can scarcely be separated from *D. longirhachis* Schubert described from a single specimen from Dem. Rep. Congo (Katanga Prov.). *Richards* 12337 (Zambia N: Lupula Distr., Mbereshi, fl. & fr. 12.i.1960) is similar but has shorter pedicels c. 1 cm long; in both the above the fruit articles are rather small, 7 × 5 mm even when ripe which agrees ± with the type of *D. longirhachis*. *D. tenuis* Schubert is very similar. Field studies are needed to determine if these "species" always have separate flowering and leafy shoots as is suggested by Schubert. It is not clear what happens in *Richards* 11461. Without much more information confusion will remain.

69. URARIA Desv.

Uraria Desv. in J. Bot. Agric. **1**: 122, t. 5, fig. 19 (1813). —Verdcourt in Kirkia **9**: 532 (1974).

Perennial herbs or subshrubs, prostrate or erect. Leaves pinnately 3–9-foliolate, less often partly or all 1-foliolate; leaflets often large and venose; stipules persistent, free, acuminate; stipels present. Inflorescences mostly terminal spike-like racemes or panicles; primary bracts ovate or lanceolate, persistent or deciduous; secondary bracts and bracteoles absent. Calyx 5-lobed; 3 lower lobes equal, and longer than the upper pair (often appearing to be lower pair due to twisting). Corolla yellowish or purplish; standard rounded or obovate, narrowed into a claw; wings oblong-falcate, adhering to the keel, which is slightly incurved and obtuse. Vexillary stamen free; anthers uniform. Ovary sessile or shortly stipitate, 2–many-ovuled; style filiform, recurved at apex; stigma terminal, capitate. Fruit folded like a concertina, mostly enclosed in the persistent calyx, subsessile, constricted between the seeds, the segments ovate, inflated, 1-seeded and indehiscent. Seeds subglobose or compressed, oblong-ellipsoid; hilum lateral, aril not developed.

A genus of about a 20 species in the Old World tropics.

Uraria picta (Jacq.) DC., Prodr. **2**: 324 (1825). —J.G. Baker in F.T.A. **2**: 169 (1871). —E.G. Baker, Legum. Trop. Africa: 340 (1929). —Robyns, Fl. Sperm. Parc Nat. Alb. **1**: 333 (1948). —Léonard in F.C.B. **5**: 232, fig. 14 (1954). —Hepper in F.W.T.A., ed. 2, **1**: 587 (1958). — Torre in C.F.A. **3**: 233 (1966). —Verdcourt in F.T.E.A., Leguminosae, Pap.: 479, fig. 67 (1971); in Kirkia **9**: 533 (1974). —Lock, Leg. Afr. Check-list: 251 (1989). TAB. 3.6.8. Type from Guinea.
    Hedysarum pictum Jacq., Collectanea **2**: 262 (1789).

Erect subshrub 20–180 cm tall. Stems scabrid with short hairs. Lower leaves 1–3-foliolate, upper 5–9-foliolate; leaflets of lower leaves 2–8 × 2–3.2 cm, ovate, of upper leaves 7–25 × 0.5–2.5(4) cm, lanceolate, subacute and mucronulate at the apex, rounded at the base, often variegated, glabrescent and shining above, scabrid and strongly reticulate beneath. Inflorescences with a rhachis 10–55 cm long; peduncles 0–5 cm long, scabrid; pedicels at first erect, later inflexed, 6–9 mm long, covered with long deciduous hairs and denser short persistent hook-tipped hairs; bracts pinkish, comose, 14–25 × 2–8 mm, ovate-lanceolate or lanceolate-caudate, deciduous. Calyx tube puberulous, 2 mm long; lobes 2–3, 4–5 mm long, covered with long, white or brownish tubercular-based hairs. Corolla pink, bluish or reddish, glabrous; standard 8–9 × 5–6 mm, obovate. Fruit 5–9 mm long, of 3–6 segments, each segment 2–3 mm wide, shining, brown to blackish then whitish-grey. Seeds yellowish-brown, compressed, 2–2.5 × 1.5–2 × 0.8 mm, oblong-ellipsoid.

Malawi. N: Karonga Distr., fl. & fr. iii.1954, *G. Jackson* 1269 (K; SRGH). S: Elephant Marsh, fr. ii.1863, *Kirk* (K). **Mozambique**. Z: Lugela Distr., Namagoa Estate, fl. 4.i., fr. 27.ii.1948, *Faulkner* Kew 223 (COI; K; SRGH).
    Widespread in tropical Africa from West Africa and Sudan to Malawi, Mozambique and Angola. Also in Asia and northern Australia. *Acacia* grassland, floodplains, old garden sites and sisal estates; 30–2400 m.
    I have seen no material collected in the Flora Zambesiaca area later than 1956.

Tab. 3.6.8. URARIA PICTA. 1, flowering branch (× 2/3), from *Grant* 666; 2, branch with inflorescences at different stages of development (× 2/3), from *Davies* 866; 3, flower (× 3); 4, calyx, opened out (× 3); 5, standard (× 3); 6, wing (× 3); 7, keel (× 3); 8, androecium (× 3); 9, upper part of androecium, spread out (× 3); 10, gynoecium (× 3), 3–10 from *Grant* 666; 11, fruit (× 3); 12, seed, side view (× 3); 13, seed, hilar view (× 3), 11–13 from *Greenway* 1400. Drawn by Derek Erasmus. From F.T.E.A.

70. ALYSICARPUS Desv.

Alysicarpus Desv. in J. Bot. Agric. **1**: 120, pl. 4, fig. 8 (1813). —Léonard in Bull. Jard. Bot. État **24**: 84 (1954). —Verdcourt in Kirkia **9**: 544 (1974). *Fabricia* Scop., Introd.: 307 (1777). —Kuntze, Revis. Gen. Pl. **1**: 181 (1891). — Taubert in Engler, Pflanzenw. Ost-Afrikas **C**: 217 (1895).

Annual or perennial, erect or decumbent herbs. Leaves 1-foliolate, less often pinnately 3-foliolate; petiole channelled, winged; stipules scarious, acuminate, persistent, free or connate; stipels present, persistent. Inflorescences terminal, axillary or leaf-opposed, falsely racemose or less often paniculate, the flowers mostly paired; bracts scarious, at length falling; bracteoles absent. Calyx scarious, persistent, appearing 4-lobed, the lobes subequal, striate, the upper lobe entire or slightly 2-fid, consisting of the 2 upper calyx teeth entirely or almost entirely connate. Corolla small, mostly pinkish or purplish; standard ovate, rounded or obovate, produced into a claw, with 2 small longitudinal folds near the base inside; wings obliquely oblong, adhering to the keel, the petals of which are often appendaged. Vexillary stamen free in the fully developed flower; free parts of the filaments alternately long and short; anthers uniform. Ovary sessile or shortly stipitate, with several to many ovules; style filiform, incurved at the apex; stigma broadly capitate. Fruit linear-oblong, at least in outline, mostly several-jointed, the margins straight or the fruit constricted between the segments (articles), which are indehiscent, compressed, subcylindrical or rounded, mostly with a raised reticulation of ridges. Seeds subglobose; hilum minute, without a rim-aril.

A genus of the Old World tropics, comprising 25–30 species, one introduced into tropical America. The species are valuable fodder plants (see Bogdan in E. Afr. Agric. For. J. **15**: 38 (1948); Tropical Pasture and Fodder Plants: 319–320 (1977)).

1. Fruit with straight margins, not constricted between the articles; calyx lobes separated by a sinus, not at all overlapping at the base ·2
 – Fruit strongly or at least definitely constricted between the articles, and the sides not at all straight; calyx lobes slightly to conspicuously overlapping at the base, or if not the sinus very obscure ·3
2. Plant perennial; inflorescences dense, the internodes mostly much shorter than the flowers ·1. *vaginalis*
 – Plant annual; inflorescences lax, the internodes mostly much longer than the flowers ·2. *ovalifolius*
3. Articles ± smooth or with the reticulation very faint; calyx lobes overlapping at the base ·3. *zeyheri*
 – Articles with rather strong to very strong sculpture; calyx lobes slightly to conspicuously overlapping at the base ·4
4. Articles with weaker more regularly reticulate sculpture; calyx lobes strongly overlapping at the base; fruit very constricted between the articles, which are transversely elliptic ·4. *rugosus* subsp. *reticulatus*
 – Articles with strong close transverse ridges but with few cross-links · · · · · · · · · · · · · · ·5
5. Calyx lobes conspicuously overlapping at the base, with white or ferruginous hairs; inflorescences mostly short; fruit usually scarcely exserted from the calyx · · · · · 4. *rugosus*
 – Calyx lobes mostly scarcely, if at all, overlapping at the base, never with ferruginous hairs; inflorescences nearly always elongate and spike-like; fruit scarcely to strongly exserted from the calyx ·5. *glumaceus*

1. **Alysicarpus vaginalis** (L.) DC., Prodr. **2**: 353 (1825). —J.G. Baker in F.T.A. **2**: 177 (1871) pro parte. —E.G. Baker, Legum. Trop. Africa: 342 (1929) pro parte. —Léonard in Bull. Jard. Bot. État **24**: 84 (1954); in F.C.B. **5**: 224, fig. 13A (1954). —Hepper in F.W.T.A., ed. 2, **1**: 587 (1958). —Torre in C.F.A. **3**: 234 (1966). —Verdcourt in F.T.E.A., Leguminosae, Pap.: 493, fig. 71A (1971). —Drummond in Kirkia **8**: 216 (1972). —Verdcourt in Kirkia **9**: 546 (1974). —Lock, Leg. Afr. Check-list: 244 (1989). TAB. 3.6.**9**, fig. 1. Type from Sri Lanka. *Hedysarum vaginale* L., Sp. Pl.: 746 (1753).

Erect or spreading procumbent perennial herb, woody at the base, 10–60 cm tall. Stems densely to sparsely pubescent or puberulous, at length ± glabrous.

Leaves 1-foliolate; leaflet 0.5–6.5 × 0.3–2.6 cm, ovate, oblong, oblong-elliptic or lanceolate, acute to emarginate and mucronulate at the apex, subcordate at the base, finely puberulous and with longer hairs, or almost glabrous on upper surface, ciliate; reticulation prominent on both surfaces; petiole 4–15 mm long; petiolule 1–2 mm long; stipules 7–17(24) mm long, lanceolate. Inflorescences terminal and leaf-opposed, 2–13 cm long, mostly dense, the internodes mostly shorter than the flowers, glabrous or usually densely pubescent; peduncle 0–5 cm long; rhachis 2–13 cm long; pedicels 0.5–2 mm long; primary bracts 4–8 × 1.5–2.5 mm, ovate-lanceolate to lanceolate, acuminate, ± glabrous, deciduous; secondary bracts similar, 2.5–3.5 mm long. Calyx puberulous or pubescent; tube 1.5–2.5 mm long; teeth 3–4 × 0.5–1 mm, narrowly triangular, acuminate, not imbricate. Standard orange, pinkish-buff or purple, often paler than other petals, 4–6 × 3–4 mm; wings mauve; keel magenta, mauve or green with purple tip. Fruit 1.2–2.5 cm long, well exserted from the calyx, not constricted between the articles; articles 4–7, 2.5–3 × 1.5–3 mm, subcylindrical, with raised reticulate ridges, puberulous with uncinulate hairs. Seeds yellowish, speckled brown or entirely yellow-brown, slightly compressed, 1.7 × 1.5 × 1 mm, ellipsoidal.

Var. **vaginalis** —Verdcourt in Kirkia **9**: 546 (1974).

Plant glabrous to densely puberulous or pubescent. Leaves mostly well over 1 cm long.

Zambia. B: Sesheke, fl. & fr. i.1925, *Borle* s.n. (PRE). **Zimbabwe**. C: Marondera Distr., Grasslands Research Station, fl. 16.iv.1961, *Corby* 997 (K; SRGH) (introduced). S: Chivi Distr., near Madzivire Dip, c. 6.4 km north of Runde (Lundi) River, fl. 16.iii.1967, *Corby* 1829 (SRGH) (presumably wild). **Mozambique**. Z: c. 32 km north of Quelimane, fl. & fr. 10.viii.1962, *Wild* 5871 (COI; K; LISC; SRGH). MS: floodplain south of Muda, fl. vi.1959, *Leach* 9190 (SRGH). GI: Inhambane, fl. & fr. 27.i.1939, *Torre* 1606 (COI; LISC). M: Maputo (Lourenço Marques), near Campo do Despontivo, fl. 1.xi.1963, *Balsinhas* 650 (LISC).
Widespread throughout the Old World tropics. Floodplain grasslands (coastal and plateau), marshland verges and sandy flats, *Acacia* savanna or *Brachystegia* woodland; 0–1580 m.

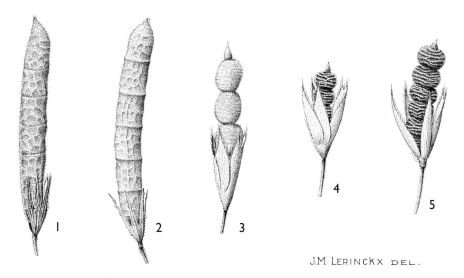

J.M. LERINCKX DEL.

Tab. 3.6.**9**. ALYSICARPUS. Fruits and calyces (× 3). 1, A. VAGINALIS, from *Flaminguy* 8154; 2, A. OVALIFOLIUS, from *Vermoesen* 1364; 3, A. ZEYHERI, from *van Meel* 7554; 4, A. RUGOSUS subsp. PERENNIRUFUS, from *G. de Witte* 2125; 5, A. GLUMACEUS, from *Bredo* 1964 bis. Drawn by J.M. Lerinckx. From Fl. Congo Belge. Reproduced with permission of Jardin Botanique National de Belgique.

Var. **parvifolius** Verdc. in Kew Bull. **27**: 443 (1972); in Kirkia **9**: 547 (1974). Type: Mozambique, Ilha de Moçambique, *A. Peter* 31267b (B; K, holotype).

Plant glabrescent or sparsely pubescent. Leaves very small, mostly well under 1 cm long, rounded oblong.

Mozambique. N: Ilha de Moçambique, fr. 31.x.1942, *Mendonça* 1163 (BM; LISC).
Also in Mauritius (and see note). Sandy dunes near sea level.
Specimens rather like this are found throughout the Asian range of the species and have sometimes been called "*A. nummulariifolius* (L.) DC." but the plant that this is based on is actually an *Indigofera*. These Ilha de Moçambique specimens are not at all like any other material from the Flora Zambesiaca area, and a varietal name has been given to them.

2. **Alysicarpus ovalifolius** (Schumach.) J. Léonard in Bull. Jard. Bot. État **24**: 88, fig. 11 (1954); in F.C.B. **5**: 226, pl. 18, fig. 13B (1954). —Hepper in F.W.T.A., ed. 2, **1**: 587 (1958). —Torre in C.F.A. **3**: 235 (1966). —Verdcourt in F.T.E.A., Leguminosae, Pap.: 493, fig. 71B (1971). —Drummond in Kirkia **8**: 216 (1972). —Verdcourt in Kirkia **9**: 547 (1974). —Gonçalves in Garcia de Orta, Sér. Bot. **5**: 63 (1982). —Lock, Leg. Afr. Check-list: 244 (1989). TAB. 3.6.**9**, fig. 2; TAB. 3.6.**10**. Type from Ghana.
   Hedysarum ovalifolium Schumach., Beskr. Guin. Pl.: 359 (1827).
   Alysicarpus vaginalis var. *paniculatus* Baker f., Legum. Trop. Africa: 342 (1929). Type from Tanzania.
   Alysicarpus vaginalis sensu auctt. mult. pro parte non (L.) DC.

Erect or spreading annual herb, sometimes woody at the base, 20–60 cm tall. Stems puberulous or pubescent, later almost glabrous. Leaves 1-foliolate; leaflet 1–10 × 0.6–3 cm, elliptic or oblong to narrowly lanceolate, acute to emarginate and mucronulate at the apex, subcordate, finely puberulous and with some hairs on the nerves beneath; reticulation obscure; petiole 2–8 mm long; petiolule 0.7–1.5 mm long; stipules 5–20 mm long, lanceolate. Inflorescences terminal or leaf-opposed, sometimes paniculate, very lax, the internodes between the 3–7 pairs of flowers long; peduncle 3–4 cm long; rhachis 3–11 cm long; pedicels 1–2 mm long; primary bracts 4–5 × 1.5–2 mm, ovate-lanceolate, acuminate, deciduous; secondary bracts 1.5–2.5 mm long. Calyx puberulous or pubescent; tube 1.5–2 mm long; teeth 3–4 × 0.5 mm, narrowly triangular, acuminate, not imbricate. Standard orange-buff to pink or reddish-violet, rarely whitish, 4–6 × 3–4 mm; wings purplish-mauve; keel pale greenish. Fruit 18–25 mm long, well exserted from the calyx, not constricted between the articles; articles 2–8, 2.5 × 2–2.5 mm, subcylindrical, with raised reticulate ridges, puberulous. Seeds brown, slightly compressed, 2.5 × 1.5 × 1.2 mm, oblong-ellipsoid.

Zambia. C: Luangwa Distr., Katondwe, fl. & fr. 23.ii.1965, *Fanshawe* 9215 (LISC; SRGH). E: Chipata Distr., Luangwa Valley, Masumba, fr. 24.iii.1963, *Verboom* 804 (K). **Zimbabwe**. N: Hurungwe Distr., south bank of River Maura (Mauora), fl. 27.ii.1958, *Phipps* 935 (K; LISC; SRGH). E: Chipinge Distr., Chivirira (Chiribira) Falls, fl. & fr. 6.vi.1950, *Wild* 3410 (K; LISC; PRE; SRGH). S: Chiredzi Distr., Runde (Lundi) River, Fishan, fl. & fr. 28.iv.1962, *Drummond* 7789 (K; PRE; SRGH). **Malawi**. N: Khondowe (Kondowe)–Karonga, fr. vii.1896, *Whyte* s.n. (K). S: Shire River, near Liwonde Ferry, fl. & fr. 13.iii.1955, *Exell, Mendonça & Wild* 842 (BM; LISC; SRGH). **Mozambique**. N: c. 10 km from Nacaroa towards Namapa, fl. & fr. 4.iv.1964, *Torre & Paiva* 11627 (LISC). Z: Lugela Distr., Namagoa Estate, fl. & fr. iii.1946, *Faulkner* Kew 136 (COI; K). T: Cahora Bassa, R. Mucangaze, fr. 14.vi.1971, *Torre & Correia* 18749 (K; LISC). MS: near Sena (Senna), fr. 9.iv.1860, *Kirk* (K).
Widespread in tropical Africa and commoner than the last species in West Africa, also in Madagascar and Asia. At lower altitudes, usually in hot low rainfall country, sandy open places, sandy river beds, often in soil pockets or cracks in granite outcrops in such places, roadsides, footpaths, savanna woodland (mopane, etc.); 65–700 m.
Verboom (L118) recorded it as a common impurity in *Lespedeza* seed but gave no localities.
I have followed Léonard in maintaining the last two as separate species. Apart from the annual or perennial habit and lax or dense inflorescence there are no proper morphological characters which can be used to separate the two, and in East Africa, particularly on the Kenya coast and in Zanzibar, the differentiation is highly unsatisfactory. In Mozambique too some specimens with condensed inflorescences appear to be annuals. Extensive hybridization may occur in some areas and field study is desirable.

J.M.LERINCKX DEL.

Tab. 3.6.**10**. ALYSICARPUS OVALIFOLIUS. 1, habit (× ¹/₂), from *Vermoesen* 1364; 2, petiole, showing stipules and base of leaflet (× 10); 3, flower pair subtended by one primary bract (turned down) and 2 secondary bracts (× 5); 4, standard (× 5); 5, wing (× 5); 6, half keel (× 5); 7, androecium (× 5), 2–7 from *M. Laurent* s.n.; 8, fruit (× 2), from *Vermoesen* 1364. Drawn by J.M. Lerinckx. From Fl. Congo Belge. Reproduced with permission of Jardin Botanique National de Belgique.

3. **Alysicarpus zeyheri** Harv. in F.C. **2**: 230 (1862). —J.G. Baker in F.T.A. **2**: 170 (1871). —E.G. Baker, Legum. Trop. Africa: 343 (1929). —Robyns, Fl. Sperm. Parc Nat. Alb. **1**: 332 (1948). —Léonard in F.C.B. **5**: 228 (1954). —Hepper in F.W.T.A., ed. 2, **1**: 587 (1958). —Torre in C.F.A. **3**: 235 (1966). —Verdcourt in F.T.E.A., Leguminosae, Pap.: 494, fig. 71C (1971). — Drummond in Kirkia **8**: 216 (1972). —Verdcourt in Kirkia **9**: 549 (1974). —Lock, Leg. Afr. Check-list: 245 (1989). TAB. 3.6.**9**, fig. 3. Type from South Africa.

Erect or somewhat spreading perennial herb 0.1–1.2 m tall, from a tough woody rootstock. Stems glabrous to pubescent. Leaves 1-foliolate; leaflet 1–10 × 0.2–1.8 cm, linear to elliptic or ovate-lanceolate, acute or obtuse and mucronulate at the apex, rounded or slightly cordate at the base, rather thick, finely puberulous and with some longer hairs on the lower surface, strongly reticulate on both surfaces; petiole 1.5–8 mm long; petiolule 0.5–1 mm long; stipules 6–17(27) mm long, lanceolate to ovate-lanceolate. Inflorescences terminal or sometimes axillary, very lax; peduncle 1–2 cm long; rhachis 2.5–45 cm long; pedicels 1–5 mm long; primary bracts 5–7 × 2–3.5 mm, ovate-lanceolate, acuminate, pubescent, ciliate, deciduous; secondary bracts 2–4 mm long, linear. Calyx hairy with brown hairs, or only pubescent or puberulous; tube 1–1.5 mm long; teeth 4–7 × 1.5 mm, elliptic to lanceolate, acuminate, very narrowly imbricate, ciliate. Standard salmon or apricot-coloured, tinged red or mauve, or sometimes cream or yellow, 6–7 × 4–4.5 mm; wings crimson-purple; keel often greenish tinged salmon at the apex. Fruit 7–15 mm long, well exserted from the calyx, much constricted between the articles; articles 2–6, 2.5–3 × 2–2.5 mm, oblong-elliptic, almost smooth, the transverse ridges very obscure, puberulous or rarely glabrous. Seeds pale yellow-brown, 1.5–2 × 1.2–1.5 × 1 mm, oblong.

Zambia. B: Masese, fl. 11.ix.1969, *Mutimushi* 3614 (K; NDO). N: Mbala Distr., Chilongowelo, path to Tasker's Deviation, fl. & fr. 21.iv.1952, *Richards* 1471 (K). W: Mwinilunga Distr., Kalenda Dambo, fl. & fr. 8.x.1937, *Milne-Redhead* 2645 (BM; K; PRE). E: Chipata, fl. & fr. 21.ix.1960, *Mutimushi* 1455 (K; SRGH). S: Choma Distr., Mapanza Mission, fl. & fr. 1.i.1953, *E.A. Robinson* 39 (K). **Zimbabwe**. N: Mwami (Miami) Experimental Farm, fl. & fr. 4.x.1946, *Wild* 1291 (K; PRE; SRGH). W: Bulawayo, fl. & fr. 1903, *Eyles* 86 (SRGH). C: Makoni Distr., near Maidstone, fl. & fr. 6.i.1931, *Norlindh & Weimarck* 4161 (COI; EA; K; LD; PRE; SRGH; UPS). E: Mutare (Umtali), fl. & fr. 20.x.1955, *Chase* 5825 (BM; COI; K; LISC; PRE; SRGH). S: Masvingo Distr., close to Masvingo (Ft. Victoria)–Birchenough Bridge road by Ndanga turn-off, fl. & fr. 19.i.1960, *Goodier* 838 (SRGH). **Malawi**. N: Karonga airstrip, fl. & fr. 5.ii.1953, *Williamson* 141 (BM; LISC; SRGH). C: Lilongwe Distr., Agricultural Research Station, fr. 30.iii.1952, *G. Jackson* 746 (K; LISC; SRGH). S: Shire Highlands, *Buchanan* 440 (K). **Mozambique**. N: Lichinga (Vila Cabral), fr. 7.v.1948, *Pedro & Pedrógão* 3644 (EA; LMA). Z: Lugela Distr., Namagoa, fl. & fr. ix.1945, *Faulkner* PRE 130 (K; PRE; SRGH). T: Angónia, Vila Ulónguè, Catsanha, fl. & fr. 5.iii.1980, *Macuácua & Mateus* 1201 (LISC; LMA). MS: Vale do R. Mussambuzi (Moçambize), região de Mavita, fr. 25.x.1933, *Mendonça* 2574 (BM; K; LISC).

Widespread in tropical Africa from Sierre Leone, Ethiopia and the Sudan to Angola and South Africa (Transvaal). Dambo and open grassland, open woodlands of *Acacia, Combretum, Brachystegia*, etc., often in recently burnt areas but also in wet places usually on sandy soil, occasionally a weed; 60–1500 m.

Several specimens have been collected which may represent an abnormal state of this species. Numerous shoots occur in tufts at ground level at the base of burnt stems, and the leaves are rather thin, elliptic, mostly under 1 cm long; these shoots sometimes bear a few atypical flowers. It probably represents the first growths after particularly severe burning. The following sheets are examples: Zambia C: near Chilanga, fl. 8.x.1957, *Angus* 1761 (K; PRE), and S: Mumbwa, *Macaulay* 1119 (K).

4. **Alysicarpus rugosus** (Willd.) DC., Prodr. **2**: 353 (1825). —J.G. Baker in F.T.A. **2**: 171 (1871) pro parte. —Léonard in Bull. Jard. Bot. État **24**: 92, fig. 12 (1954); in F.C.B. **5**: 229 (1954). —Hepper in F.W.T.A., ed. 2, **1**: 587 (1958). —Torre in C.F.A. **3**: 236 (1966). —Schreiber in Merxmüller, Prodr. Fl. SW. Afrika, fam. 60: 14 (1970). —Verdcourt in F.T.E.A., Leguminosae, Pap.: 495 (1971); in Kirkia **9**: 550 (1974). —Lock, Leg. Afr. Check-list: 244 (1989). Type from Ghana.
  Hedysarum rugosum Willd., Sp. Pl. **3**: 1172 (1802).
  Alysicarpus violaceus (Forssk.) Schindl. in Repert. Spec. Nov. Regni Veg. **21**: 13 (1925) pro parte non *H. violaceum* Forssk. nec *H. violaceum* L.
  Alysicarpus violaceus var. *pilosus* Schindl., loc. cit. Type from Ethiopia.

Erect, prostrate or ascending, robust annual or perennial somewhat suffruticose herb 0.3–1(2) m tall. Stems pubescent, pilose or practically glabrous. Leaves 1-foliolate or less often pinnately 3-foliolate; leaflets 1.5–11 × 0.2–2.3 cm, oblong, ovate-lanceolate, linear-lanceolate or linear, acute and mucronulate at the apex, rounded or subcordate at the base, finely puberulous and with longer hairs beneath, slightly reticulate; petiole 2–17 mm long; petiolules 0.5–1.5 mm long; stipules 0.5–3 cm long, lanceolate. Inflorescences axillary, terminal or leaf-opposed, dense to lax; peduncle 1–3 cm long; rhachis 3–20 cm long; pedicels 2–5 mm long; primary bracts 5–10 × 3 mm, elliptic, acuminate, glabrous or pubescent, ciliate, deciduous; secondary bracts 1–2.5 mm long, linear, or sometimes lacking. Calyx glabrous to densely pubescent; tube 1–1.5 mm long; teeth 5–8(9.5) × 1.5–2.5 mm, lanceolate, acuminate, imbricate, noticeably rounded or subcordate at the base, ciliate with white or brownish hairs. Standard whitish, pinkish-buff, reddish-purple or bluish, 6–7 × 3 mm, obovate; keel often greenish. Fruit 5–10 mm long, scarcely exserted from the calyx, much constricted between the articles; articles 3–6, 1–1.5 × 2–2.5 mm, transversely elliptic, with strong, close transverse ridges which join up here and there, or with a raised more regular reticulation, glabrous or pubescent. Seeds olive, compressed, 1.5 × 0.8 mm, squarish.

1. Articles of the fruit with a raised more regular reticulation not particularly orientated in one direction · iii) subsp. *reticulatus*
 – Articles of the fruit very strongly closely transversely ribbed with few cross-links · · · · · · ·2
2. Plants annual; calyx lobes with white cilia · i) subsp. *rugosus*
 – Plants mostly perennial but annual in some areas of Tanzania; calyx lobes with ferruginous cilia · ii) subsp. *perennirufus*

i) Subsp. **rugosus** —Verdcourt in F.T.E.A., Leguminosae, Pap.: 496 (1971). —Drummond in Kirkia **8**: 216 (1972). —Verdcourt in Kirkia **9**: 551 (1974). —Lock, Leg. Afr. Check-list: 244 (1989).

Annual, mostly erect herbs. Lobes of calyx with white cilia. Articles with strong close transverse ridges with limited anastomosing.

Botswana. N: Mababe Depression, fl. 14.vi.1978, *P.A. Smith* 2443 (K; SRGH). **Zambia**. N: Kaputa Distr., Mweru Wantipa, Bulaya, fl. 13.iv.1957, *Richards* 9194 (K; SRGH). W: Kitwe Distr., Mwekera, fl. & fr. 20.iv.1962, *Fanshawe* 6763 (PRE; SRGH). E: Chipata Distr., Lukuzye (Rukuzi) River, fl. iii.1962, *Verboom* 603 (LISC; SRGH). S: Mumbwa Distr., Kafue National Park, Kafue Hook Pontoon, fl. 27.iii.1963, *B.L. Mitchell* 18/46 (SRGH). **Zimbabwe**. N: c. 16 km from Gokwe, on bank of River Siatengwe, fl. & fr. 25.iii.1963, *Bingham* 563 (K; LISC; SRGH). W: Hwange Distr., Kazungula, iv.1955, *Davies* 1110 (K; SRGH). **Malawi**. N: Rumphi Distr., escarpment road above Chiweta, fr. 4.vi.1989, *Brummitt* 18323 (K). S: Blantyre Distr., Ndirande Forest, fr. 13.v.1968, *Banda* 1034 (K). **Mozambique**. N: Cuamba, fr. 13.v.1948, *Pedro & Pedrógão* 3354 (EA; LMA). MS: Gorongosa National Park, Chitengo Area, fl. & fr. viii.1970, *Tinley* 1990 (K; LISC; SRGH). GI: Xai-Xai Distr., Barra do Limpopo, Zongoene, fr. 21.xii.1979, *de Koning* 7829 (K; LMU).
Also in Mali and Eritrea to Mozambique and Namibia, and in Madagascar. Seasonally waterlogged grassland, *Acacia* and mopane wooded grassland, sandy river banks, rice paddy fallows, sandy beaches, etc. mostly on black clays; 10–1280 m.
Several plants occur in India which are only variants of this, but I have made no detailed assessment of their status. One plant cultivated in Harare (fl. & fr. 17.ii.1932, *Stent* in *GHS* 5576) is var. *heyneanus* (Wight & Arn.) Baker, distinguished by the spreading hairs on the stem.

ii) Subsp. **perennirufus** J. Léonard in Bull. Jard. Bot. État **24**: 95 (1954); in F.C.B. **5**: 230, fig. 13D (1954). —Torre in C.F.A. **3**: 236 (1966). —Verdcourt in F.T.E.A., Leguminosae, Pap.: 496, fig. 71D (1971). —Drummond in Kirkia **8**: 216 (1972). —Verdcourt in Kirkia **9**: 552 (1974). —Lock, Leg. Afr. Check-list: 244 (1989). TAB. **3.6.9**, fig. 4. Type from Dem. Rep. Congo.

Perennial often decumbent herb with a woody base. Calyx lobes with brown cilia. Articles with strong close transverse ridges with limited anastomosing.

Zambia. N: Mbala Distr., Lumi Marsh, fl. & fr. 9.ii.1955, *Richards* 4391 (K). **Zimbabwe**. C: Harare (Salisbury), fl. 29.x.1953, *Kerr* in *GHS* 44234 (PRE; SRGH). E: Nyanga (Inyanga), above Cheshire, fl. & fr. 4.ii.1931, *Norlindh & Weimarck* 4802 (K; LD; SRGH). **Malawi**. N: Karonga to Khondowe (Kondowe), fl. & fr. vii.1896, *Whyte* (K). C: Kasungu, Ntchisi (Nchisi) Mt., fr. 6.v.1963, *Verboom* 847 (SRGH). **Mozambique**. N: Nampula, fl. & fr. 11.v.1937, *Torre* 1401 (COI; LISC). Z: Morrumbala, fl. & fr. 30.xii.1858, *Kirk* (K). M: Marracuene, between Matola and Umbelúzi, fr. 29.iv.1947, *Pedro & Pedrógão* 830 (LMA).

Also in Central African Republic, Dem. Rep. Congo and Eritrea to Angola and South Africa. Swampy grassland and dambos, termite mounds, sometimes in cultivations; 900–1800 m.

iii) Subsp. **reticulatus** Verdc. in Kew Bull. **24**: 67 (1970); in F.T.E.A., Leguminosae, Pap.: 497 (1971); in Kirkia **9**: 552 (1974). —Lock, Leg. Afr. Check-list: 244 (1989). Type from Tanzania.

Annual (?or perennial) erect or decumbent herb. Lobes of calyx with white cilia. Articles with raised reticulation of veins not prominently orientated in any direction.

Zambia. N: Mbala Distr., near Kawimbe, Nkali (Kali) Dambo, fl. & fr. 16.v.1955, *Richards* 5718 (K; SRGH). **Malawi**. N: Chitipa Distr., Kaseye Mission, fl. & fr. 19.iv.1975, *Pawek* 9408 (K; MAL; MO).

Also in southern Tanzania. Dambo and short grassland, on sand and termite mounds, and in rock crevices; 1050–1680 m.

Although the fruit is undoubtedly very different from that of other variants of *A. rugosus* and from that of *A. glumaceus* further evidence is required. The species of *Alysicarpus* with non-constricted fruits have a faint reticulate sculpture and there is a possibility that this variant has arisen through a complicated hybridization.

5. **Alysicarpus glumaceus** (Vahl) DC., Prodr. **2**: 353 (1825). —Léonard in Bull. Jard. Bot. État **24**: 98, fig. 13 (1954); in F.C.B. **5**: 231, fig. 13E (1954). —Hepper in F.W.T.A., ed. 2, **1**: 587 (1958). —Torre in C.F.A. **3**: 236 (1966). —Verdcourt in F.T.E.A., Leguminosae, Pap.: 497 (1971); in Kirkia **9**: 553 (1974). —Lock, Leg. Afr. Check-list: 243 (1989). TAB. 3.6.**9**, fig. 5. Type from Yemen.

    Hedysarum violaceum Forssk., Fl. Aegypt.-Arab.: 136 (1775) non L. Type as for *Alysicarpus glumaceus* nom. illegit.

    Hedysarum glumaceum Vahl, Symb. Bot. **2**, Add. & Corrig.: 106 (1791).

    Alysicarpus hochstetteri A. Rich., Tent. Fl. Abyss. **1**: 209 (1847). Type from Ethiopia.

    Alysicarpus violaceus (Forssk.) Schindl. in Repert. Spec. Nov. Regni Veg. **21**: 13 (1925) pro parte. —É.G. Baker, Legum. Trop. Africa: 342 (1929) pro parte.

Erect annual herb, sometimes procumbent, often somewhat woody at the base, 0.15–1.5 m tall. Stems pubescent, sometimes only in narrow longitudinal lines, or densely covered with spreading hairs. Leaves 1-foliolate; leaflet 0.65–12.5 × 0.2–1.1 cm, oblong, linear or linear-lanceolate, acute and mucronulate at the apex, rounded or slightly cordate at the base, finely puberulous and with longer hairs on lower surface, slightly reticulate; petiole 2–6 mm long; petiolule 0.5–1 mm long; stipules 4–16 mm long, lanceolate. Inflorescences terminal and leaf-opposed; peduncle 3 cm long; rhachis 3–30 cm long; pedicels 2–5(9) mm long; primary bracts 4–6 × 2 mm, elliptic, acuminate, deciduous; secondary bracts 1.5–2 mm long, linear. Calyx puberulous or pubescent; tube 1–1.5 mm long; teeth 4–6 × 0.7–1.5 mm, lanceolate or narrowly triangular, at first slightly imbricate at the base, ciliate with white hairs or glabrous. Standard salmon-coloured, red or mauve, 5–6 × 3–5 mm, obovate; wings and keel often pale, but margined and tipped with purple. Fruit 7–15 mm long, much constricted between the articles; articles 4–7, 1.5–3 × 2–3 mm, transversely elliptic, with strong close transverse ridges which join up here and there, shortly pubescent or puberulous. Seeds reddish-brown or olive, compressed, 1.8–2.2 × 1.8 × 1–1.2 mm, rounded-rhomboid or ellipsoid.

Subsp. **glumaceus** —Verdcourt in F.T.E.A., Leguminosae, Pap.: 498 (1971); in Kirkia **9**: 553 (1974). —Lock, Leg. Afr. Check-list: 243 (1989).

Stems with appressed hairs mostly in longitudinal lines. Leaflets oblong or oblong-lanceolate to linear-lanceolate.

Var. **glumaceus** —Verdcourt in F.T.E.A., Leguminosae, Pap.: 498, fig. 71E (1971); in Kirkia **9**: 554 (1974).

Calyx lobes narrow, 4–6 × 1 mm, scarcely if at all overlapping at the base. Fruit well exserted from the calyx.

Zimbabwe. E: Save (Sabi) Valley, by pool at Honde Dip, fl. & fr. 26.ix.1947, *Whellan* 247 (SRGH). **Mozambique**. M: Namaacha Distr., between Changalane and Mazeminhama (near the Cabeça de Elefante), fl. & fr. 3.v.1960, *Myre* 3913 (LMA; LMU).

Also in Arabia and widely distributed throughout tropical Africa from Senegal to Angola and Ethiopia to South Africa (KwaZulu-Natal). Grassland, grassland with scattered trees, *Acacia–Combretum* woodland with *Themeda*, and in maize fields on black cotton soil; 5–360 m.

Var. **intermedius** Verdc. in Kew Bull. **24**: 68 (1970); in F.T.E.A., Leguminosae, Pap.: 498 (1971); in Kirkia **9**: 554 (1974). Type from Kenya.

Calyx lobes slightly broader, 5.5–7.5 × 1–2.5 mm, very slightly overlapping at the base. Fruit scarcely or only slightly exserted.

Malawi. N: Lake Malawi (Nyasa), Karonga, fr. xi.1893, *Scott-Elliot* 8424 (K) (very poor specimen — record dubious). S: Nsanje, fr. 26.ix.1956, *G. Jackson* 2060 (K; SRGH). **Mozambique**. N: Meconta Distr., c. 19 km from Corrane towards Liúpo, fl. & fr. 18.i.1964, *Torre & Paiva* 10061 (LISC). Z: between Mopeia and Régulo Changalaze, 43.6 km from Mopeia, fl. & fr. 2.viii.1949, *Barbosa & Carvalho* 3806 (K; LISC; LMA). MS: Zambezi, below Mazaro (Mazzaro), fl. & fr. 21.iii.1860, *Kirk* s.n. (K). GI: Chibuto Distr., Maniquenique, fr. 17.vii.1947, *Pedro & Pedrógão* 1486 (K; LMA; SRGH). M: environs of Matutuíne (Bela Vista), fl. & fr. 26.iv.1948, *Torre* 7727 (BM; K; LISC).
Also in East Africa. Floodplain or swampy grassland, wooded grassland and woodland (e.g. of *Sterculia, Milletia, Androstachys*) on sandy or alluvial soils, also on dunes with *Cyperaceae* and *Sporobolus* and occasionally in plantations; 0–210(538) m.

Tribe 10. **PSORALEEAE**

by B. Verdcourt

Psoraleeae (Benth.) Rydb. in North Amer. Fl. **24**: 1 (1919).
Subtribe Psoraleineae Benth. in Bentham & Hooker f., Gen. Pl. **1**: 443 (1865).

Herbs, shrubs or small trees with most parts glandular-punctate. Leaves (1)3(5)-foliolate, or sometimes reduced to scales; leaflets entire or toothed; stipules adnate to the petiole; stipels absent. Flowers basically 3 per node, in fascicles, heads or spike-like or racemose inflorescences, axillary or less often also terminal. Calyx campanulate with upper lobes joined for most of their length. Corolla mostly white, blue or purple; standard without appendages. Filaments all joined, or vexillary filament free; anthers uniform, alternately basifixed and dorsifixed. Ovary with single ovule; style slender, curved, glabrous or penicillate. Fruit indehiscent; pericarp thin or tough and glandular-verrucose. Seeds oblong-reniform with a small lateral hilum.

1. Flower pedicel subtended by a lobed cupulum of 2 joined bracts; leaflets linear to linear-oblong; erect woody herbs to shrubs* · 73. **Psoralea**
   – Flower pedicel not subtended by a lobed cupulum, but ordinary bracts present; leaflets not linear · 2
2. Fruit without black glandular warts; leaflets oblanceolate; erect shrubs · · 72. **Otholobium**
   – Fruit with conspicuous black glandular warts; leaflets obovate to elliptic, slightly dentate to crenate-lobulate; many spreading stems from a woody rootstock · · · · · · · · · · 71. **Cullen**

71. CULLEN Medik.

Cullen Medik. in Vorles. Churpfälz. Phys.-Öcon. Ges. **2**: 381 (1787). —J.W. Grimes in Austr. Syst. Bot. **10**: 565–648 (1997).

Annual or perennial herbs or shrubs. Leaves petiolate, pinnately or palmately 1–3(5)-foliolate, the leaflets very variable in shape, entire or often toothed or

---

* The leaflet and habit characters in this key refer only to the species growing in the Flora Zambesiaca area.

undulate; stipules very variable, triangular to linear, usually attenuate at the base or somewhat decurrent, occasionally amplexicaul. Inflorescences pseudoracemose or spike-like, axillary, fasciculate or pedunculate, each triad of flowers subtended by a bract of very variable shape. Flowers sessile or subsessile. Calyx tube cylindric or campanulate with teeth unequal, the lowest the longest. Standard obovate, narrowing into the claw, not or scarcely auriculate; wings longer or shorter than the keel. Stamens monadelphous in part at first, the vexillary filament free below, later becoming entirely free. Ovary very shortly stipitate; stigma capitate. Fruit 1-seeded, indehiscent, ± ellipsoid to round; epicarp glandular and often hairy; seed adhering to the pericarp.

A genus of 33 species, extending from Australia (the centre of the genus), through southern Asia and Asia Minor into southern Europe and southwards through Africa to South Africa. Only one species occurs in the Flora Zambesiaca area.

Cullen tomentosum (Thunb.) J.W. Grimes in Austr. Syst. Bot. **10**: 589 (1997). TAB. 3.6.**11**. Type from South Africa.
  Trigonella tomentosa Thunb., Prodr. Pl. Cap., pars post.: 137 (1800); Fl. Cap., ed. 2, **2**: 611 (1823).
  Psoralea obtusifolia DC., Prodr. **2**: 221 (1825). —J.G. Baker in F.T.A. **2**: 64 (1871). —Hiern, Cat. Afr. Pl. Welw. **1**: 206 (1896). —Harms in Engler, Pflanzenw. Afrikas [Veg. Erde 9] **3**: 584 (1915). —Forbes in Bothalia **3**: 128 (1930). —Schreiber in Merxmüller, Prodr. Fl. SW. Afrika, fam. 60: 96 (1970). —Torre in C.F.A. **3**: 83 (1962). —Drummond in Kirkia **8**: 225 (1972).
  Cullen obtusifolium (DC.) C.H. Stirt. in Bothalia **13**: 317 (1981) as "*obtusifolia*".

Herb up to 60 cm tall, often mat- or cushion-forming, white strigose or grey hairy but eventually ± glabrous; branches many, pale, up to 75 cm long (up to 9 m fide Forbes), widely spreading prostrate to ascending from a much branched rootstock. Leaves pinnately 3-foliolate; leaflets 3–30 × 2–20 mm, obovate or elliptic, obtuse at the apex, plicate, with margin slightly dentate to lobulate-crenate, glandular, thinly grey pubescent; petioles 0.5–7 cm long; stipules 2.5–3.5 × 1.25–2 mm, triangular to ovate, ciliate and grey pubescent. Inflorescences spicate or subcapitate with 3–8 flowers; peduncle 5–28 mm long; rhachis 3–27 mm long with 1–6 nodes; pedicels very short. Calyx tube 1.5–3 mm long; lobes unequal, lanceolate to ovate, ciliate, hairy outside, pilose inside, the upper 1–2 × 0.5–1 mm, the lowest, 2–3 × 1–1.5 mm. Corolla white and blue or pink; standard 3–5(7) × 3–4.5 mm with claw 1–2 mm long, ± obovate; wings 3–4.5 × 1–2 mm, with a linear claw 1.5–4 mm long; keel 2–3 × 1.5 mm with linear claw 2–2.5 mm long. Fruit 3.8–5 × 2.5–3 × 1.75–2.5 mm, oblong-reniform, densely white hairy and glandular.

Botswana. N: Groot Laagte (East), fossil river valley at 20°56.8'S, 21°25.75'E, st. 15.iii.1980, *P.A. Smith* 3194 (K; SRGH). SW: Masetlheng (Masetleng) Pan, south side, 23°41'S, 20°51'E, fl. 12.xi.1976, *Skarpe S* 86 (K). SE: near Gaborone, fl. ix.1967, *Lambrecht* 317 (K; SRGH). **Zimbabwe.** S: Beitbridge Distr., Lindi Valley, Tuli Circle, fl. 18.ix.1967, *Cleghorn* 1810 (K; SRGH). **Mozambique.** GI: 31 km from Mabalane (Vila Pinto Teixeira) to Balule, 1969, *Correia & Marques* 1240 (WAG) (cited by Grimes as *Correier & Marquas*).
  Also in Angola, Namibia and South Africa. Grassy areas, edges of pans and rivers, hot dry wooded grassland and scrubland on sandy soil; 90–1000 m.
  Grimes, in his synonymy, tentatively includes *Psoralea holubii* Burtt Davy (Fl. Pl. Ferns Transvaal, pt. 2: xxix & 374 (1932), type from South Africa). Further material seen is identical with the type material, and the distinctive facies persuades me it is best left distinct. Stirton has published the combination needed, *Cullen holubii* (Burtt Davy) C.H. Stirt.
  Forbes' "up to 9 m" is probably incorrect but numerical data in most field notes is sparse. Some of the qualitative data certainly suggests the branches are longer than 75 cm.
  Grimes includes in his synonymy *Psoralea tomentosa* Thunb. (Prodr. Fl. Cap. pars: 135) as "(Thunb.) Thunb." but the description does not agree and I think Stirton is correct in referring it to *Otholobium sericeum* (Poir.) C.H. Stirt.

Tab. 3.6.**11**. CULLEN TOMENTOSUM. 1, habit (× ²/₃), from *Welwitsch* 1988; 2, flower (× 4); 3, standard (× 6); 4, wing (× 6); 5, keel (× 6); 6, keel, spread out (× 6); 7, androecium (× 6); 8, gynoecium (× 6), 2–8 from *W. Giess* 56; 9, fruiting calyx (× 4); 10, fruit (× 6), 9 & 10 from *Lambrecht* 317. Drawn by Pat Halliday.

72. OTHOLOBIUM C.H. Stirt.

Otholobium C.H. Stirt. in Adv. Leg. Syst.: 341 (1981); in S. African J. Bot. **52**: 1–6 (1986); A revision of Otholobium C.H. Stirton (Papilionoideae, Leguminosae) Thesis, Univ. Cape Town, 2 vols. (1989). —J.W. Grimes in Mem. New York Bot. Gard. **61**: 16–32 (1990).

Small trees, shrubs, subshrubs, or herbs. Leaves 1-foliolate or 3-foliolate, stipulate, subsessile or petiolate; leaflets entire, black- or pellucid-dotted, mostly recurved-mucronate; stipules evident, striate, adnate to the base of the petiole. Flowers sessile to pedicellate, bracteate, aggregated in 1 or 5–50 triplets (rarely doublets) in axillary or apparently terminal fascicles, imperfect racemes or lax pseudo-spikes, each triplet subtended by a solitary bract; cupulum absent. Calyx campanulate, 5-lobed, the lowest lobe usually longer than the rest; vexillary lobes generally fused to some degree above the tube. Standard scarcely reflexed vertically, with auricles weakly developed and appendages absent; wings with a lamellate sculpturing pattern, distal edges characteristically overlapping and rounded; keel petals purple-tipped on inner face of apices, fused along lower edges, free along upper edges. Vexillary stamen free or variously fused to androecial sheath; anthers uniform, alternately basifixed and medifixed. Ovary uniovulate; style bent; stigma capitate. Fruit swollen, indehiscent, pubescent, papery or cartilaginous. Seeds compressed, longer than broad.

A genus of 61 species, all but one of the Old World species restricted to South Africa, the extralimital one extending to Zimbabwe, Malawi, Tanzania and Kenya; 8 species also in South America (Andes) from Columbia and Venezuela to Chile and Argentina.

Otholobium foliosum (Oliv.) C.H. Stirt. in South African J. Bot. **52**: 2 (1986); A revision of *Otholobium*, Thesis, Univ. Cape Town: 333, fig. 9.120 (1989). —Lock, Leg. Afr. Check-list: 455 (1989). TAB. 3.6.**12**. Type from Kenya.
    Psoralea foliosa Oliv. in J. Linn. Soc., Bot. **21**: 399 (1885). —E.G. Baker, Legum. Trop. Africa: 94 (1926). —Brenan, Check-list For. Trees Shrubs Tang. Terr.: 436 (1949). —Dale & Greenway, Kenya Trees & Shrubs: 374 (1961). —White, F.F.N.R.: 161 (1962). —Gillett in F.T.E.A., Leguminosae, Pap.: 1011 (1971). —Drummond in Kirkia **8**: 225 (1972). —Beentje, Kenya Tree Shrubs & Lianes: 307 (1994).

Branched erect shrub 1–3.5 m tall with densely pubescent branchlets. Leaves digitately 3-foliolate, mostly crowded; leaflets 7–35 × 2.5–14 mm, obovate to oblanceolate, acute, rounded or very slightly emarginate at the apex and recurved-mucronate, cuneate at the base, undulate, pubescent or glabrescent, densely gland-dotted, the glands smaller and more numerous on upper surface; petiole 1–2 mm long; petiolules c. 1 mm long; stipules 2–4 mm long, oblong-falcate. Inflorescences dense, subcapitate, mostly in uppermost axils and appearing terminal, comprised of 3–15 triplets of flowers; pedicels 2–3 mm long; bracts subtending each triplet 1.5–5 mm long, ovate or obovate, persistent, glandular, ciliate; bracts subtending flowers 1.5–3 mm long, linear to lanceolate. Calyx tube 3 mm long; lobes unequal, the lowest ± lanceolate, 6 mm long, the rest triangular-ovate, all pubescent and sparsely glandular. Standard white or pale blue with purple veins or darker purple centre, 9–10 × 7–8 mm with claw 1–3 mm long; wings longer than the keel, 9–10 × 2 mm with claws 2.5–3 mm long; keel purple with yellow glands at the tip, petals 6.5–10 × 2 with claws 3.5–4 mm long. Androecium 6–7 mm long, the vexillary filament lightly adherent. Ovary 1–2 mm long; stipe 1–1.5 mm long; style with straight part 3–5 mm long and apical upwardly curved part c. 1 mm long. Fruit 5–7 × 3–4 mm, papery, reticulate, puberulous. Seeds chestnut-brown to dark brown, 3.5–4.5 × 2.5–2.7 mm, oblong-reniform with central hilum.

Subsp. **foliosum**

Leaflets mostly acute at the apex. Fruit surface more strongly reticulate particularly when immature.

Zambia. E: Chama Distr., Nyika Plateau, upper slopes of Kangampande Mt., fl. 4.v.1952, *White* 2552 (FHO; K). **Malawi**. N: Rumphi Distr., 11.2 km north of entrance to Nyika National Park, fl. 1.x.1969, *Pawek* 2880 (K).

Tab. 3.6.**12**. OTHOLOBIUM FOLIOSUM. 1, flowering branch (× 1); 2, detail of leaflet upper surface (× 10); 3, flower (× 6); 4, calyx (× 6); 5, calyx, opened out (× 6); 6, standard (× 6); 7, wing (× 7); 8, keel (× 6); 9, androecium (× 6); 10, androecial sheath spread out (× 6); 11, gynoecium (× 6), 1–11 from *Milne-Redhead & Taylor* 11007; 12, fruit (× 6); 13, seeds (× 6), 12 & 13 from *Lynes* D.73. Drawn by Christine Grey-Wilson. From F.T.E.A.

Also in Tanzania and Kenya. Montane grassland, upland evergreen forest margins, lake and stream edges; 1980–2135 m.

Subsp. **gazense** (Baker f.) Verdc. comb. et stat. nov. Type: Zimbabwe, Chimanimani (Melsetter), *Swynnerton* 1417 (BM, holotype; K).
  Psoralea foliosa var. *gazensis* Baker f. in J. Linn. Soc., Bot. **40**: 52 (1911). —Eyles in Trans. Roy. Soc. South. Africa **5**: 375 (1916). —E.G. Baker, Legum. Trop. Africa: 94 (1926).
  Otholobium gazense (Baker f.) C.H. Stirt., A revision of *Otholobium*, Thesis, Univ. Cape Town: 316 (1989).

Leaflets rounded to slightly emarginate at tip. Fruit surface less strongly reticulate to almost smooth.

Zimbabwe. E: Chimanimani (Melsetter), fl. 25.iv.1947, *Wild* 1957 (K; SRGH). **Mozambique.** Shown as occurring just over the border on Stirton's map fig. 9.112 but no material is cited.
  Also in South Africa (Transvaal). Forest edges, *Acacia abyssinica* woodland, bushland, *Loudetia–Themeda* grassland, streambeds; (900)1200–2100 m.
  Stirton maintains these two taxa at specific rank but the differences seem slight. Some material from South Africa (northern Transvaal) may form a third subspecies.

73. PSORALEA L.

Psoralea L., Sp. Pl.: 762 (1753); Gen. Pl., ed. 5: 336 (1754). —Forbes in Bothalia **3**: 116–136 (1930) pro parte. —Hutchinson, Gen. Fl. Pl. **1**: 418 (1964) pro parte. — C.H. Stirton in Polhill & Raven, Adv. Leg. Syst.: 341 (1981).

Small trees, shrubs or suffrutices, covered in prominent blackish-red glands, especially on the calyx. Leaves digitately 3-foliolate, 3–5-pinnate, 1-foliolate, or rarely reduced to scales; leaflets entire; stipules embracing the stem by the broad base, fused to petiole near the base. Flowers axillary, fasciculate, 1–5, each with or without a bract but always subtended by a lobed cupulum which itself is subtended by 2 free bracts; bracteoles absent. Calyx lobes equal, the upper 2 mostly connate; inner face of lobes invested with stubby black hairs. Corolla blue; standard round, appendaged above the claw with inflexed auricles; wings longer than keel, distinctly heeled above claw; keel incurved, shortly clawed, somewhat falcate, with darker patch at tip. Vexillary stamen free or joined only near base, arching over ovary chamber; anthers uniform, alternately basifixed and versatile. Ovary distinctly stipitate, 1-ovulate, glabrous with a few scattered recurved club-headed glands. Style glabrous, dilated at the base, incurved in upper part; stigma penicillate. Fruit enclosed by the calyx at maturity, ovate, indehiscent; pericarp reticulately veined, fragile. Seeds black, shortly funiculate.

A genus of c. 20 species, mostly restricted to the Cape Province of South Africa but two extending to Swaziland and southern Mozambique.

Leaflets 1.5–4(6) mm wide; species from above 500 m · 1. *arborea*
Leaflets up to 1 mm wide; species from near sea-level · · · · · · · · · · · · · · · · · · · 2. *glabra*

1. **Psoralea arborea** Sims in Bot. Mag. **46**: t. 2090 (1819). —Lock, Leg. Afr. Check-list: 458 (1989). TAB. 3.6.**13**. Type grown in Kensington, England from seed sent from the Cape.
  Psoralea pinnata var. *latifolia* Harv. in F.C. **2**: 145 (1862). —Burtt Davy, Fl. Pl. Ferns Transvaal, pt. 2: 375 (1932). —Verdoorn in Fl. Pl. Africa **26**: t. 1029 (1947).
  Psoralea affinis sensu Hutchinson in Bot. Mag. **136**: t. 8331 (1910) non Eckl. & Zeyh.

Woody shrub or subshrub 1–3.5 m tall; branches striate, hairy or glabrous. Leaves 5–7(9)-foliolate; leaflets 10–50 × 1.5–4(6) mm, linear-oblong to linear-lanceolate, apiculate at the apex, cuneate to narrowly rounded at the base, densely gland-dotted, glabrous to sparsely pubescent; stipules separate, thickened, 2.5–4 × 2 mm, ovate-lanceolate, acuminate. Peduncles 15–25 mm long, 1-flowered, fasciculate, 1–3(6) together, axillary in upper leaves and terminal, pubescent with black hairs; 2 bracts at the apex united to form a cup around the peduncle, c. 3

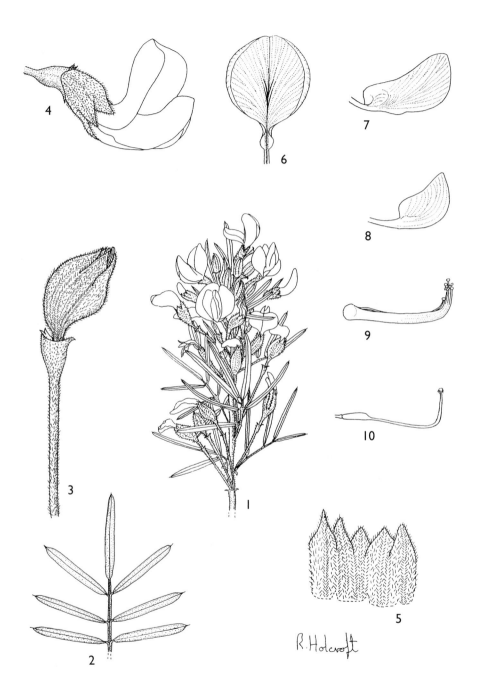

Tab. 3.6.**13**. PSORALEA ARBOREA. 1, flowering branch (× ²/₃); 2, single leaf (× ²/₃); 3, peduncle with bud, bracts united to form a cap; 4, flower (× ¹/₂); 5, calyx, opened out (× ¹/₂); 6, standard (× ¹/₂); 7, wing (× ¹/₂); 8, keel (× ¹/₂); 9, androecium (× ¹/₂); 10, gynoecium (× ¹/₂). Drawn by R. Holcroft.

mm long, split on one side, at first close to calyx base but later distant when the true pedicel elongates, bristly pubescent with black hairs. Calyx tube 6 mm long, lobes 4–5 × 3 mm, oblong-ovate, subequal, subacute or rounded at the apex, the lowest the largest, both tube and lobes glabrous to pubescent on the outer surface (and lobes within) with characteristic black hairs; margins of lobes ciliate. Standard dark violet or blue, paler at the margins, 13–15 × 12–15 mm, broadly obovate with white claw 4 mm long; wings white below and pale violet in the upper half, 12.5–13 × 4–6 mm with claw 7 mm long; keel about as long, white, apically suffused with dark violet-purple. Vexillary filament shortly united with staminal sheath at the base. Ovary c. 3 mm long with stipe 1 mm long; style straight for c. 7 mm then curved upwards for 4 mm. Fruit 6 × 3.8 mm, ellipsoid, wrinkled; seed dark brown, 4.2 × 3 mm, ellipsoid; hilum 1.2 mm long.

Mozambique. M: Namaacha Distr., Mte M'Ponduine (Ponduini), fl. 25.vii.1980, *Schäfer* 7197 (BM; K; LMU).
Also in South Africa. Rocky places, near waterfalls; 700–800 m.

2. **Psoralea glabra** E. Mey., Comment. Pl. Afr. Austr. **1**: 83 (1836). —Walpers, Repert. Bot. Syst. **1**: 656 (1842). —Lock, Leg. Afr. Check-list: 458 (1989). Syntypes from South Africa.
    Psoralea pinnata var. *glabra* (E.Mey.) Harv. in F.C. **2**: 145 (1862). —Burtt Davy, Fl. Pl. Ferns Transvaal, pt. 2: 375 (1932).

Woody herb or subshrub 0.6–1.5 m tall; branches striate, glabrous or with a few short hairs on youngest parts. Leaves 5–9-foliolate; leaflets 10–25 × 0.7–1.1 mm, linear, apiculate, glabrous, gland-dotted; stipules 1–2 × c. 0.5 mm, lanceolate, acute. Peduncles 1–2 cm long, 1-flowered, fasciculate, 1–4 together, axillary in upper leaves and terminal, glabrous or sparsely pubescent; 2 bracts at apex united to form a cup around the peduncle, 2.5 mm long, triangular; pedicels 2–3 mm long. Calyx tube 3–4.5 mm long, ribbed and gland-dotted; lobes (1)2–4.5 × 1.2–2 mm, ovate-triangular, subequal; tube and lobes sometimes densely dark pubescent within and appearing ciliate. Standard blue, 10–12 × 9–11 mm, obovate with claw 2 mm long; wings 9 × 3 mm, narrowly oblong-oblanceolate; keel dark blue-purple at apex, 5 × 2.8 mm, claw 5 mm long. Vexillary filament free. Ovary 1.3 mm long; style straight for 6 mm then curved upwards for 2.5 mm. Fruit 5 × 2.5 mm, ± oblong, wrinkled. Seed dark brown, 4.2 × 2.2 mm, oblong-reniform; hilum 1.2 mm long.

Mozambique. M: Matutuíne Distr., Ponta do Ouro, 4 km along the road to Maputo, fl. 9.i.1980, *de Koning* 7895 (K; LMU).
Also in South Africa. Damp vlei on sandy soil; c. 15 m.
The single specimen seen from Mozambique matches material from northern KwaZulu-Natal but further material is needed to confirm the identification.

Tribe 11. **AESCHYNOMENEAE**

by B. Verdcourt

Aeschynomeneae (Benth.) Hutch., Gen. Fl. Pl. **1**: 470 (1964).
Subtribe Aeschynomeninae Benth. in Bentham & Hooker f., Gen. Pl. **1**: 448 (1865), as "*Aeschynomeneae*".
    Stylosantheae (Benth.) Hutch., Gen. Fl. Pl. **1**: 435 (1964).

Shrubs or herbs, sometimes with tubercular-based glandular hairs. Leaves paripinnate or imparipinnate, sometimes digitately 2- or 4-foliolate, sometimes pinnately 3-foliolate, rarely 1-foliolate or simple; stipules often striate, sometimes prolonged basally or rarely joined to the petiole; stipels usually absent. Inflorescences axillary; pedicels inserted singly; bracts and bracteoles often conspicuous and striate. Hypanthium generally apparent, sometimes extended as a narrow receptacular tube resembling a pedicel. Calyx usually 2-lipped, sometimes 5-toothed or the 2 or 4 upper lobes joined higher than the lower ones. Staminal tube not closed, the filaments in a sheath or 2 bundles, rarely with the vexillary one free;

anthers uniform. Style tapered to a small terminal stigma. Fruits transversely jointed, the articles usually indehiscent, sometimes only 1, occasionally buried underground by the elongated receptacle. Seeds reniform, with a rather small radicular lobe and small lateral hilum.

About 25 genera in tropical and warm temperate regions.

The New World genera formerly referred to the subtribe Bryinae B.G. Schub. are now placed in the tribe Aeschynomeneae.

Brya ebenus (L.) DC., West Indian Ebony, native of the West Indies, is mentioned as an ornamental by Biegel, Check List Ornam. Pl. Rhod. Parks & Gard.: 31 (1977). Shrub or small tree 1–2(8) m tall, with simple obovate leaves clustered on short shoots in the axils of small recurved spines; racemes of orange-yellow flowers (standard 10 mm long) and fruits of generally 1 article c. 13 mm across on a distinct stipe.

1. Leaves in clusters in the axil of a reduced leaf with its petiole modified as a small recurved spine; ornamental shrub or small tree with racemes of orange-yellow flowers and stipitate 1-seeded fruits (see above) ·**Brya**
   – Leaves not so modified; other features not combined · 2
2. Leaflets digitately 2 or 4 (occasionally a few leaves with 3 leaflets), often glandular-punctate · 80. **Zornia**
   – Leaflets 3–numerous (sometimes in *Aeschynomene* undeveloped at flowering time), pinnate, the leaflets not all arising at the same point · 3
3. Leaves 4-foliolate; fruits developing underground; receptacle (extended hypanthium) long and filiform and easily mistaken for a pedicel but flowers are actually sessile; cultivated for seeds (peanuts) but sometimes found as an escape · · · · · · · · · · · · · · · · · · 82. **Arachis**
   – Leaves not all 4-foliolate; fruits not developing underground (or if so, in *Aeschynomene nematopoda*, leaflets very narrow); receptacle not developed except in *Stylosanthes* · · · · 4
4. Leaves pinnately 3-foliolate; calyx extended basally as a pedicel-like receptacle; fruits with 1–2 articles, the apical article narrowed into a distinct hooked beak · · · · 81. **Stylosanthes**
   – Leaves paripinnate or imparipinnate with more leaflets; calyx not so extended; fruits rarely as above · 5
5. Calyx subequally 5-toothed; shrubs with hard woody stems; articles of the fruit elliptic or oblong, veined, often bristly or papillate; leaves imparipinnate, the leaflets usually ± alternate · 74. **Ormocarpum**
   – Calyx 2-lipped, or rarely 5-toothed but then tiny herbs; shrubs or herbs; articles rectangular or semicircular, less often elliptic; leaves usually paripinnate or rarely absent · · · · · · · · 6
6. Fruit curved into a ring, the articles dehiscent, leaving the persistent sutures; annual herb · 79. **Cyclocarpa**
   – Fruit not as above, if curved then plant not an annual · 7
7. Inflorescence mostly lax; fruit straight, curved or rarely coiled, of 1–∞ articles, mostly well exserted from the calyx; bracts mostly small (save in sect. *Rubrofarinaceae*), not scarious and not masking the flowers and fruits; bracteoles often deciduous · · · · · · 75. **Aeschynomene**
   – Inflorescence mostly dense and ± scorpioid, often strobilate; fruit not visible, enclosed within the calyx or bracts, folded like a concertina, of 1–9 articles · · · · · · · · · · · · · · · 8
8. Fruit not included in the calyx, of 1–2 articles; bracts membranous, larger than the flowers and hiding the fruit, persistent; bracteoles and calyx membranous · · · · · · 78. **Humularia**
   – Fruit included in the calyx, of 2–9 articles; bracts scarious, smaller than the flowers; bracteoles and calyx scarious · 9
9. Stipules not spurred; leaflets alternate, with 2–7 basal nerves; inflorescence distichous, usually dense and strobilate; bracts persistent; lateral appendages of keel petals short or lacking · 76. **Kotschya**
   – Stipules spurred; leaflets opposite, with only 1 main nerve; inflorescence subumbellate (in the Flora Zambesiaca area); bracts deciduous; lateral appendages of keel petals nearly as long as the claws · 77. **Smithia**

74. ORMOCARPUM P. Beauv.

Ormocarpum P. Beauv., Fl. Owar. & Ben. **1**: 96 (1806) *nom. conserv.* —Gillett in Kew Bull. **20**: 323 (1966). —Verdcourt in Kirkia **9**: 361 (1974).
Diphaca Lour., Fl. Cochinch.: 457 (1790).
Saldania Sim, For. Fl. Port. E. Afr.: 42, fig. 33 (1909).

Shrubs or small trees with usually two kinds of indumentum, weak white deciduous hairs and stiffer tubercular-based hairs which are often aromatic and glandular and frequently persist as conical tuberculars on the older stems. Leaves often fasciculate on short shoots, 1-foliolate (not in the Flora Zambesiaca area) or more usually imparipinnate; stipules striate, persistent; stipels absent; leaflets usually ± alternate, entire or minutely notched, mostly glabrous and with minute black dots above. Flowers in axillary racemes or solitary, rarely in panicles; bracts persistent, sometimes tripartite; bracteoles persistent, opposite, at the base or just below the base of the receptacle (hypanthium). Calyx usually strongly veined; lobes mostly longer than the tube, the lowest the longest, the two upper united for half their length and connivent at the tips. Corolla glabrous or wings and standard pubescent at their apices and margins, often strongly veined; blade of standard forming an angle with the claw and bearing two variable ridges or appendages at the base. Staminal arrangement various, monadelphous or variously divided, 5 + 5, 1 + 9 or 1 + 4 + 5, sometimes varying in a single specimen; anthers uniform, or less often alternate anthers slightly longer. Ovary stipitate, 3–9-ovulate, often curving; style curved near its centre with the upper side concave, thin, cylindric, glabrous or sparsely hispid near the base; stigma terminal, minute. Pod breaking transversely into 1–6 indehiscent articles. Seeds flattened, asymmetrically ellipsoid; hilum towards one end, round or elliptic, c. 0.5 mm long, devoid of any appendages.

A genus of c. 20 species ranging from Senegal to South Africa and Madagascar, and through southern Asia to the Philippines, northern Australia and Fiji.

1. Corolla persistent · 2
 – Corolla deciduous, falling off soon after fertilization · 3
2. Leaflets with revolute margins, the venation beneath usually not drying dark; widespread
 · 4. *kirkii*
 – Leaflets with plain margins, the venation beneath drying blackish; mid-Zambezi valley · · ·
 · 5. *zambesianum*
3. Swollen-based hairs on the ovary and pod copious, many of them over 3 mm long · · · · ·
 · 3. *trichocarpum*
 – Swollen-based hairs on the ovary and pod short, under 3 mm long* · · · · · · · · · · · · · 4
4. Pod up to 6 mm wide, usually c. 3 mm thick, glabrous or with bristle-tipped prickles but without ordinary hairs; leaflets nearly or quite glabrous beneath; lateral nerves not noticeably dark nor margins strongly revolute · 1. *sennoides*
 – Pod more than 6 mm wide, c. 2 mm thick, pubescent with ordinary hairs and also with swollen-based hairs; leaflets appressed-puberulous beneath, at least on the midrib and with dark lateral veins and revolute margins · 2. *schliebenii*

1. **Ormocarpum sennoides** (Willd.) DC., Prodr. **2**: 315 (1825). —J.G. Baker in F.T.A. **2**: 143 (1871) pro parte. —Harms in Engler, Pflanzenw. Afrikas [Veg. Erde 9] **3** (1): 608 (1915). —Léonard in Bull. Jard. Bot. État **24**: 103 (1954). —Gillett in Kew Bull. **20**: 332 (1966); in F.T.E.A., Leguminosae, Pap.: 355 (1971). —Verdcourt in Kirkia **9**: 362 (1974). —Lock, Leg. Afr. Check-list: 119 (1989). Syntypes from India.
 Hedysarum sennoides Willd., Sp. Pl. **3**: 1207 (1802).

Shrub, scrambler or small tree 1–5 m tall, mostly devoid of white hairs; branchlets reddish-purple, glabrous or with viscid swollen-based hairs. Leaves sometimes fasciculate; leaflets 6–17, 0.6–5 × 0.4–2.2 cm, elliptic or oblong-elliptic, shortly mucronate, somewhat narrowed to the base, the margins flat, glabrous; petiole 4–7 mm long; rhachis 2.5–6.5 cm long; stipules up to 4 × 1.5 mm, ovate or lanceolate, acuminate with subulate tips. Racemes glandular-setose, sticky, 2–13-flowered, 0.5–7 cm long; peduncles 0.5–2 cm long; pedicels 3–9 mm long; bracts brown, scarious, 1.5–2 mm long, ovate, acuminate or 3-pointed, setose on the margins; bracteoles

---

* *Ormocarpum* sp. of Verdcourt in Kirkia **9**: 368 (1974) would key out here. Although also recognised in this account it has not been formally described, or keyed out, because there is as yet insufficient material.

1–2 mm below the receptacle, 2–3 mm long, ovate-acuminate. Calyx glabrous or with a few swollen-based hairs; receptacle cup (hypanthium) 3 mm long; calyx tube c. 2.5 mm long, the upper lobes c. 4 mm long, lateral lobes 4.5–5 mm long, triangular or lanceolate, and lower lobe 6–7 mm long, very acute, lanceolate-boat-shaped. Corolla deciduous, 9–16 mm long; standard yellow, red-veined, or dark purple-red almost black, ovate, 7–9 × 6–8 mm. Stamens monadelphous or diadelphous. Ovary 5–8-ovulate, glabrous or with swollen-based hairs on only the sutures or all over the faces also. Fruit dark blackish-green, longitudinally ribbed; stipe 3–8 mm long; articles 9–20 × 5–6 mm, 3 mm thick, glabrous or covered with backwardly-curved prickles 0.5–1.5 mm long.

Subsp. **hispidum** (Willd.) Brenan & J. Léonard in Bull. Jard. Bot. État **24**: 103, fig. 14, 2 (1954); in F.C.B. **5**: 244 (1954). —Hepper in F.W.T.A., ed. 2, **1**: 576 (1958). —White, F.F.N.R.: 161 (1962). —Gillett in Kew Bull. **20**: 334 (1966). —Torre in C.F.A. **3**: 191 (1966). —Verdcourt in Kirkia **9**: 363 (1974). —Lock, Leg. Afr. Check-list: 119 (1989). Type probably from Ghana.

    Cytisus hispidus Willd., Sp. Pl. **3**: 1121 (1802).

    Robinia guineensis Willd., Enum. Pl. Hort. Berol. **2**: 769 (1809). Type as for *Cytisus hispidus* Willd.

    Ormocarpum guineense (Willd.) Hutch. & Dalziel ex Baker f. in J. Bot. **66**, suppl. 1: 109 (1928). —E.G. Baker, Legum. Trop. Africa: 278 (1929).

Corolla 9–14 mm long. Ovary glabrous or with swollen-based hairs on the sutures only; style 6–8 mm long.

Zambia. W: Mwinilunga Distr., in forest fringing Zambezi River, 6.4 km north of Kalene Hill Mission, fl. 20.ix.1952, *White* 3298 (FHO; K).

Also in Senegal to Central African Republic, Angola and perhaps in Sri Lanka. Fringing forest and evergreen forest remnants; 1500 m.

A list of synonyms not concerned with the Flora Zambesiaca area is given by Gillett in Kew Bull. **20**: 334 (1966).

Subsp. *sennoides* is confined to peninsular India and Sri Lanka, while subsp. *zanzibarianus* Brenan & J.B. Gillett, distinguished mainly by its verrucose pods, occurs in Kenya and Tanzania including Zanzibar.

2. **Ormocarpum schliebenii** Harms in Notizbl. Bot. Gart. Berlin-Dahlem **13**: 418 (1936). —Brenan, Check-list For. Trees Shrubs Tang. Terr.: 435 (1949). —Gillett in Kew Bull. **20**: 336 (1966); in F.T.E.A., Leguminosae, Pap.: 357, fig. 52/3, 4 (1971). —Verdcourt in Kirkia **9**: 363 (1974). —Lock, Leg. Afr. Check-list: 119 (1989). Type from Tanzania.

    Pterocarpus sp. of Pires de Lima in Brotéria, Sér. Bot. **19**: 126 (1921). —E.G. Baker, Legum. Trop. Africa: 279 (1929).

Shrub 1–3 m tall; branchlets at first purplish, later whitish, sparsely pubescent with both white and swollen-based hairs; bark on older stems eventually splitting longitudinally. Leaves single or fasciculate; leaflets 9–17, (3)8–20 × (1.5)3–9 mm, elliptic-oblong, shortly mucronate, slightly cordate at the base, the margins revolute, sparsely appressed white pubescent beneath; lateral nerves and midrib dark purplish or brownish beneath; petiole 2–5 mm long; rhachis 2–5.5 cm long, pubescent and with swollen-based hairs; stipules up to 7 mm long, narrowly triangular, long drawn out to a fine tip. Racemes c. 10–15 mm long, mostly 2-flowered; peduncles very short; pedicels 4–15 mm long, glabrous or pubescent; bracts c. 5 per inflorescence, 2–3 mm long, ovate-lanceolate, acuminate; bracteoles similar, c. 4 mm long, 2–4 mm below the receptacle. Calyx glabrous save at the margin, receptacle cup (hypanthium) 2 mm long; calyx tube c. 2–3 mm long, the upper lobes 3 mm long, lateral lobes 3–4 mm long, triangular, lower lobe folded, 5 mm long, lanceolate, acuminate. Corolla deciduous; standard lemon-yellow, 8–13 × 8–10 mm, obovate, with claw 1.5–2 mm long, the blade with a thickened somewhat lobed callous protuberance at its base. Stamens diadelphous. Ovary 4–5-ovulate, pubescent. Fruit brownish, usually of 3 articles and 3.5 cm long, longitudinally and reticulately ribbed; stipe 2–3 mm long or appearing much longer when basal article does not develop, articles up to 16 × 8 mm, 2 mm thick, shortly and rather sparsely pubescent with stiff hairs c. 0.3 mm long, many of them with minutely tubercular bases and glandular. Seeds buff, 5.5 × 3 × 1.2 mm, very compressed ellipsoid with almost keeled margin.

Mozambique. N: between Memba and Cavá, fl. & fr. 28.x.1948, *Barbosa* 2622 (BM; K; LISC). Also in southern Tanzania. *Brachystegia–Isoberlinia, Combretum, Albizia* woodland and coastal bushland; 0–100 m.

3. **Ormocarpum trichocarpum** (Taub.) Engl. in Sitzungsber. Königl. Preuss. Akad. Wiss. **10**: 385 (1904). —Harms in Mildbraed, Wiss. Ergebn. Deutsch. Zentr.-Afrika Exped., Bot. **2**: 259 (1911); in Engler, Pflanzenw. Afrikas [Veg. Erde 9] **3** (1): 609 (1915). —E.G. Baker, Legum. Trop. Africa: 280 (1929). —Steedman, Trees, Shrubs & Lianes S. Rhod.: 26, pl. 21 (1933). —Brenan, Check-list For. Trees Shrubs Tang. Terr.: 435 (1949). —O.B. Miller in J. S. African Bot. **18**: 34 (1952). —Dale & Eggeling, Indig. Trees Uganda Prot., ed. 2: 309 (1952). —Léonard in F.C.B. **5**: 247 (1954). —F.W. Andrews, Fl. Pl. Anglo-Egypt. Sudan **2**: 244 (1955). —Pardy in Rhodesia Agric. J. **53**: 210, fig. (1956). —Dale & Greenway, Kenya Trees & Shrubs: 373 (1961). —Gillett in Kew Bull. **20**: 342 (1966); in F.T.E.A., Leguminosae, Pap.: 357, fig. 52/5, 6 (1971). —Drummond in Kirkia **8**: 224 (1972). —Verdcourt in Kirkia **9**: 364 (1974). —Gonçalves in Garcia de Orta, Sér. Bot. **5**: 100 (1982). —Lock, Leg. Afr. Check-list: 120 (1989). Type from Tanzania.

Diphaca trichocarpum Taub. in Engler, Pflanzenw. Ost-Afrikas **C**: 213 (1895). —E.G. Baker in J. Linn. Soc., Bot. **37**: 143 (1905).

Saldania acanthocarpa Sim, For. Fl. Port. E. Afr.: 42, fig. 33 (1909). Type: Mozambique, Maputo Prov., frequent in the Umbelúzi Valley and thence to the Lebombo Range, *Sim*, not located.

Ormocarpum setosum Burtt Davy, Fl. Pl. Ferns Transvaal, pt. 2: xxviii, 425 (1932). Type: Mozambique, Chimanimani Mts., Moribane, *W.H.Johnson* 221 (K, holotype).

Shrub or small tree 1–5 m tall, with white hairs and swollen-based hairs on young stems, leaf rhachis and inflorescence; old stems with fissured corky bark; branchlets often white. Leaves mostly fasciculate; leaflets 7–15, rather thick, 3–11(14?) × 1.5–3.5(6.5?) mm, elliptic-oblong, rounded or retuse and shortly mucronate, rounded or subcordate at the base, the margins flat, pubescent beneath at least on midrib and margins with white hairs, or glabrous, usually densely minutely black-dotted; petiole 2–6 mm long; rhachis 1–3.5 cm long; stipules up to 3 mm long, narrowly triangular, acute, striate. Racemes short, up to 2.5 cm long but often the flowers appearing fascicled, 1–3-flowered; peduncle 0–5 mm long; pedicels 7–20 mm long, with mixed hairs; bracts and bracteoles 2–3 mm long, ovate-lanceolate to lanceolate, the latter 2–4 mm below the receptacle. Calyx usually glabrous except at the margin or sometimes sparsely white-pubescent but always without swollen-based hairs; receptacle cup (hypanthium) 2 mm long; calyx tube 3–5 mm long, the upper lobes 2–3.5 mm long, obtuse, lateral lobes 3–4 mm long, oblong or triangular, sometimes slightly falcate, obtuse, lower lobe 4–6 mm long, triangular, acute. Corolla deciduous; standard cream-coloured or bluish with deep blue-purple venation, the blade 10–14 × 10–17 mm, ovate, truncate or very broadly cuneate at the base, with two protuberances forming an inverted V, the claw 2.5–3 mm long. Stamens diadelphous in 2 bundles of 5. Ovary 5–8-ovulate, densely covered with appressed swollen-based hairs c. 2 mm long. Fruit ± straight or curved, up to 5(6) cm long, all densely covered with stiff yellowish swollen-based hairs up to 7 mm long; stipe 4–7 mm long; articles excluding bristles 8–12 × 6–8 mm, 2 mm thick. Seeds pinkish-brown, compressed, 5 × 3 × 1.2 mm, oblong-ellipsoid.

Botswana. N: between Francistown and Bisole (Bosoli), fl. 7.iv.1931, *Pole Evans* 3235 (BM; K; PRE). SE: 80 km NW of Francistown, 3.ii.1966, *Drummond* 8500 (LISC; SRGH). **Zimbabwe**. W: Bulawayo, fl. & fr. 11.xi.1920, *Borle* 1 (K; PRE). C: Gweru (Gwelo), fl. 8.ii.1949, *West* 2858 (K; PRE; SRGH). E: Mutare (Umtali), Circular Drive, fr. 7.iii.1966, *Corby* 1567 (K; SRGH). S: Lower Save (Sabi), west bank, fr. 2.ii.1948, *Wild* 2482 (B; K; SRGH). **Malawi**. S: Machinga Distr., Sitola to Milanje Village, 14.ii.1964, *Salubeni* 247 (SRGH). **Mozambique**. T: Chicoa to Chinhanda, c. 5 km from Chicoa, fr. 27.ii.1972, *Macêdo* 4933 (K; LMA). MS: between R. Revué and R. Munhinga, fr. 27.iv.1948, *Barbosa* 1607 (BM; LISC). GI: between Inharrime and Nhacoongo (Inhacoongo), 27.6 km from Inharrime, fl. 16.iii.1952, *Barbosa & Balsinhas* 4923 (BM; LISC). M: between Moamba and Ressano Garcia, fl. & fr. 17.ii.1948, *Torre* 7341 (BM; LISC).

Also in Sudan, Ethiopia, Rwanda, Uganda, Kenya, Tanzania, South Africa (Transvaal, KwaZulu-Natal) and Swaziland. In *Brachystegia, Acacia* or *Colophospermum* woodland and savanna; 50–1400 m.

Certain specimens from central and eastern Zimbabwe are apparently intermediate with *O. kirkii* and Gillett in Kew Bull. **20**: 350 (1966) has suggested that the two may hybridize. Unfortunately the specimens are not in ripe fruit. The small dotted leaflets, small flowers, and straight hairy ovary seem to me to suggest *O. trichocarpum* but most have been annotated as *O. kirkii* since the corolla does seem to be persistent and purple. The following are examples: Zimbabwe C: Chirumanzu Distr., 2.4 km from Mvuma (Umvuma), fl. i.1956, *Barrett* 56/56 (K; LISC; PRE; SRGH); E: Mutare Distr., commonage south of Cross Hill, fl. 1.ii.1957, *Chase* 6312 (COI; SRGH); Mukuni Purchase area, i.1960, *Davies* 2677 (K; LISC; SRGH). If ripe fruits were available the problem would probably disappear. In *Barrett* 51/56 (K; SRGH), from Zimbabwe C: Chikomba Distr., 19.2 km south of Chivhu (Enkeldoorn) on Mvuma (Umvuma) road, the petals are small but apparently persistent, the pod is twisted and with short tubercular-based hairs longer than usual for *O. kirkii*, and the general facies is that of *O. trichocarpum*. Field work is needed to solve this problem.

4. **Ormocarpum kirkii** S. Moore in J. Bot. **15**: 290 (1877). —Vatke in Oesterr. Bot. Z. **28**: 215 (1878). —Harms in Engler, Pflanzenw. Afrikas [Veg. Erde 9] **3** (1): 609 (1915). —E.G. Baker, Legum. Trop. Africa: 279 (1929). —Sprague & Milne-Redhead in Bull. Misc. Inform., Kew **1934**: 42 (1934). —Brenan, Check-list For. Trees Shrubs Tang. Terr.: 435 (1949). —Pardy in Rhodesia Agric. J. **53**: 210 (1956). —Dale & Greenway, Kenya Trees & Shrubs: 372 (1961). —Gillett in Kew Bull. **20**: 348 (1966). —Schreiber in Merxmüller, Prodr. Fl. SW. Afrika, fam. 60: 93 (1970). —Gillett in F.T.E.A., Leguminosae, Pap.: 362, fig. 53/5–7 (1971). —Drummond in Kirkia **8**: 224 (1972). —Verdcourt in Kirkia **9**: 366 (1974). —Gonçalves in Garcia de Orta, Sér. Bot. **5**: 99 (1982). —Lock, Leg. Afr. Check-list: 119 (1989). TAB. **3.6.14**. Syntypes from Somali Republic (S) and Kenya.

   Ormocarpum mimosoides S. Moore in J. Bot. **15**: 291 (1877). —Harms in Engler, Pflanzenw. Afrikas [Veg. Erde 9] **3** (1): 609 (1915). —E.G. Baker, Legum. Trop. Africa: 279 (1929) pro parte. —Burtt Davy, Fl. Pl. Ferns Transvaal, pt. 2: 425 (1932). —Brenan, Check-list For. Trees Shrubs Tang. Terr.: 435 (1949) pro parte. Type: Malawi, Manganja Hills, *Waller* (K, holotype).
   Ormocarpum discolor Vatke in Oesterr. Bot. Z. **29**: 223 (1879). Type from Kenya.
   Diphaca kirkii (S. Moore) Taub. in Engler & Prantl, Nat. Pflanzenfam. **3** (3): 319 (1894); in Engler, Pflanzenw. Ost-Afrikas C: 213 (1895).
   Ormocarpum affine De Wild. in Ann. Mus. Congo, Sér. IV, Bot. [Études Fl. Katanga] **1**: 197 (1903); in Contrib. Fl. Katanga: 84 (1921). —E.G. Baker, Legum. Trop. Africa: 279 (1929). Type from Dem. Rep. Congo.
   Ormocarpum bibracteatum sensu J. Léonard in F.C.B. **5**: 248 (1954). —White, F.F.N.R.: 160 (1962) non *Ormocarpum bibracteatum* (A. Rich.) Baker.

Shrub or small tree 2–9 m tall with mostly whitish branchlets, sparsely white-pubescent at first and with some persistent swollen-based hairs; bark rough, corky, longitudinally furrowed; older stems rough with a covering of old short shoots. Leaves single and spaced out on long young shoots but later fasciculate on short shoots; leaflets 7–13(17), 6–12(24) × 2–7(12) mm, elliptic-oblong, or the terminal one obovate, truncate, emarginate or rounded and mucronate at the apex, rounded or slightly subcordate at the base, the margins revolute, glabrous to sparsely to ± densely appressed white-pubescent beneath; lateral veins mostly scarcely visible, rarely dark; petiole 5–13 mm long; rhachis 2–4(7) cm long with white and swollen-based hairs; stipules up to 6 mm long, lanceolate, striate. Racemes 1–3-flowered, up to 15 mm long but usually short, white pubescent; peduncle 0–10 mm long; pedicels up to 17 mm long; bracts and bracteoles c. 2–3.5 mm long, ovate, the latter 1–5 mm below the receptacle. Calyx glabrous save on the margins, or rarely some white pubescence on the lobes but never with swollen-based hairs; receptacle 1 mm long; calyx tube 3–4 mm long, the upper lobes 3–6 mm long, rounded, joined for half their length, lateral lobes 5–8 mm long, broadly triangular, obtuse, lower lobe lanceolate, boat-shaped, acute. Corolla persistent, pinkish or purple, rather obscurely veined or pinkish-yellow with purple veins; standard blade 12–17(26) × 15–20(29) mm, broadly ovate or transversely elliptic, emarginate, cordate, with two rounded appendages at the base, usually with rather cottony hairs around the margins; claw 3–4 mm long. Stamens diadelphous in 2 bundles of 5, or rarely monadelphous. Ovary 5–6-ovulate, densely pubescent with swollen-based hairs. Pod rarely maturing, possibly due to insect attack, coiled within the persistent corolla, mostly only 1-seeded, pubescent, glandular and minutely tuberculate; articles 6–8 × 3–5 mm, 2–3 mm thick. Seeds straw-coloured, 3–4 × 2–2.5 × 1.5 mm, compressed-ellipsoid, smooth; seeds produced sparsely.

Zambia. N: Mpika, fl. & fr. 12.xi.1955, *Fanshawe* 2057 (K). C: Mt. Makulu Research Station, near Chilanga, fl. & fr. 10.vi.1957, *Angus* 1619 (FHO; K). S: Gwembe Distr., 4.8 km south of Sinazeze (Sinazezi), fl. & fr. 4.iv.1952, *White* 2607 (FHO; K). **Zimbabwe**. N: Gokwe South Distr.,

Tab. 3.6.**14**. ORMOCARPUM KIRKII. 1, flowering branch (× ²/₃); 2, standard (× 1); 3, wing (× 1); 4, keel (× 1); 5, half androecium (× 1); 6, gynoecium (× 1), 1–6 from *Brummitt* 11778; 7, fruit (× 1½), from *Faulkner* 399. Drawn by Pat Halliday.

on Gokwe–Charama road c. 25 km from Gokwe, fl. 15.iii.1963, *Bingham* 100 (K; SRGH). W: Shangani Reserve, fl. 1.iv.1951, *West* 3148 (K; SRGH). C: Chikomba Distr., 11.2 km north of Chivhu (Enkeldoorn) on the Harare (Salisbury) road, fl. & fr. i.1956, *Barrett* 58/56 (K; LISC; PRE; SRGH). E: Mutare Distr., Battery Spruit, fl. 20.i.1945, *Hopkins* in *GHS* 13229 (K; SRGH). S: Masvingo (Fort Victoria), fl. 28.x.1951, *Chase* 4152/4153 (BM; SRGH). **Malawi**. N: Karonga Distr., fl. 24.iii.1954, *G. Jackson* 1275 (BM; FHO; K). C: 16 km north of River Dwangwa (Dwanga), 8.vi.1938, *Pole Evans & Erens* 618 (PRE). S: Mulanje (Mlanje) Mt., fr. 6.viii.1957, *Chapman* 401 (BM; FHO). **Mozambique**. N: Nampula, fl. 19.ix.1936, *Torre* 677 (COI; LISC). Z: Lugela Distr., Namagoa, fl. 19.iii.1949, *Faulkner* Kew 399 (COI; K; SRGH). T: between Fíngoè and the frontier, 83.5 km from Fíngoè, fl. 28.vi.1949, *Barbosa & Carvalho* in *Barbosa* 3368 (K; LISC; LM; SRGH). GI: Inhambane, *Sim* 21186 (PRE) (poor specimen probably belonging here).

Also in Dem. Rep. Congo (Katanga Province), Somali Republic (S), Kenya, Tanzania, Namibia and South Africa (Transvaal). In *Brachystegia–Julbernardia*, *Combretum* or *Colophospermum* woodland, riparian woodland or thicket, or less often in open grassland, sometimes on termite mounds; 100–1500 m.

See note at end of species 3 for the possible occurrence of hybrids in east and central Zimbabwe. *Blackmore & Hargreaves* 653 (Malawi, Mangochi Distr., Namaso Bay, fl. 17.iii.1979 (BM) is unusual in having large leaflets up to 22 × 12 mm with venation very dark beneath; unfortunately no fruits are present.

5. **Ormocarpum zambesianum** Verdc. sp. nov.* Type: Mozambique, Tete Prov., Zumbo, *Macêdo* 5496 (LISC, holotype).
   Ormocarpum trachycarpum sensu Gonçalves in Garcia de Orta **5**: 100 (1982) non (Taub.) Harms.
   Ormocarpum sp. of Verdcourt in Kirkia **9**: 369 (1974).

Small shrub 1.5–2(4) m tall; branchlets blackish-grey covered with short tubercular-based hairs when young, but later glabrous, lacking white pubescence on young shoots. Leaves single or in tufts; leaflets 9–13, 5–12 × 3–6 mm, obovate to oblong, rounded at the apex with midrib projecting as short mucro, rounded or minutely subcordate at the base, glabrous on faces or with few hairs on midrib but with short tubercular-based hairs on margins which are entire or slightly crenate, the venation drying blackish-purple beneath and very conspicuous; petiole and rhachis combined 2.5–4 cm long with short tubercular-based hairs but no pubescence; stipules 5–7 mm long, triangular-lanceolate. Inflorescence 1–2-flowered; peduncle up to 10 mm long; pedicels 7–13 mm long, the bracteoles 2–3 mm long, ciliate, c. 2 mm below calyx base. Calyx tube 5 mm long, glabrous; upper lobes c. 3 mm long, oblong-falcate, lateral lobes c. 6 mm long, oblong-triangular, lowest lobe 7–8.5 mm long, all glabrous save apex sometimes ciliate. Corolla rose-salmon in colour, glabrous, the petals with dark venation when dry; standard up to 13 × 14 mm, oblate, broadly cuneate to cordate at the base; wings with blade 12 × 8 mm, semicircular; keel with blade 11.5 × 6.5 mm, similar in shape but more narrowed to the apex. Ovary c. 6 mm long, faces appressed-pubescent and margins with tubercular-based hairs. Fruits not seen.

Zimbabwe. N: Binga, fl. 11.i.1963, *Whellan* 2017 (SRGH). W: Hwange Distr., Victoria Falls, area east of Big Tree, fl. 14.ii.1979, *Ncube* 49 (K; SRGH). **Mozambique**. T: c. 8 km from Zumbo on road from Miruro, fl. 19.iv.1972, *Macêdo* 5223 (LISC).

Not known elsewhere. Along the middle reaches of the Zambezi R., usually on the valley floor in rocky, yellow or reddish sandy places; 358–880 m.

6. **Ormocarpum sp**. —Verdcourt in Kirkia **9**: 368 (1974).

Rhizomatous shrub c. 1.5 m tall; branches covered with short glandular swollen-based hairs or the remains of these. Leaves not fasciculate; leaflets 15–17, 7–25 × 6–16 mm, elliptic-oblong, very rounded to slightly retuse, shortly mucronulate, rounded to almost truncate at the base, deep green but paler beneath, margins not

---

* **Ormocarpum zambesianum** Verdc. sp. nov. *Ormocarpum trachycarpi* (Taub.) Harms valde affinis sed ramulis juvenilibus nigrogriseis tuberculatis haud albopubescentibus, calycis lobo infimo 7–8.5 mm longo, vexillo basi late cuneato usque cordato; superficiebus ovarii breviter appresse pubescentibus differt. Typus: Mozambique, Tete Prov., Zumbo, *Macêdo* 5496 (LISC, holotypus).

revolute, glabrous but very minutely punctate; lateral nerves and midrib not or only very slightly darkened; petiole just over 1 cm long, with swollen-based hairs; rhachis 5–10 cm long, mostly glabrous; stipules c. 6 mm long, narrowly triangular, acuminate, ciliate-dentate. Racemes in upper and lower axils, 2.5–3.5 cm long, 3–4-flowered; peduncles 1–2 cm long, densely covered with swollen-based hairs; pedicels 6–7 mm long, similarly hairy; bracts c. 12 per inflorescence, 2–3-fid or toothed, similarly hairy; bracteoles 2.5 mm long, ovate, c. 1 mm below the receptacle. Calyx with scattered swollen-based hairs, particularly on the lower lip; receptacle cup (hypanthium) c. 2 mm long; calyx tube 2 mm long, the upper lobes 3 mm long, rather obtuse, lateral lobes 4 mm long, triangular-ovate, lower lobe 5 mm long, narrowly triangular, acute. Standard colour not known, perhaps bluish, c. 8.5 mm long and wide including a short claw 2 mm long, almost round, the blade with two callous protuberances at its base. Stamens in 2 bundles of 5. Ovary margined with short swollen-based hairs, at least 5-ovulate.

Mozambique. Z: Pebane, fl. 9.iii.1966, *Torre & Correia* 15110 (LISC).
Not known elsewhere. *Craibia, Cassia, Mimusops, Ozoroa* thicket on dunes near ancient beach; 10 m.
This species is related to *O. sennoides* and *O. schliebenii* but is not identical with either. Until fruiting material is available it would be premature to describe it. So little information is available that it is mentioned only in a footnote in the key.

75. AESCHYNOMENE L.

Aeschynomene L., Sp. Pl.: 713 (1753); Gen. Pl., ed. 5: 319 (1754). —Verdcourt in Kirkia **9**: 370 (1974).
Herminiera Guill. & Perr. in Guillemin, Perrottet & Richard, Fl. Seneg. Tent.: 201 (1832).
Rueppellia A. Rich., Tent. Fl. Abyss. **1**: 203 (1847).

Erect or prostrate shrubs, subshrubs or annual or perennial herbs or in very few cases climbers, mostly covered with tubercular-based hairs. Leaves alternate or subfasciculate on short lateral branches, paripinnately 2–many-foliolate (one species is aphyllous, and one Brazilian species is unifoliolate); stipules truncate or produced below the point of attachment, membranous to leaf-like in texture, persistent or deciduous; stipels absent. Inflorescences axillary or less often terminal or leaf-opposed, falsely racemose or paniculate, rarely umbellate, or sometimes flowers solitary to ternate or in fascicles; bracts entire or 2–3-fid, persistent or deciduous, rarely wanting; bracteoles inserted below the calyx, mostly deciduous. Flowers small to fairly large, mostly yellow, often lined with purple. Calyx either 2-lipped, the lips* practically free, the upper lip entire or 2-fid, the lower entire or 3-fid, or campanulate with subequal lobes. Standard mostly rounded or pandurate, often emarginate, shortly clawed, often with thickenings at the base of the limb and tiny appendages at the base of the claw; wings straight or falcate, oblong or obovate with lateral spur and a series of small pockets; keel petals either partly joined or adhering only by means of the marginal fimbriations when present. Stamens monadelphous or rarely diadelphous, sometimes in 2 joined groups of 5 or the vexillary filament free in a few species; anthers uniform (see below). Ovary linear, usually stipitate, 2–28-ovuled; style inflexed, mostly glabrous; stigma terminal. Fruit linear or elliptic, compressed, straight, slightly curved or in one case rolled into a spiral, shortly to distinctly stipitate, mostly exserted from the calyx, the sutural nerve entire, crenulate, 1–28-jointed; articles disarticulating, round, elliptic or rectangular, flat or slightly convex, smooth or tuberculate, sometimes papery, nearly always indehiscent. Seeds oblong, reniform or lunate, sometimes slightly beaked; hilum small, often eccentric; rim-aril not developed or rarely slightly so.

---

* Throughout this account lips are measured from the base of the calyx, not just the true free lips.

A large genus of c. 150 species (sometimes estimated as high as 250), about half in Central and South America, the rest in Africa save for a few in Asia. Léonard's subgenus *Bakerophyton* (occurring in Angola and Dem. Rep. Congo) has been raised to generic rank by Hutchinson (Gen. Fl. Pl. I: 474 (1964)); in this the anthers are alternately large and very small and the fruits 1–2-jointed, enclosed in the bracteoles and calyx, but the general habit is so similar that I would prefer to follow Léonard and consider it a subgenus. Out of the 59 species of *Aeschynomene* known from Central Africa 5 are unfortunately too poorly known to describe and much work remains to be done.

Three subgenera are represented, *Aeschynomene* sensu stricto, *Rueppellia* (A. Rich.) J. Léonard and *Ochopodium* (Vogel) J. Léonard, distinguished in the conspectus below. Subgen. *Aeschynomene* was not subdivided by Léonard, but some may care to follow E.G. Baker in putting *A. elaphroxylon* and *A. pfundii* in a sect. *Herminiera* (Guill. & Perr.) Baker f. Subgen. *Rueppellia* can be divided into several sections as shown.

Subgen. **Aeschynomene**. Stipules spurred beneath or unilaterally auriculate, generally membranous and not leaf-like; ovules and pod-joints 3–28; pod with sutures crenulate; flowers small or large; keel petals with margins partially laciniate (save in species 2 and 3); wings free. *Species 1–13.*

Subgen. **Rueppellia** (A. Rich.) J. Léonard. Stipules unilaterally or bilaterally, generally obliquely, appendaged or unappendaged, usually leaf-like as in the bracts; ovules 1–2(4) and pod-joints 1–2(3) separated by a narrow isthmus; flowers usually small; keel petals not laciniate; wings free or adhering.
Sect. LIBERAE J. Léonard. Stipules not or slightly appendaged. Inflorescences axillary; bracts deciduous, rarely persistent. Standard not panduriform, mostly without calli at the base; wings free. Ovules 2–6; articles 1–7, not thin and papery. *Species 14–18.*
Sect. MARGINULATAE (Harms) J. Léonard. Stipules not appendaged. Inflorescences axillary and terminal; bracts deciduous. Standard not panduriform, without calli at the base; wings free. Ovules 1–2; articles rather thin, margined. *Species 19.*
Sect. BASIADHAERENTES J. Léonard. Stipules appendaged or not. Inflorescences axillary or axillary and terminal, rarely only terminal; bracts persistent or deciduous or absent. Standard panduriform or obpanduriform, mostly with 2 basal calli; wings usually joined by the lateral appendages. Ovules 2–3; articles not thin. *Species 20–53;* some of the species do not fit well, but are left here at present.
Sect. SAMAROIDEAE J. Léonard. Stipules not appendaged. Inflorescences terminal and sometimes also axillary; bracts deciduous; standard not panduriform, with 2 calli at base; wings joined briefly above the appendages. Ovules 2–3; articles thin and samara-like, not margined. *Species 54–56.*
Sect. RUBROFARINACEAE (P.A. Duvign.) Verdc. Bark breaking down to form a soft ferruginous floury covering*. Bracts large and papery but not hiding the flowers and fruit. *Species 57, 58.*

Subgen. **Ochopodium** (Vogel) J. Léonard. Stipules not spurred mostly leaf-like; calyx campanulate with 5 subequal lobes; ovules 2–8 and pod-joints 1–8 separated by a narrow isthmus; flowers usually small; keel petals not laciniate; wings free (*species 59* and *A. falcata*).

Key 1

1. Fruit with straight margins or at the most the lower one undulate, only slightly narrowed between the articles**; articles and ovules 3–28; stipules mostly membranous, spurred or appendaged; keel petals laciniate along the lower edges save in *spp. 2, 3* and usually *sp. 5* ··· **Key 2** (Subgen. *Aeschynomene*)
  – Fruit with 1–3(4, less often 5–9) articles joined by narrow or very narrow necks; stipules usually leaf-like in texture, appendaged or not at the base; keel petals not laciniate ···· 2
2. Stipules striate, strongly spurred, not leaf-like; fruit with upper margin straight, lower deeply incised between the 3–9 articles ························· 13. *americana*
  – Without the above combination of characters ····························· 3

---

* Only species 44 and 50 in sect. *Basiadhaerentes* share this character and neither has the large papery bracts.
** In rare instances alternate seeds do not develop and the fruit appears moniliform but in this case there are numerous articles and the necks are centrally, not marginally placed.

3. Calyx campanulate with 5 almost subequal lobes; prostrate herb with mostly odd number of leaflets (5–11) some of which are not opposite, the venation openly and prominently reticulate beneath; standard c. 7 mm long, sparsely pubescent outside; fruit comprised of 1–4 pubescent articles · 59. *micrantha**
   – Calyx distinctly 2-lipped; lips entire or 2–3-fid; standard glabrous, wings free or adhering · **4**
4. Bracts persistent, equally 3-partite often to the base, the lobes mostly small and lanceolate; fruit stipitate, the stipe 3–15(80) mm long; slender erect to prostrate annual or perennial plants with corolla 3–9 mm long · **Key 3**
   – Bracts persistent or deciduous, 2-partite or sometimes one of the lobes again unequally 2-fid, or if truly 3-partite then fruit never so distinctly stipitate; perennial herbs or subshrubs · **5**
5. Bark on all but very young stems soon breaking down to form a dark reddish-brown floury powder and bracts large, persistent and papery, bilobed, 4–17 × 30–35 mm · · · · · · · · · **6**
   – Bark on older stems not powdery, or if so then bracts not large and papery · · · · · · · · · **7**
6. Bracts widely spaced, not overlapping, with acute lobes; leaflets 6–23 × 2–10 mm in (6)9–13 pairs · 57. *pararubrofarinacea*
   – Bracts crowded, overlapping with rounded lobes; leaflets 5–17 × 2.5–8 mm in (3)4–8 pairs · 58. *rubrofarinacea*
7. Wings free** · **Key 4**
   – Wings adhering by basal appendages** · **8**
8. Flowers solitary; bracts absent · **9**
   – Flowers in several-flowered inflorescences; bracts present but sometimes very deciduous · · **10**
9. Erect herb or subshrub with tufted glabrous or glabrescent stems; leaflets 2.5–6 × 0.6–1.5 mm; standard 7.5–10 mm long · 23. *solitariiflora*
   – Prostrate herb with densely hairy stems; leaflets 1.5–3 × 0.5–0.8 mm; standard 7 mm long · 24. *laterÍticola*
10. Articles of the fruit 12–27 × 7–15 mm, elliptic, thin and papery; standard more ovate or elliptic; stipules not appendaged; wings adhering above the appendages · see last 3 species in **Key 6**
   – Articles 3–13 × 3–8 mm, not thin and papery; standard often violin-shaped or at least narrowest near the middle; stipules with or without appendages, or if articles thin and up to 16 × 9 mm then stipules conspicuously appendaged · **11**
11. Bracts persistent or only falling very tardily, often 2–3-fid · · · · · · · · · · · · · · · · · **Key 5**
   – Bracts soon deciduous · **Key 6**

Key 2

1. Herb with long prostrate jointed stems floating and spreading out over the water; inflorescences axillary, 1-flowered · 4. *fluitans*
   – Herbs without long prostrate floating stems, or shrubs; inflorescences axillary or leaf-opposed, 1–8-flowered · **2**
2. Robust shrub with pithy spiny stems, growing by lakes and often ± floating; fruit spirally coiled; corolla 3–4.7 cm long · 12. *elaphroxylon*
   – Herbs or subshrubs, branches not spiny; fruit mostly quite straight, never spirally coiled · · **3**
3. Fruit oblong-elliptic, 6–7-jointed, 3–5 × 1.1–1.9 cm; small shrub similar in habit to *A. elaphroxylon* but stems not spiny; leaflets serrulate towards the apex · · · · · · · · 11. *pfundii*
   – Fruit linear or linear-oblong · **4**
4. Keel petals not laciniate along lower margin · **5**
   – Keel petals laciniate along lower margin · **7**
5. Stipe of pod 1–2 mm long; pod margins straight; plant not drying black; articles of the fruit smooth · 5. *mediocris*
   – Stipe of pod 5–9 mm long; lower pod margin distinctly to slightly undulate; plant drying blackish or greenish · **6**
6. Plants (at least stems and fruit) drying blackish; articles smooth when adult · · 2. *sensitiva*

---

* See also note on *Aeschynomene falcata*.
** This character may be difficult in certain cases – both **Key 4** and couplet 8 should be tried.

– Plants drying greenish; articles strongly tuberculate when adult · · · · · · · · · · · 3. *indica*
7. Flowers 1–3, placed singly in the axils; bracts absent · · · · · · · · · · · · · · · · · 1. *uniflora*
– Flowers in 1–8-flowered racemes or umbellate inflorescences; bracts present but sometimes soon falling · 8
8. Leaves 1–8(11.5) cm long; inflorescences 1- rarely 2–4-flowered; bracteoles ovate-lanceolate, not closely appressed to the calyx, 3–7 × 1–2 mm (contrast with *sp. 10*); calyx lips usually entire or rarely 2–3-toothed; fruit with straight margins; articles 4–28, each 3–6 mm long, smooth when adult; standard 8–33 mm long, glabrous · · · · · · · · · · · 6. *schimperi*
– Leaves 4–20 cm long; inflorescences (1)2–8-flowered; calyx lips always 2–3-lobed; fruit with margins slightly constricted between the articles; articles 4–11, each 8–11 mm long · · · 9
9. Fruit with margins verrucose and articles muriculate over the entire surface when adult; calyx lips equal, 5–9 mm long; standard 9–12 mm long, glabrous; wings c. 7 mm long; keel 9–11 mm long · 7. *afraspera*
– Fruit with margins not verrucose; calyx lips mostly unequal, 7–12 and 7–16 mm long; standard 10–33 mm long and wide; keel 12–32 mm long · 10
10. Standard glabrous; wings 6–14 mm long; articles of the fruit 6–7 mm wide sparsely finely verrucose when adult · 8. *nilotica**
– Standard hairy outside towards the apex; wings 15–25 mm long; articles 7–10 mm wide, smooth or with small spine-like tubercles when adult; bracteoles ovate, closely appressed to the calyx, 4–6 × 2–4 mm · 10. *cristata*

Key 3

1. Erect or more rarely decumbent annual herbs; stipules 4.5–16(18) × (5)2–3(5) mm; bracts small and narrow, 2–3(4.5) mm long, corolla 3–4.5(5.5) mm long; articles of the fruit small, 2–3.5 × 1.5–3 mm, the stipe rather short, c. 3 mm long · · 26. *minutiflora* subsp. *minutiflora*
– Mostly prostrate annual or perennial herbs · 2
2. Stipules mostly broader, 8–28 × 1.2–8 mm; bracts mostly rather larger and broader, c. (2)4–6 mm long; corolla 6–10 mm long; articles 2.7–3.5 × 2.5–3 mm with stipe 3–4(6) mm long; perennial herb · 31. *rhodesiaca*
– Stipules narrower, 4–25 × 1–5 mm; bracts often small and narrow as in *A. minutiflora* and without other characters combined · 3
3. Fruit developing below ground, the stipe elongating and pushing the fruit into the ground · 30. *nematopoda*
– Fruit not developing below ground · 4
4. Articles of the fruit mostly larger, up to 5 × 4.5 mm with a distinctly elongated stipe 7–8(15) mm long; corolla 4–6.5 mm long; stipules larger; annual or perennial herbs · 28. *mossambicensis*
– Articles mostly smaller, under 4 mm long or wide; stipe 3–5 mm long; stipules mostly distinctly smaller · 5
5. Inflorescence lax; stipules 6–10 × 1–2 mm; corolla (6.5)8–9 mm long; articles 3.5 × 3.5 mm, sometimes with a very distinct median spine; stipe 3–4 mm long · 26. *minutiflora* subsp. *grandiflora*
– Inflorescence condensed; stipules 2–5 × 0.5–1.2 mm; corolla 4.5–6 mm long; articles 3–3.5 × 2.5–3 mm, without a median spine; stipe c. 3.5–5 mm long · · · · · · · · · · · · · 29. *sp. B*

Key 4**

1. Prostrate perennial herb; leaves 8–36-foliolate; leaflets 4–12 × 1.5–4 mm, oblong; articles of the fruit 6 × 3.5 mm · 49. *stipulosa*
– Erect herbs or shrubs · 2
2. Leaflets larger, 2.5–35 × 2–22 mm · 3
– Leaflets smaller, 3–5 × 0.5–1.2 mm, or leaves absent · 8

---

* 9. *Aeschynomene sp. A* will key out here and differs from *Aeschynomene nilotica* in having a larger standard and keel.
** If the specimen being keyed out is not found in **Key 4** return to **Key 1**, couplet 7 "wings adhering", since there may be some difficulty with this character.

3. Leaflets with marginal gland dots but sometimes not very easy to see, slightly crenulate; pedicels 10–27 mm long; calyx lips obviously lobed; articles of the fruit thin, 14–25 × 8–14 mm with a narrow but distinctive smooth wing-like margin around the central reticulate portion; stems not appearing jointed ···························· 19. *baumii*
– Leaflets without gland dots around the margins, quite entire; pedicels 2–16 mm long; calyx lips not or obscurely lobed; articles without marked wing-like margins; stems mostly appearing jointed where marked by stipule scars ························· 4
4. Standard shorter than the keel ··································· 5
– Standard and keel about the same length ····························· 6
5. Fruit with 3–7 articles, each article 5–8 × 4–6 mm, pubescent or rarely glabrous; leaflets mostly glabrous; pedicels 9–16 mm long ····················· 14. *nodulosa*
– Fruit with 1–2 articles, each article 10–20 × 6–10 mm, almost entirely glabrous; leaflets mostly hairy; pedicels 5–6 mm long ························ 15. *semilunaris*
6. Standard 17–22 × 14–18 mm; fruit with 4–8 articles; stipules 17–28 × 5–8 mm; leaflets 10–35 × 7–22 mm ·· 16. *megalophylla*
– Standard shorter; fruit with 2–3 articles ···························· 7
7. Leaflets 12–30 × 6–20 mm, glabrous; stipules 15–37 × 13–30 mm; standard 11–13 × 7–8 mm ··· 17. *grandistipulata*
– Leaflets 6–15 × 3–6 mm, hairy; stipules 6–19 × 5–7 mm; standard 11 × 8 mm ······· ··· 18. *chimanimaniensis*
8. Leaves normally developed, 14–24-foliolate; leaflets 1–4(5) × 0.6–1.2 mm, elliptic to oblong; standard 10–12 mm long; articles of the fruit 4–5 × 3.5–4 mm ············ 22. *gazensis*
– Leaves absent or only rarely a few present at the base of the stem; leaves and petioles represented by a vestige, but stipules well-developed and congested ············· 9
9. Stipules 2.5–5(7.5) × 1–1.5 mm, with margins only minutely setulose; leaf reduced to a linear vestige 2–4.5 mm long (or rarely leaves developed) ··· ·· 20. *inyangensis*
– Stipules 6–16 × 1.5–3 mm, with margins densely setulose; articles reticulate; leaf reduced to a mere scale less than 0.5 mm long ···························· 21. *aphylla*

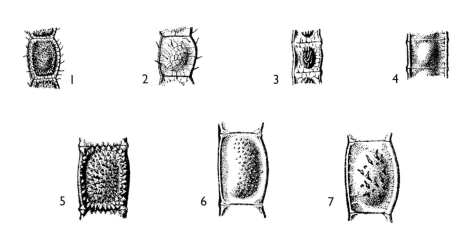

Tab. 3.6.**15**. AESCHYNOMENE. Articles of fruits (× 2). 1, A. UNIFLORA, from *Dewèvre* 23; 2, A. SENSITIVA, from *Louis* 10763; 3, A. INDICA, from *van Meel* in *G. de Witte* 5267; 4, A. SCHIMPERI, from *Lebrun* 5379; 5, A. AFRASPERA, from *Callens* 3368; 6, A. NILOTICA, from *Louis* 7559; 7, A. CRISTATA, from *Kesler* 1022. Drawn by J.M. Lerinckx. From Fl. Congo Belge. Reproduced with permission of Jardin Botanique National de Belgique.

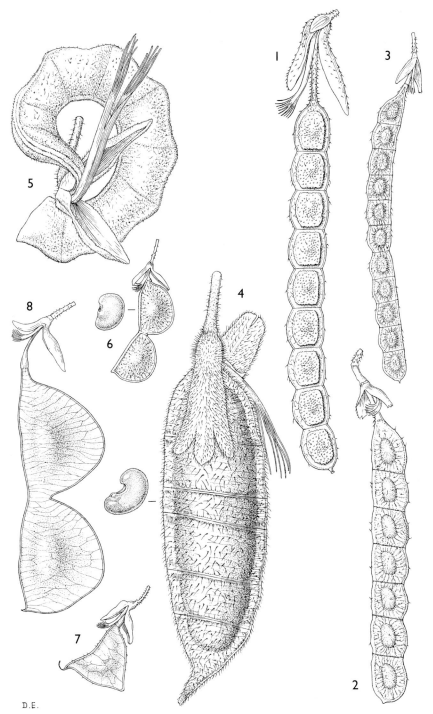

Tab. 3.6.**16**. AESCHYNOMENE.   Fruit (× 2).   1, A. UNIFLORA, from *Faulkner* 662; 2, A. SENSITIVA, from *Jackson* 575/65; 3, A. INDICA, from *Maitland* 176; 4, A. PFUNDII, from *Vesey-FitzGerald* 443; 5, A. ELAPHROXYLON, from *Schweinfurth* 1869; 6, A. ABYSSINICA, from *Chandler* 390; 7, A. TRIGONOCARPA, from *Richards* 19800; 8, A. NYASSANA, from *Jefford & Newbould* 2786.   Drawn by Derek Erasmus.   From F.T.E.A.

Key 5

1. Articles of the fruit sparsely hairy · 2
 – Articles glabrous* · 5
2. Leaflets in 3–5 pairs, elliptic-oblong or obovate-oblong, 3–10 × 1.5–5.5 mm; articles of the
 fruit 6.5–9 × 5–6 mm; standard 10–13 mm long · 3
 – Leaflets more numerous, or if few then narrower and mostly differently shaped and articles
 smaller · 4
3. Straggling subshrub 15–30 cm tall, with glabrescent stems · · · · · · · · · · · · · 32. *sparsiflora*
 – Small tufted herb of fireswept areas, c. 10 cm tall, with glandular pubescent stems · · 33. *sp. C*
4. Leaflets in 5–12 pairs, oblong, obovate-oblong or almost linear, 1–7 × 0.5–2 mm; stipules
 3–9 × 0.6–2 mm; articles of the fruit 3.5 × 3 mm; standard 8–10 mm long · · · · · 37. *stolzii*
 – Leaflets in 4–7 pairs, oblanceolate or oblong-oblanceolate, 4–15 × 1–3.5 mm; stipules 8–28
 × 1.2–8 mm; articles 2.7–3.5 × 2.5–3 mm; standard 6–10 mm long · · · · · · · 31. *rhodesiaca*
5. Stipules not appendaged or scarcely so · 6
 – Stipules appendaged at the base · 8
6. Lobes of bracts joined; leaflets 2–4.5 mm wide; articles of the fruit 8.5–9.5 × 6–6.5 mm · ·
 · 48. *pseudoglabrescens*
 – Lobes of bracts free; leaflets 0.3–7 mm wide; articles 4–5(7) × 3.5–4.5(6) mm · · · · · · · · 7
7. Leaflets 0.3–1.5(2.5) mm, typically 2–3-nerved rarely up to 6-nerved at base, not glaucous;
 branchlets viscid-hairy · 25. *bracteosa*
 – Leaflets up to 7 mm wide, 5–7-nerved at the base, ± glaucous; branchlets sparsely pilose or
 glabrous, not viscid-hairy · 27. *pawekiae*
8. Leaves 2–4-foliolate; petiole and rhachis mostly well under 1 cm long; seeds with a distinct
 rim-aril · 34. *oligophylla*
 – Leaves (4)6–32-foliolate; petiole and rhachis 0.5–10 cm long; seeds, where known, without
 a distinct rim-aril · 9
9. Straggling diffuse or prostrate subshrubs, sometimes scrambling over vegetation · · · · 10
 – Herbs suffrutescent with few to numerous erect stems from a woody rootstock, not in any
 way diffuse or scrambling · 11
10. Mat-forming herb or small bushy plant; leaflets predominantly 3–4 mm long · · 37. *stolzii***
 – More robust straggling plant; leaflets predominantly 7–8 mm long, usually ciliolate when
 young; stipules 6–25 mm long; pedicels 3–13 mm long; standard 7–11 mm long · · · · · ·
 · 38. *heurckeana*
11. Stems numerous; standard 7–8 mm long; articles of the fruit 4–7 × 3.5–5 mm*** · 35. *glauca*
 – Stems 1–5; standard 11 mm long; articles 8.5–9.5 × 5–6.5 mm · · · · · · · · · · · · · · · · 12
12. Stems and inflorescences glabrous; stipules 3–6.5 mm wide · · · · · · · · 48. *pseudoglabrescens*
 – Stems and inflorescences densely pubescent with tubercular-based hairs (in Flora
 Zambesiaca area); stipules 1.3–3 mm wide · 51. *sp. E*

Key 6

1. Leaves 4-foliolate; upper leaflets very asymmetrical with lateral petiolules; lower leaflets
 more symmetrical, slightly subcordate; stipules large and leaf-like · · · · · · · · · 39. *venulosa*
 – Leaves more than 4-foliolate and leaflets shaped differently · 2
2. Articles of the fruit 12–27 × 8–15 mm, elliptic, thin and papery; standard more ovate or
 elliptic; stipules not appendaged; wings adhering above the appendages · · · · · · · · · · · 3
 – Articles 4–13 × 3.8–8 mm, not thin and papery; standard often violin-shaped or at least
 narrowest near the middle; stipules with or without appendages; wings adhering by their
 appendages; or if articles thin, 16 × 9 mm then stipules conspicuously appendaged · · · 5
3. Leaflets with lateral nerves reaching the margin and joining to form a completely marginal
 nerve · 56. *glabrescens*
 – Leaflets with lateral nerves ascending and anastomosing in areas near the margin but not
 actually forming a marginal nerve · 4

---

* Save for one variant of *Aeschynomene glauca*.
** Included here since ovary and articles are sometimes practically glabrous.
*** 36. *Aeschynomene sp. D* will key near here but the habit is not evident; the standard is small,
6–7 mm long; see description. A further imperfectly known species of this group is mentioned
in a note after *sp. 34*.

4. Wings about the same length as the keel; standard 4.5–7 mm wide; staminal sheath often as long as or longer than the fruit stipe which measures 7–13 mm long · · · · · 54. *leptophylla*
– Wings longer than the keel; standard 2–5 mm wide; staminal sheath mostly shorter than the fruit stipe which measures 4–5(10) mm · 55. *nyassana*
5. Bark breaking down and forming a reddish powdery coating on the older shoots; stipules appendaged · 6
– Bark not breaking down to form a powdery coating · 7
6. Powdering effect usually very noticeable; shoots mostly thick; inflorescences mostly short; articles of the fruit with rounded lower margin, glabrous or slightly pubescent at the edges · 44. *schliebenii*
– Powdering effect less noticeable; shoots slender with longer more slender inflorescences; articles distinctly triangular, glabrous or almost so · · · · · · · · · · · · · · · · 50. *trigonocarpa*
7. Stipules not appendaged at the base · 8
– Stipules appendaged at base, or some appendaged and others not · · · · · · · · · · · · · · · 15
8. Articles of the fruit hairy on the faces · 9
– Articles glabrous or with only a few marginal hairs · 11
9. Leaflets 4–5-nerved from the base, lateral nerves not evident; leaflets 1.5–7 × 0.3–1.8 mm, mostly very narrow and falcate; articles of the fruit 5–8 × 4.5–6 mm · · · · · · · · · · · · · ·
· 41. *pygmaea* var. *hebecarpa*
– Leaflets with lateral nerves present; leaflets not falcate · 10
10. Leaflets larger, 4–17 × 1.5–7 mm, ciliate; articles 5.5–10 × 4.5–7.5 mm · · · · · · · · · · · ·
· 47. *mossoensis* var. *pubescens*
– Leaflets smaller, 3–8 × 1–2 mm, glabrous; articles 7–9 × 6–7 mm · · · · · · · · · · · · · · · · ·
· 52. *tenuirama* var. *hebecarpa*
11. Branched shrub or undershrub 0.6–3.6 m tall; standard 16–19 mm long; leaflets up to 12 × 3 mm; articles of the fruit 8–12 × 5–8 mm (very rarely there are slight traces of stipular appendages) · 40. *fulgida*
– Herbs or shrubs; standard 5–12(15) mm long; articles 4.5–10 × 4–7 mm (if dimensions of standard and articles overlap with those in the last statement then leaflets 1.5–3.5 × 0.5–0.7 mm) · 12
12. Main nerve of leaflets marginal or submarginal; leaves short, mostly fasciculate; petiole and rhachis together 5–30 mm long; leaflets 1.5–7 × 0.5–1.5 mm; standard 7–8 mm long; subshrub 0.3–2.7 m tall · 42. *mimosifolia*
– Main nerve of leaflets central or nearly so or, if submarginal, then standard over 10 mm long · 13
13. Leaves mostly distinctly fasciculate · 14
– Leaves alternate, well-spaced or sometimes practically absent · · · · · · · · · · · · · · · · 15*
14. Leaflets 2–12 × 0.5–3 mm; standard 5–13 × 3.5–6 mm; articles of the fruit 5–9 × 4–6 mm; stems mostly sparsely pubescent or glabrous · 45. *abyssinica*
– Leaflets 1.5–3.5 × 0.5–0.7 mm; standard 12–15 × 5–11 mm; articles 7.5–10 × 5.6–7 mm; stems mostly distinctly hairy** · 43. *nyikensis*
15. Subshrub with numerous erect stems from a very woody rootstock, mostly flowering after annual burn · 46. *multicaulis*
– Subshrub or herb with single or few stems from the rootstock · · · · · · · · · · · · · · · · · 16
16. Leaflets 4–13 × 2–4.5 mm; stipules 6–20 × 3–6.5 mm · · · · · · · · · · · 48. *pseudoglabrescens*
– Leaflets 2–8 × 0.5–2.5 mm; stipules 5–16 × 0.5–3 mm · 17
17. Articles of the fruit triangular, distinctly pointed at the base; stems leafy; wings c. 10 mm long · 50. *trigonocarpa****
– Articles of the fruit semicircular, rounded at the base · 18
18. Standard broadest at apex; stems leafy · 51. *sp. E*
– Standard broadest near the base · 19
19. Leaves short, 5–18 mm long, the stems often practically leafless; leaflets with secondary venation obscure, obtuse to acute and mucronulate at the apex, 0.5–1 mm wide; wings 8–12 mm long · 53. *katangensis*

---

* Since some of the remaining species in this group can rarely have all stipules unappendaged continue with couplet 15.
** The vast majority of specimens do not overlap as much as the couplet suggests and in practice distinction is easy.
*** Included here since young portions do not show the floury bark.

– Leaves mostly longer, 7–55 mm long, but sometimes almost absent from flowering shoots; leaflets with secondary venation obscure or very marked, retuse to rounded or rarely acute and mucronulate at the apex, 0.7–2.5 mm wide; wings 6–8 mm long · · · · · 52. *tenuirama**

1. **Aeschynomene uniflora** E. Mey., Comment. Pl. Afr. Austr. **1**, 1: 123 (1836). —J.G. Baker in F.T.Á. **2**: 146 (1871). —E.G. Baker, Legum. Trop. Africa: 286 (1929). —Brenan, Check-list For. Trees Shrubs Tang. Terr.: 407 (1949). —Léonard in F.C.B. **5**: 256, fig. 18A (1954). — Hepper in F.W.T.A., ed. 2, **1**: 580 (1958). —Rudd in Reinwardtia **5**: 31 (1959). —White, F.F.N.R.: 141 (1962). —Torre in C.F.A. **3**: 194 (1966). —Verdcourt in F.T.E.A., Leguminosae, Pap.: 372, fig. 54/3 (1971); in Kirkia **9**: 380 (1974). —Lock, Leg. Afr. Check-list: 108 (1989). TAB. 3.6.**15**, fig. 1; TAB. 3.6.**16**, fig. 1. Type from South Africa (KwaZulu-Natal).

Erect shrubby herb, often an annual, 0.5–2(?4.5) m tall. Stems covered with tubercular-based hairs. Leaves sensitive, 18–56-foliolate; leaflets 3–16 × 1–2.5 mm, linear-oblong, rounded at the apex, obliquely rounded at the base, finely serrulate, ± ciliate but otherwise glabrous; reticulation scarcely evident; petiole and rhachis together 2.5–8 cm long, hairy like the stems; petiolules 0.5 mm long; stipules straight, spurred, 3–8 × 1–2 mm, ovate-lanceolate, finally deciduous. Flowers solitary or ternate, axillary; pedicels 5–10(27) mm long, pubescent with tubercular-based hairs; bracts absent; bracteoles 3–4 × 1.5 mm, ovate, acuminate, ciliate. Calyx with tubercular-based hairs, 2-lipped; lips 6–7(12) × 2.5–4 mm, one wider than the other. Standard white to yellow, sometimes with dark veins, 6–22 × 4–18 mm, obovate, emarginate; keel petals paler greenish-yellow with margins partly laciniate. Fruit stipitate, 3–5.5 cm long (excluding stipe), linear, straight or slightly curved, 4–9-jointed, the suture somewhat constricted between the articles; articles 4–5.5 mm long and wide, oblong, mostly compressed, sparsely pilose, a central oblong-elliptic part slightly thickened, densely and finely tuberculate when adult or sometimes this area inflated and longitudinally crested. Seeds chestnut-brown, shiny, compressed, 3.5 × 2.2 × 1.2 mm, oblong-reniform, the end nearest the eccentric hilum somewhat canoe-shaped.

Var. **uniflora** —Verdcourt in Kirkia **9**: 381 (1974).

Standard up to 13 × 9 mm.

Zambia. N: Ncheta Island, Lake Bangweulu, 1.ix.1969, *Verboom* 2804 (K). W: Kitwe, fr. 14.iii.1966, *Mutimushi* 1284 (SRGH). C: South Luangwa National Park, c. 3.2 km north of Big Lagoon Game Camp, fl. & fr. 15.iii.1967, *Prince* 376 (K; SRGH). E: Nsefu Game Camp, Luangwa River, fl. & fr. 15.x.1958, *Robson* 129 (BM; K). S: Livingstone, fl. & fr. iv.1909, *Rogers* 7062 (K). **Malawi**. S: Liwonde National Park, near office, fl. & fr. 17.iv.1980, *Blackmore, Brummitt & Banda* 1259 (K; MAL). **Mozambique**. N: Mocímboa da Praia, 5 km from latter towards Diaca, 14.iv.1964, *Torre & Paiva* 11930 (LISC). Z: Namacurra Distr., Macuze, fl. & fr. 28.viii.1949, *Barbosa & Carvalho* in *Barbosa* 3861 (K; LMA). M: Maputo, Costa do Sol, fl. & fr. 4.v.1968, *Balsinhas* 1254 (LMA).

Widespread in tropical Africa from Senegal and Sudan to South Africa (KwaZulu-Natal) and Angola; also in Madagascar, Mascarene Islands and Comoro Islands, formerly cultivated in Jawa; rather rare in Central Africa. In wet places, sometimes even in water, on dambo fringes and sandy river beds, *Echinochloa* grassland, etc.; 20–1170 m.

Var. *grandiflora* Verdc., from southern Tanzania, has a larger standard, up to 22 × 18 mm.

2. **Aeschynomene sensitiva** Sw., Nov. Gen. Sp. Pl. Prodr.: 107 (1788). —J.G. Baker in F.T.A. **2**: 147 (1871). —E.G. Baker, Legum. Trop. Africa: 286 (1929). —Léonard in F.C.B. **5**: 258, fig. 18B (1954). —Hepper in F.W.T.A., ed. 2, **1**: 580 (1958). —Lind & Tallantire, Fl. Pl. Uganda: 83, fig. 35 (1962). —Verdcourt in F.T.E.A., Leguminosae, Pap.: 373, fig. 54/10 (1971); in Kirkia **9**: 381 (1974). —Lock, Leg. Afr. Check-list: 108 (1989). TAB. 3.6.**15**, fig. 2; TAB. 3.6.**16**, fig. 2. Type from the West Indies.

Erect much branched sometimes slightly viscid subshrub, 1–2 m tall; mostly blackening on drying. Stems pubescent with tubercular-based hairs or, in some areas, glabrescent. Leaves sensitive, 10–40-foliolate; leaflets 6–15 × 1–4 mm, linear-

---

* 52. *tenuirama* and 53. *katangensis* are mostly readily distinguishable but to cover all cases the ranges of measurements in the couplet statements have had to overlap.

oblong, rounded but mucronulate at the apex, obliquely rounded at the base, entire, glabrous; venation obscure; petiole and rhachis together 1–6 cm long, with tubercular-based hairs; petiolules 0.5 mm long; stipules 6–16 × 1–2.5 mm, oblong-lanceolate, straight, spurred, soon falling.   Inflorescences leaf-opposed, 3–7-flowered; peduncle 1–15 mm long; rhachis 1.5–8 cm long; pedicels 2–5 mm long; bracts 4–5 × 1–1.5 mm, deciduous; bracteoles 2.5–3.5 × 1.5 mm, ovate-elliptic, ciliate. Calyx glabrous or ciliate, 2-lipped; lips 3.5–4 × 2.5–3 mm, broadly oblong, one emarginate, the other 3-toothed at the apex.   Standard yellow, often flushed purplish-red or apricot outside, 7–8 × 5–6 mm, rounded, emarginate; keel petals yellow, not laciniate. Fruit stipitate, 3–4.5 cm long, linear, straight or slightly curved, (4)6–9-jointed, the suture somewhat constricted between the articles; articles 4–6 × 4–5 mm, ± oblong, compressed, very sparsely pubescent or at length glabrous, raised in the middle but smooth save for a slightly raised venation.   Seeds dark chestnut-brown, shiny, compressed, 3.2 × 2.2 × 1.8 mm, oblong-ellipsoid, almost beaked at one end of the hilum.

Malawi. C: Nkhotakota (Kota-Kota), fl. & fr. 2.v.1963, *Verboom* 891 (K; LISC; PRE; SRGH).
Mozambique. N: Nametil (Namatil), fr. 12.vii.1948, *Pedro & Pedrógão* 4442 (EA; LMA).
Also in West Africa, Dem. Rep. Congo, Sudan, Madagascar, Mascarene Islands, West Indies and tropical America. Lake shores; 170–500 m.

3. **Aeschynomene indica** L., Sp. Pl.: 713 (1753). —J.G. Baker in F.T.A. **2**: 147 (1871). —E.G. Baker, Legum. Trop. Africa: 286 (1929). —Léonard in F.C.B. **5**: 259, fig. 18C (1954). —Hepper in F.W.T.A., ed. 2, **1**: 580 (1958). —Rudd in Reinwardtia **5**: 30 (1959). —Torre in C.F.A. **3**: 195 (1966). —Schreiber in Merxmüller, Prodr. Fl. SW. Afrika, fam. 60: 12 (1970). —Verdcourt in F.T.E.A., Leguminosae, Pap.: 373, fig. 54/4 (1971). —Drummond in Kirkia **8**: 216 (1972). —Verdcourt in Kirkia **9**: 382 (1974). —Gonçalves in Garcia de Orta, Sér. Bot. **5**: 61 (1982). —Lock, Leg. Afr. Check-list: 104 (1989). TAB. 3.6.**15**, fig. 3; TAB. 3.6.**16**, fig. 3. Type from India.
   Aeschynomene oligantha Welw. ex Baker in F.T.A. **2**: 146 (1871). —Hiern, Cat. Afr. Pl. Welw. **1**: 234 (1896). Type from Angola.

Erect subshrubby annual or perennial herb, 0.3–2.5 m tall. Stems mostly slender, c. 5 mm wide at the base but sometimes thick and spongy up to 2.5 cm wide, pubescent with mostly rather sparse tubercular-based sometimes glandular hairs. Leaves sometimes sensitive, 16–50(70)-foliolate; leaflets 3–13 × 1–3 mm, linear-oblong, rounded and mucronulate at the apex, obliquely rounded at the base, entire or very finely serrulate, glabrous; venation obscure; petiole and rhachis together 1.2–10 cm long, with tubercular-based hairs; petiolules 0.3 mm long; stipules 3–15 × 1–3.5 mm, elliptic-lanceolate, straight, spurred, deciduous.   Inflorescences leaf-opposed or axillary, 1–6-flowered; peduncle 8–21 mm long; rhachis 1.5–6 cm long; pedicels 1–2 and finally up to 8 mm long; bracts 3.5–6 × 1.5–2.5 mm, mostly ovate, often lacerate or toothed, deciduous; bracteoles 2.5–4.5 × 1–2 mm, ovate-lanceolate to lanceolate, acuminate, ciliolate. Calyx glabrous, 2-lipped; lips 4–6 × 2–3 mm, oblong, one 2-fid, the other 3-fid.   Standard yellow or whitish, mostly lined and suffused with red outside or purplish, 7–10 × 4–7 mm, elliptic; wings and keel greenish-white or pale yellow, the petals of the latter not laciniate. Fruit 2.4–4.8 cm long excluding the 6–9 mm stipe, linear, straight or slightly curved, 5–13-jointed, one suture ± straight, the other slightly constricted between the articles; articles 3–5 mm long and wide, oblong, compressed, with sparse short tubercular-based hairs, central part raised and venulose or sometimes rugose, surrounding lower area at length rather coarsely rugose.   Seeds dark olive, black or brownish, 2.8 × 2.5 × 1.3–1.8 mm, oblong, slightly beaked near the eccentric small hilum.

Botswana. N: Khwebe (Kwebe) Hills, fl. & fr. 12.iii.1898, *Mrs. Lugard* 219 (K). SE: Sefare (Sephare), Mahalapye, fl. & fr. 25.ii.1958, *de Beer* 702 (K; SRGH). **Zambia**. B: Barotse Floodplain, fl. & fr. 19.iii.1964, *Verboom* 1203 (K; SRGH). N: Kaputa Distr., Mweru Wantipa (Mweru-wa-Ntipa), Mwawe River, fl. & fr. 6.iv.1957, *Richards* 9063 (K). W: Kitwe, fl. & fr. 28.ii.1954, *Fanshawe* 884 (K). C: Kabwe (Broken Hill), fl. & fr. v.1909, *Rogers* 8126 (K). E: Lundazi Distr., fl. & fr. i.1962, *Verboom* 443, 443a (K). S: Namwala Distr., Kafue National Park, Musa–Kafue River confluence, fl. & fr. 23.iii.1964, *B.L. Mitchell* 25/08 (K; LISC; SRGH). **Zimbabwe**. W: Hwange National Park, Ngamo Pans, fl. & fr. 17.iv.1972, *Grosvenor* 712 (K; LISC; SRGH). C: Gweru (Gwelo), fr. 10.v.1955, *Fitt* in *GHS* 51552 (K; LISC; SRGH). E: Chipinge

Distr., Save (Sabi) Escarpment behind Dotts Drift near Dakata, 9.iv.1959, *Savory* 402 (K; PRE; SRGH). S: Mwenezi (Nuanetsi), fl. & fr. 6.v.1958, *Drummond* 5600 (COI; K; LISC; PRE; SRGH). **Malawi**. N: Karonga Distr., Sangilo Hill, fl. & fr. 25.ii.1978, *Pawek* 13906 (K; MAL; MO; SRGH; UC). C: Lilongwe Nature Sanctuary, Zone A, fr. 28.i.1985, *Patel & Banda* 1986 (K; MAL). S: Shire Highlands, banks of Likangala R., fl. & fr., *Buchanan* 336 (K). **Mozambique**. N: Mutuáli, fl. & fr. 12.v.1948, *Pedro & Pedrógão* 3343 (EA; LMA). Z: Quelimane, fr., *Sim* 20769 (PRE). T: Lupata, iii.1859, *Kirk* (K). MS: Gorongosa Nat. Park, Urema floodplain, fl. & fr. ii.1971, *Tinley* 2034 (K; LISC; SRGH). GI: between Guijá and Inchobane (Bilene), fl. & fr. 11.v.1944, *Torre* 6624 (LISC; PRE). M: Maputo, fr. 17.iv.1947, *Hornby* 2645 (LMA).

Widespread in tropical Africa, from Senegal, Ethiopia and N Somali Republic to South Africa (Transvaal) and Namibia; also in São Tomé, Madagascar, tropical and subtropical Asia, Australia and N America. Grassland, savanna, scrub mopane usually in very wet places, swamps, pond and river edges, floodplains, pans, etc., often growing in standing water; on sandy or black clay soils, sometimes a weed; 30–1200 m.

4. **Aeschynomene fluitans** Peter in Abh. Königl. Ges. Wiss. Göttingen, Math.-Phys. Kl. **13** (2): 82, t. 11 (1928). —E.G. Baker, Legum. Trop. Africa: 883 (1930). —Léonard in F.C.B. **5**: 260 (1954). —Torre in C.F.A. **3**: 195 (1966). —Schreiber in Merxmüller, Prodr. Fl. SW. Afrika, fam. 60: 12 (1970). —Verdcourt in F.T.E.A., Leguminosae, Pap.: 374 (1971); in Kirkia **9**: 384 (1974). —Lock, Leg. Afr. Check-list: 103 (1989). Type from Tanzania.

   Aeschynomene schlechteri Harms ex Baker f., Legum. Trop. Africa: 289 (1929). Syntypes from Dem. Rep. Congo and Zambia, Barotseland, Sesheke, *Macaulay* (*Gairdner*) (K).

Perennial herb, 1–4.5 m long, floating on water surface. Stems hollow, spongy, thick, 0.5–1 cm in diameter, densely covered with adventitious roots. Leaves 16–26-foliolate; leaflets 9–25 × 2–6.5 mm, oblong to linear-oblong, rounded and mostly mucronulate at the apex, obliquely rounded or truncate at the base, entire or very finely serrulate, glabrous and glaucous; venation obscure; petiole and rhachis together 2.5–8 cm long, with a line of bristly hairs on the upper side; petiolules 0.5–1 mm long; stipules 13–21 × 3–5 mm, ovate-lanceolate, straight, spurred, at length deciduous. Flowers axillary, solitary; peduncle and pedicel together 3.5–10 cm long; bracts 2, one of which subtends the pedicel, 7–8 × 2–3 mm, ovate, appendiculate and amplexicaul, persistent; bracteoles 5–9 × 3–4.5 mm, ovate, ciliolate. Calyx glabrous, 2-lipped; lips oblong, one 1–11 × 3–9 mm, ± entire, and the other 1–15 × 4–6 mm, 3-lobed at the apex. Standard yellow, 18–25(30) × 18–20(30) mm, rounded; wings and keel yellow, the petals of the latter laciniate along their lower margins. Fruit 1.5–5 cm long including the short 5–7 mm long stipe, irregularly linear-oblong in outline, straight or slightly curved, 1–4(5)-jointed; suture slightly broadly constricted between the articles, and mostly with a raised corrugated or tuberculate ridge beside it forming a margin all round the fruit; articles 1–14 × 8–9.5 mm, irregularly oblong, mostly curved, compressed, glabrous, slightly venulose but otherwise smooth or only slightly tuberculate apart from the marginal ridge. Seeds dark purplish-brown, 8.5 × 5 × 2 mm, reniform, with a long central hilum.

   Botswana. N: Mborogha (Mboroga) River, 19°11'S, 23°08'E, fl. 31.iii.1975, *P.A. Smith* 1336 (K; SRGH). **Zambia**. B: Mongu, fl. 14.iii.1966, *E.A. Robinson* 6880 (K; SRGH). N: Kawambwa Distr., Kalungwishi (Kalungwisi) River, fl. 18.iv.1957, *Richards* 9243 (K). S: Namwala Distr., Bambwe (Baambwe), fl. & fr. 17.iv.1963, *van Rensburg* 2008 (K; SRGH).
   Also in Dem. Rep. Congo, western Tanzania, Angola and Namibia. Floating on water of slow rivers and ponds, and in floodplains, river and pond margins; 1000–1500 m.

5. **Aeschynomene mediocris** Verdc. in Kew Bull. **28**: 429 (1974); in Kirkia **9**: 385 (1974). —Lock, Leg. Afr. Check-list: 105 (1989). TAB. 3.6.**17**. Type: Zimbabwe, Zvimba Distr., Raffingora, *Corby* 2222 (BM; BR; COL; EA; K, holotype; LISC; PRE; SRGH; US).

Weak annual herb to 0.9 m tall, sometimes forming a spreading mat to 45 cm in diameter from the taproot. Main stem somewhat woody at the base; young branches glabrous to densely covered with tubercular-based yellowish bristly hairs. Leaves 14–46-foliolate; leaflets 2–4(7) × 0.8–1.2(2) mm, linear-oblong, rounded and mucronulate at the apex, obliquely rounded at the base, entire or very slightly serrulate near the apex, glabrous or with a few marginal setae; midrib and venation often purplish but the latter usually obscure; petiole and rhachis together 1–3(5) cm long, glabrous or with tubercular-based hairs; petiolules 0.3 mm long; stipules 3–10 × 1–1.5 mm, lanceolate, straight, spurred, persistent. Inflorescences axillary,

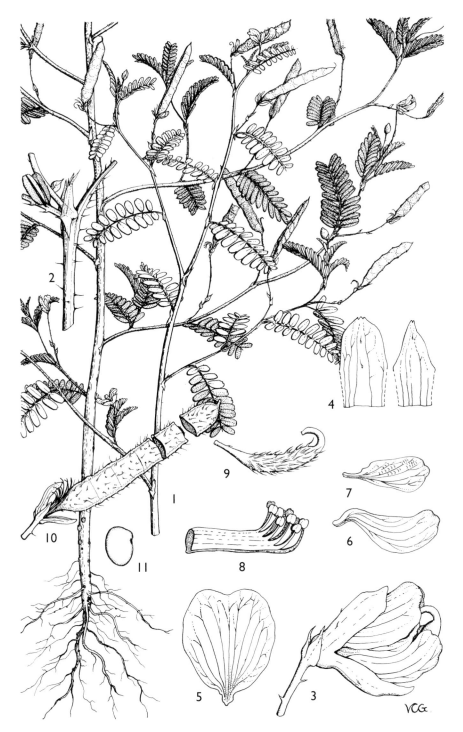

Tab. 3.6.**17**. AESCHYNOMENE MEDIOCRIS. 1, habit (× 1), from *Corby* 2222; 2, detail of stem showing stipule (× 4); 3, flower (× 6); 4, calyx, opened out (× 6); 5, standard (× 6); 6, keel (× 6); 7, wing (× 6); 8, androecium (× 6); 9, gynoecium (× 6); 10, fruit (× 2); 11, seed (× 4), 2–11 from *Robinson* 914. Drawn by Victoria Friis. From Kew Bull.

1-flowered; peduncle 5–10 mm long, pedicels 3–10 mm long with scattered short setae; bracts 2, 1–3 × 1–2 mm, rounded-ovate, clasping, acuminate; bracteoles 1.5–3 × 0.7 mm, narrowly lanceolate, acuminate, with very sparse marginal hairs. Calyx glabrous or with a few bristly hairs, 2-lipped; lips 4 × 1–2 mm, oblong, one entire the other 3-fid. Standard yellow, probably with darker veins (or flowers described as pale violet-pink), 5–6 × 3.5 mm, obovate or oblong; wings 2–4 mm long, oblong or rounded triangular, short; keel petals not or slightly lacerate beneath. Fruit 8–30 × 3–3.8 mm excluding the 1.5–2 mm stipe, linear, straight or very slightly curved, 3–11-jointed, both sutures almost straight; articles 2.5–3 × 3–3.8 mm, oblong, compressed, with sparse tubercular-based hairs, smooth including the part raised over the seed. Seeds pale brown, 2.2 × 1.6 × 1 mm, compressed ellipsoid, minutely pitted.

Zambia. B: Kaoma Distr., Kasempa Road, fl. & fr. 3.iv.1964, *Verboom* 1209 (K; SRGH). W: Mwekera, 16 km NE of Kitwe, fl. & fr. 24.iv.1962, *Fanshawe* 6781 (K). S: Muckle Neuk, 19.2 km north of Choma, fl. & fr. 11.x.1954, *E.A. Robinson* 914 (K; PRE; SRGH). **Zimbabwe**. N: Zvimba Distr., Raffingora, Mafuta Farm, fl. & fr. 8.iv.1972, *Corby* 2222 (BM; BR; COL; EA; K; LISC; PRE; SRGH; US).

Not known outside the Flora Zambesiaca area. Dambo grassland and swampy grassland by streams, also in gravel bed of drying stream; 1230–1280 m.

This seems to be related to both *A. indica* and *A. schimperi* and particularly has the facies of the latter but has much smaller corollas and fruit. The three cited sheets are not quite identical. *Fanshawe* 10559 (Zambia, Kitwe, fl. & fr. 20.iii.1969) resembles *Fanshawe* 6781 in having glabrescent fruits and slightly lacerate keel petals.

6. **Aeschynomene schimperi** Hochst. ex A. Rich., Tent. Fl. Abyss. **1**: 202 (1847). —J.G. Baker in F.T.A. **2**: 146 (1871). —E.G. Baker, Legum. Trop. Africa: 287 (1929). —Léonard in F.C.B. **5**: 262, fig. 18D (1954). —Hepper in F.W.T.A., ed. 2, **1**: 580 (1958). —White, F.F.N.R.: 141 (1962). —Verdcourt in F.T.E.A., Leguminosae, Pap.: 376, fig. 55 (1971). —Drummond in Kirkia **8**: 216 (1972). —Verdcourt in Kirkia **9**: 386 (1974). —Lock, Leg. Afr. Check-list: 107 (1989). TAB. 3.6.**15**, fig. 4; TAB. 3.6.**18**. Syntypes from Ethiopia.

  Aeschynomene telekii Schweinf. in van Hohnel, Zum Rud. See und Steph. See, Anhang: 866 (1892). —E.G. Baker, Legum. Trop. Africa: 268 (1929). —Brenan, Check-list For. Trees Shrubs Tang. Terr.: 407 (1949). Syntypes from Sudan and Kenya.

  Aeschynomene paludicola Harms in Bot. Jahrb. Syst. **30**: 329 (1901). —E.G. Baker, Legum. Trop. Africa: 287 (1929). Syntypes from Tanzania.

  Aeschynomene mearnsii De Wild. in Rev. Zool. Bot. Africaines **13**: B6 (1925). —E.G. Baker, Legum. Trop. Africa: 289 (1929). Syntypes from Kenya.

Shrubby aromatic herb or small shrub, 0.6–3 m tall, usually erect but rarely straggling. Stems soft at the base, pubescent to hispid with yellow often viscid tubercular-based hairs, sometimes glabrous, often reddish. Leaves sometimes sensitive, (10)18–68-foliolate; leaflets 2–12 × 1–2.5 mm, linear to linear-oblong, rounded to slightly emarginate and with a deciduous bristle at the apex, obliquely rounded at the base, subentire or very finely serrulate, margins often ciliolate, glabrous; venation often obscure; petiole and rhachis 1–8(11.5) cm long, glabrescent to hairy; petiolules obsolete; stipules 6–20 × 2–3 mm, lanceolate, straight, spurred, at length deciduous, veined. Inflorescences axillary, 1 (rarely 2–4)-flowered; peduncle 0.5–5 cm long; rhachis 5–6 cm long; pedicels 0.7–1.5 (3 in fruit) cm long; upper bract sterile, 3–6 mm long, ovate, acuminate; lower bract or bracts 2–4 mm long, ovate, sometimes acuminate, amplexicaul; bracteoles 3–7 × 1–2 mm, ovate-lanceolate, acuminate, mostly ciliate. Calyx glabrescent to hairy, 2-lipped, 6–18 × 3–8 mm, oblong, subequal, one lip entire or rarely 2-fid, the other entire or rarely 2–3-fid. Standard orange-yellow to orange, sometimes streaked with crimson, 8–33 mm long and wide, rounded, glabrous; wings and keel greenish, speckled or lined with purple, the petals of the latter bristly and laciniate along their lower margins. Fruit 3–10 cm long excluding the 4–12 mm stipe, linear, straight or rarely curved, (4)6–28-jointed, glabrous to densely hairy, the margins straight; articles 3–6 × 4–7 mm, oblong, compressed, smooth. Seeds brown or purplish-brown, 3.8–4.5 × 2.8–3 × 1.5–1.7 mm, irregularly oblong, produced slightly near the small eccentric hilum.

Zambia. N: near Mbala (Abercorn), old road to Pans, fl. & fr. 1.iii.1955, *Richards* 4727 (K). W: Ndola, fl. & fr. 30.iii.1954, *Fanshawe* 1038 (K; SRGH). C: Ndola–Kapiri Mposhi road, fl. &

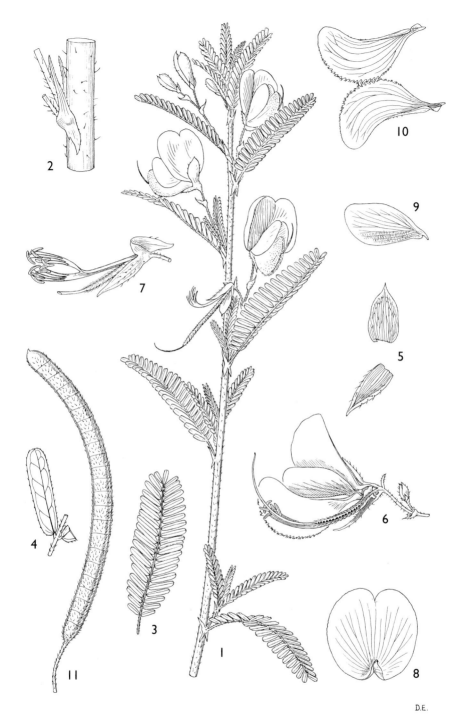

Tab. 3.6.**18**. AESCHYNOMENE SCHIMPERI. 1, flowering branch (× ²⁄₃), from *Pielou* 90; 2, detail of stem showing stipule (× 1), from *Lynes* P.r.62; 3, leaf (× ²⁄₃); 4, leaflet (× 2), 3 & 4 from *Kanure bin Kabue*; 5, calyx lips (× 1); 6, section of flower (× 1); 7, flower with petals removed (× 1); 8, standard (× 1); 9, wing (× 1); 10, keel, spread out (× 1), 5–10 from *Gichuru* 22; 11, fruit (× 1), from *Grote* in *Herb. Amani* 5614. Drawn by Derek Erasmus. From F.T.E.A.

fr. 19.iv.1961, *Verboom* LK 262 (K; SRGH). E: Petauke Distr., Chilongozi area, fl. & fr. ii.1963, *Verboom* 687 (K; SRGH). **Zimbabwe**. C: Chishawasha, fl. & fr. iii.1911, *Kolbe* 4307 (BOL; K). E: Mutare (Umtali) Commonage, fl. & fr. iii.1947, *Chase* 354 (BM; K; SRGH). **Malawi**. N: Mzimba Distr., Mbawa, fl. & fr. 14.vii.1952, *G. Jackson* 972 (K). C: Nambuma, fl. 10.ii.1959, *Robson* 1544 (BM; K). S: Blantyre, Nkolakote Village, bank of Maone Stream, 3 km north of Limbe, fl. & fr. 3.v.1970, *Brummitt* 10333 (K). **Mozambique**. N: Maniamba, Serra Jéci (Géci), fl. & fr. 29.v.1948, *Pedro & Pedrógão* 4062 (EA; LMA). Z: Ile (Errego), fl. & fr. 26.vi.1943, *Torre* 5591 (LISC). MS: Sussundenga Distr., Rotanda prox. du fronteira, fl. & fr. 5.iii.1948, *Pedro* 3641 (LMA).

Widespread throughout tropical Africa from Senegal and Ethiopia to Zimbabwe; also in Madagascar. Swampy places, stream banks, lake-sides, flooded grassland or wooded areas, marshes, dambos, etc.; 400–1500 m.

An exceptionally variable species, particularly in the size of the flowers and fruit. The wings are particularly prone to variation which does not seem to be in any way clearly geographically correlated. In some specimens they are almost as long as the keel but in others they are very much smaller, well under a half or even a third of its length. In the Flora Zambesiaca area these mostly occur in NW Zambia but are to be found in E Africa and as far away as N Uganda. An investigation of this genetically would be interesting — it may be a simple Mendelian variant. *Mutimushi* 3524 (Zambia W: Mufulira, fl. & fr. 14.viii.1969) is an exceptionally small-flowered variant with the petals c. 7 mm long.

7. **Aeschynomene afraspera** J. Léonard in Bull. Jard. Bot. État **24**: 64 (1954); in F.C.B. **5**: 264, fig.18E (1954). —Hepper in F.W.T.A., ed. 2, **1**: 579 (1958). —Torre in C.F.A. **3**: 196 (1966). —Verdcourt in Kirkia **9**: 387 (1974). —Lock, Leg. Afr. Check-list: 101 (1989). TAB. 3.6.**15**, fig. 5. Type from Dem. Rep. Congo.

Aeschynomene aspera sensu Baker in F.T.A. **2**: 147 (1871). —Hiern, Cat. Afr. Pl. Welw. **1**: 234 (1896). —E.G. Baker, Legum. Trop. Africa: 288 (1929) non L.

Shrub or shrubby herb 1–3 m tall. Stems hollow, thick and with abundant pith at the base, glabrous. Leaves sensitive, (20)36–100-foliolate; leaflets (4)8–22 × (1.2)5–3.5 mm, linear to linear-oblong, rounded or retuse and mucronulate at the apex, obliquely rounded at the base, entire or very finely denticulate, obscurely ciliolate; venation visible, sometimes dark; petiole and rhachis together (1.5)6–19(20) cm long, with sparse, tubercular-based hairs; petiolules c. 0.5 mm long; stipules 6–23 × 1.5–4 mm, lanceolate, straight, spurred, eventually sometimes deciduous. Inflorescence axillary, (1)2–4(6)-flowered; peduncle 1.3–4 cm long; pedicels 5–10 mm long, with short stiff hairs; bracts 4–8 × 2–3 mm, ovate, ciliate, but otherwise glabrous; bracteoles 3–7 × 1.6–3 mm, ovate or lanceolate, ciliolate. Calyx glabrous or pubescent, 2-lipped; lips (5)6–9 × 3–4.5 mm, oblong, one slightly 2-fid the other 3-fid. Standard yellow or greenish, 9–12 × (6)7–10 mm, elliptic, obovate or round; wings and keel yellow, the petals of the latter laciniate along their lower margins and pubescent outside. Fruit dark, 5–9 cm long excluding the 0.7–1.4 cm long stipe, linear, straight or slightly curved, 6–10-jointed, the margins ± straight or distinctly wavy, thickened and verrucose when adult; articles (7)8–12 × (6)7–8.5 mm, oblong, flat, veined when immature, then finely muricate over the entire surface when quite adult. Seeds pale to blackish-brown, 6 × 4 × 2 mm, compressed oblong-ellipsoid, beaked beyond the hilum at one end so as to resemble the prow of a canoe.

Zambia. E: Chipata Distr., Luangwa Valley, fr. 24.iii.1963, *Verboom* 690 (K). **Malawi**. N: Mwenembwe (Mwanemba), fl. & fr. iii.1903, *McClounie* 119 (K). C: Nkhotakota (Kota Kota), near jetty along bank of Lake Malawi (Nyasa), fl. & fr. 16.i.1964, *Salubeni* 197 (K; SRGH). S: Shire River, Elephant Marsh, *Kirk* (K). **Mozambique**. MS: Beira, fr. v.1947, *Pedro* 3146 (LMA).

Widespread in West Africa from Gambia to Nigeria, also Mali, Sudan, Angola and South Africa (Transvaal). Growing in water, river floodplains and lagoons, river banks and other wet habitats; 0–1050 m.

Verboom refers to this as an annual.

8. **Aeschynomene nilotica** Taub. in Bot. Jahrb. Syst. **23**: 189 (1896). —E.G. Baker, Legum. Trop. Africa: 287 (1929). —Léonard in Bull. Jard. Bot. État **24**: 68 (1954); in F.C.B. **5**: 265, fig. 18F (1954). —Hepper in F.W.T.A., ed. 2, **1**: 579 (1958). —White, F.F.N.R.: 141 (1962). — Schreiber in Merxmüller, Prodr. Fl. SW. Afrika, fam. 60: 13 (1970). —Verdcourt in F.T.E.A., Leguminosae, Pap.: 378 (1971). —Drummond in Kirkia **8**: 216 (1972). —Verdcourt in Kirkia **9**: 389 (1974). —Lock, Leg. Afr. Check-list: 106 (1989). TAB. 3.6.**15**, fig. 6. Type from Sudan.

Shrubby herb or small shrub, 1–2.5 m tall, floating in deep water but rooting in the soil when in shallow water, the lower stems thick, hollow, covered densely with adventitious roots and black tubercles. Young stems slightly pubescent or glabrescent, but soon becoming almost or quite glabrous. Leaves 24–80-foliolate; leaflets 5–25 × 1–5 mm, linear or linear-oblong, rounded and mucronulate at the apex, oblique at the base, entire or very finely serrulate, margins often ciliolate, glabrous; venation obscure; petiole and rhachis 4–20 cm long, slightly pubescent with tubercular-based hairs; petiolules 0.5 mm long; stipules 5–20 × 1–4 mm, lanceolate, straight, spurred, at length deciduous. Inflorescences axillary, 3–8-flowered, peduncle up to 2 cm long, pubescent with tubercular-based hairs; rhachis 1–8 cm long; pedicels 0.5–1(2.3 in fruit) cm long; bracts 3–6 × 2–3 mm, ovate; bracteoles 3–8 × 2–4.5 mm, ovate-elliptic, ciliolate. Calyx glabrescent or somewhat pubescent with tubercular-based hairs, 2-lipped; lips 7–13 × 3–6 mm, oblong, somewhat unequal, one entire or slightly emarginate, the other 3-lobed. Standard yellow or greenish-yellow, 11–25 × 1–28 mm, rounded, emarginate, crenellate, glabrous; wings greenish-yellow, irregularly crenellate; keel petals greenish or brown, laciniate along their lower margins. Fruit 2–8.5 cm long excluding the 0.8–1.5(2) cm stipe, linear, straight, 4–9-jointed, glabrous, the margins straight or slightly constricted between the joints; articles 8–10 × 6–7 mm, rounded-oblong, compressed, at first venulose, later often sparsely and finely verrucose. Seeds dark reddish-brown, 7 × 4 × 1 mm, compressed ellipsoid-reniform; hilum 3 mm long, oblong, the seed-end canoe-shaped beyond it.

Botswana. N: R. Okavango, upstream of Kaokwe (Xaokwe), fl. & fr. 22.ii.1984, *P.A. Smith* 4373 (K). **Zambia**. B: Sesheke, fl. & fr. iii.1910, *Macaulay* 444 (K). N: Kaputa Distr., Mweru Wantipa, Chishela (Chishyela) Swamp, fl. & fr. 6.iv.1957, *Richards* 9066 (K). S: Mumbwa Distr., between Stambi(?) and Shikatende, fl. & fr. 5.iii.1964, *van Rensburg* 2830 (K; SRGH). **Zimbabwe**. W: Hwange Distr., Kazungula, iv.1955, *Davies* 1095 (SRGH).

Also from Mali to Sudan, Cameroon and Dem. Rep. Congo, northwestern Tanzania and Namibia. Swamps, marshes, floodplains, around water holes, etc. often standing in water; c. 850–1050 m.

9. **Aeschynomene sp. A**
   Aeschynomene sp. B of Verdcourt in Kirkia **9**: 390 (1974).

Herb to 1.2 m tall. Stems glabrous or with some tubercular-based hairs near the nodes and on some of the youngest parts. Leaves 28–36-foliolate; leaflets 3–6.5 × 1–1.5 mm, narrowly oblong, rounded and minutely mucronulate at the apex, obliquely rounded at the base, glabrous, midnerve ± median, purplish, venation mostly blackish beneath; petiole and rhachis together 3.5–5.5 cm long, petioles with hairs from black tubercles near the base; petiolules very short; stipules 8 mm long, lanceolate, straight, spurred, soon deciduous. Inflorescences axillary 2(?several)-flowered; peduncle 5–15 mm long, with some tubercular-based hairs; pedicels 7–10 mm long; bracts 4 × 2.5 mm, ovate; bracteoles 4–4.5 × 2–3 mm, ovate to ovate-lanceolate, minutely ciliolate. Calyx glabrous save for ciliolate margins, 2-lipped; lips elliptic, the upper 8 mm long and wide, entire, acuminate, the lower 14 × 8 mm, deeply 3-lobed. Standard yellow, 16 × 17 mm, rounded-obovate, glabrous; wings and keel yellow, the petals of the latter laciniate along their lower margins and bristly hairy outside. Fruit 2.5–5(?) cm long excluding the 1.5–1.8 cm long stipe, linear, straight, 2–8-jointed the upper margin ± straight, the lower slightly undulate, somewhat thickened; articles 9 × 6.5 mm, oblong, flat, muricate with fine conical tubercles over the part covering the seed but the rest smooth and veined. Seeds dark reddish-brown, 5.5 × 3.5 × 1.2 mm, elongate-reniform, ± canoe-shaped with one end raised over the long hilum.

Mozambique. MS: Nhamatanda (Vila Machado), Beira to Dondo road, near Dondo, fl. & fr. 12.x.1935, *Lea* 67 (PRE).

Not known elsewhere. Sandy soil near ditches; 15 m.

This plant had been referred to as "probably *A. nilotica*" but although the standard is glabrous it and the keel are far too large. It shows a resemblance to *A. cristata* var. *pubescens* but that has a densely hairy standard. The possibility of a hybrid should not be overlooked but further material is needed. It also shows resemblances to *A. aspera* and *A. afraspera*.

10. **Aeschynomene cristata** Vatke in Oesterr. Bot. Z. **28**: 215 (1878). —E.G. Baker, Legum. Trop. Africa: 288 (1929). —Brenan, Check-list For. Trees Shrubs Tang. Terr.: 406 (1949). — Léonard in F.C.B. **5**: 266, pl. 20, fig. 18G (1954). —Hepper in F.W.T.A., ed. 2, **1**: 579 (1958). —White, F.F.N.R.: 141 (1962). —Verdcourt in F.T.E.A., Leguminosae, Pap.: 378 (1971). — Drummond in Kirkia **8**: 216 (1972). —Verdcourt in Kirkia **9**: 390 (1974). —Lock, Leg. Afr. Check-list: 102 (1989). TAB. 3.6.**15**, fig. 7. Type from Zanzibar.

Shrub or herb forming a dense coppice-like growth 0.9–3(6) m tall, with immersed procumbent rooting stems covered with prominent lenticels and erect branches. Branches thick and soft at the base, glabrous to densely covered with sticky tubercular-based golden-brown hairs. Leaves 40–86-foliolate; leaflets 6–22 × 1–4.5 mm, linear to oblong, rounded to emarginate and mucronulate at the apex, obliquely emarginate at the base, entire or finely serrate, glabrous but with ciliolate margins; venation sometimes obvious and purplish beneath; petiole and rhachis together 4–19 cm long, hairy like the branches; petiolules 0.5 mm long; stipules 7–20 × 1.5–3 mm, ovate-lanceolate, straight, spurred, glabrous or hairy, eventually deciduous. Inflorescences axillary, (1)2–8-flowered; peduncle 3 cm long, glabrous or pubescent; rhachis 6 cm long; pedicels 8–16(22) mm long; bracts 4–7 × 2–3.5 mm, ovate; bracteoles 4–6 × 2–4 mm, ovate, ciliate. Calyx pubescent or hairy, 2-lipped; lips oblong, unequal, one 8–12 × 5–8 mm, emarginate, the other 11–16 × 6–8 mm, 3-lobed. Standard deep yellow to orange-yellow or sometimes brownish-purple, 1.8–3.3 cm long and wide, round, pubescent above outside; wings deep yellow to orange-yellow; keel petals yellow or brownish, laciniate along their lower margins, hairy. Fruit 5–11 cm long excluding the 2 cm stipe, linear, straight or slightly curved, 6–12-jointed, glabrous or sparsely hairy, the margins slightly thickened, ± straight or slightly constricted between the joints; articles 8–12 × 7–10 mm, oblong, compressed, at first venulose, later often with the central part ornamented with spine-like tubercles up to 1 mm long, but mostly quite smooth. Seeds purplish-brown or pale chestnut-brown, shiny, up to 7 × 3.5 × 1.8 mm, narrowly oblong-reniform, curved and canoe-shaped at one end; hilum linear-oblong.

Var. **cristata** —Verdcourt in Kirkia **9**: 391 (1974). TAB. 3.6.**19**.

Erect branches mostly densely hairy. Stipules hairy; petiole and rhachis hairy. Inflorescence rhachis mostly hairy; pedicels 8–22 mm long.

Caprivi Strip. Katima Mulilo area, c. 7 km from Katima on road to Lisikili, fl. 24.xii.1958, *Killick & Leistner* 3075 (K; PRE). **Botswana**. N: Okavango Swamp, 16 km south of Seronga, fl. & fr. 29.ix.1954, *Story* 4791 (K; PRE). **Zambia**. B: Senanga, fl. & fr. 17.x.1961, *Fanshawe* 6741 (K). N: Samfya Distr., Lake Bangweulu, southern part, 10 km south of Lake Chali, Mboyaluambe Island, fl. & fr. 6.iii.1996, *Renvoize* 5757 (K). S: Livingstone Distr., River Zambezi, fl. & fr. v.1909, *Rogers* 7155 (K). **Zimbabwe**. W: Hwange Distr., Kazungula, fl. & fr. 15.iv.1955, *Davies* 1127 (SRGH). **Malawi**. C: Nkhotakota (Kota Kota) Dambo, fl. & fr. 16.ii.1944, *Benson* 298 (PRE). S: without locality, fl. & fr., *Buchanan* 1160 (K). **Mozambique**. N: Muatua to Angoche (António Enes), fl. & fr. 7.xi.1936, *Torre* 993 (COI; LISC). Z: between Pebane and Regone, fl. & fr. 30.x.1942, *Torre* 4728 (LISC). MS: Sussundenga Distr., entre Dombe e Matarara do Lucite, junto ao Rio Mucombe, fl. & fr. 19.iv.1948, *Barbosa* 1500 (LISC) (intermediate with var. *pubescens*). GI: Inhambane, between Inharrime and Chacane, fl. & fr. 28.i.1941, *Torre* 2577 (LISC).

Widespread in tropical Africa from the Sudan to east and central Africa and the Dem. Rep. Congo. River and lake margins, permanent and seasonal swamps in grassland, etc.; 5–1160 m.

Var. **pubescens** J. Léonard in Bull. Jard. Bot. État **24**: 67 (1954); in F.C.B. **5**: 269 (1954). — Verdcourt in F.T.E.A., Leguminosae, Pap.: 379 (1971); in Kirkia **9**: 392 (1974). Type from Tanzania.

Tab. 3.6.**19**. AESCHYNOMENE CRISTATA var. CRISTATA. 1, flowering branch (× ¹/₂); 2, detail of stem showing stipule (× 2); 3, flower, longitudinal section (× 1); 4, bracteoles and lips of calyx (× 1); 5, standard, external face (× 1); 6, wing (× 1); 7, keel, spread out, internal face (× 1); 8, detail of the joined margins of part of the keel (× 2); 9, androecium and portion of calyx (× 1), 1–9 from *Ghesquière* 61; 10, fruit (× ¹/₂); 11, article with spines (× 2), 10 & 11 from *Kesler* 1022. Drawn by J.M. Lerinckx. From Fl. Congo Belge. Reproduced with permission of Jardin Botanique National de Belgique.

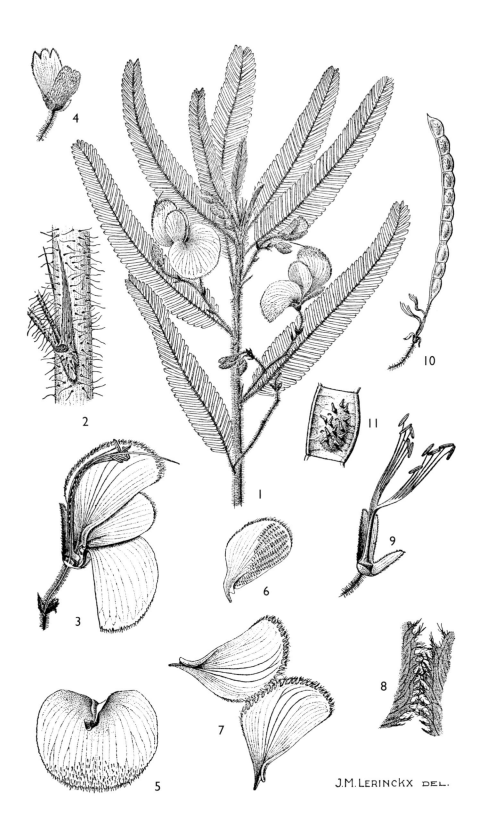

J.M. LERINCKX DEL.

Branches glabrous. Stipules glabrous; petiole and rhachis sparsely pubescent. Inflorescence rhachis pubescent or glabrous; pedicels 6–12 mm long.

Zambia. N: Kaputa Distr., Lake Mweru, below Kapindi, fl. 10.iv.1957, *Richards* 9135 (K). **Malawi**. S: Lower Shire Valley, Elephant Marsh, fl. & fr. ix.1956, *V.C. Robertson* 11 (K).

Widely distributed, from Benin to Angola and Dem. Rep. Congo, and from the Sudan to east Africa. Habitat as for typical variety; 50–1050 m.

11. **Aeschynomene pfundii** Taub. in Engler, Pflanzenw. Ost-Afrikas **C**: 215 (1895). —E.G. Baker, Legum. Trop. Africa: 289 (1929). —Brenan, Check-list For. Trees Shrubs Tang. Terr.: 406 (1949). —White, F.F.N.R.: 141 (1962). —Verdcourt in F.T.E.A., Leguminosae, Pap.: 380, fig. 54/7 (1971); in Kirkia **9**: 392 (1974). —Lock, Leg. Afr. Check-list: 106 (1989). TAB. 3.6.**16**, fig. 4. Syntypes from Tanzania.

A small shrub 1.2–4.5 m tall, often with swollen pithy floating stems similar to *A. elaphroxylon* but without spines. Stems densely hispid with golden-brown tubercular-based glandular hairs. Leaves 16–32(40)-foliolate; leaflets 1–3.5 × 0.4–1.2 cm, oblong, rounded and mucronulate at the apex, obliquely rounded at the base, ciliate on margins and main nerve beneath, ciliate-serrulate towards the apex, otherwise glabrous; petiole and rhachis together 7–14(22) cm long; petiolules 1 mm long; stipules 8–10 × 3.2–5 mm, ovate or elliptic, acute, not appendaged, densely ciliate, soon falling. Inflorescences axillary, 3–7-flowered; peduncle 2.5–9 cm long; rhachis 2–9 cm long; pedicels 1–2 mm long, glandular hairy; bracts often 2–3-fid, 8–11 × 4.5–9 mm, ovate, hairy, soon falling; bracteoles 6–8.5 × 1.5–2.5 mm, lanceolate, soon falling. Calyx hairy, 2-lipped; lips 13–18 × 6–12 mm, oblong, one emarginate, the other 3-lobed. Standard orange-yellow, 2–4 × 2.5–3 cm, rounded, ciliolate at the apex; wings orange-yellow; keel paler orange-yellow, the petals laciniate. Fruit 5 cm long, oblong-elliptic, sessile, straight, beaked, 6–7-jointed, the suture straight or smoothly curved, not constricted between the articles; articles 3–12 × 11–17(19) mm, irregularly oblong, compressed, hispid with yellow tubercular-based hairs, venulose. Seeds dark brown or black, 6–7 × 4.5–5 × 2.8 mm, truly reniform with strongly curved outer edge and marked sinus containing the curved elliptic hilum.

Zambia. N: Kaputa Distr., Lake Mweru Wantipa, below Kampinda (Kapindi), fl. 10.iv.1957, *Richards* 9140 (K). S: Mazabuka River, fl. 30.iii.1964, *van Rensburg* 2892 (K; SRGH). **Malawi**. S: Upper Shire Valley, fl. & fr. iv.1859, *Kirk* (K). **Mozambique**. GI: Chigubo Distr., Parque Nacional de Banhine (Banhine National Park), st. x.1973, *Tinley* 2968 (K; LISC; SRGH).

Also in Sudan, Kenya and Tanzania. Riverside fringing swamps and swampy grassland, ditches by lake edges, etc.; 300–1050 m.

12. **Aeschynomene elaphroxylon** (Guill. & Perr.) Taub. in Engler and Prantl, Nat. Pflanzenfam. **3** (3): 319, fig. 124A–C (1894). —Harms in Engler, Pflanzenw. Afrikas [Veg. Erde 9] **3** (1): 611, fig. 298/A–C (1915). —E.G. Baker, Legum. Trop. Africa: 289 (1929). —Brenan, Check-list For. Trees Shrubs Tang. Terr.: 405 (1949). —Dale & Eggeling, Indig. Trees Uganda Prot., ed. 2: 295 (1952). —Léonard in F.C.B. **5**: 261 (1954). —Hepper in F.W.T.A., ed. 2, **1**: 578 fig. 168 (1958). —Rudd in Reinwardtia **5**: 28 (1959). —Dale & Greenway, Kenya Trees & Shrubs: 354 (1961). —Lind & Tallantire, Fl. Pl. Uganda: 83, fig. 34 (1962). —White, F.F.N.R.: 141 (1962). —Torre in C.F.A. **3**: 196 (1966). —Verdcourt in F.T.E.A., Leguminosae, Pap.: 375, fig. 54/1 (1971); in Kirkia **9**: 393 (1974). —Lock, Leg. Afr. Check-list: 103 (1989). TAB. 3.6.**16**, fig. 5. Type from Senegal.

   Herminiera elaphroxylon Guill. & Perr. in Guillemin, Perrottet & Richard, Fl. Seneg. Tent.: 201, t. 51 (1832). —J.G. Baker in F.T.A. **2**: 144 (1871). —Hutchinson & Dalziel, F.W.T.A. **1**: 415, fig. 149 (1928). —Brenan in Mem. New York Bot. Gard. **8**: 253 (1953). Type as above.

Shrub or small tree, 2–9(12) m tall, with a swollen pithy often almost conical stem acting as a float. Stems with short and long sticky hairs and nearly always with short sharp spines 2–15 mm long. Leaves 20–40-foliolate; leaflets 8–26.5 × 4–10 mm, oblong, truncate or slightly emarginate at the apex, obliquely rounded at the base, entire, somewhat glaucous beneath, glabrescent above, pubescent with slightly tubercular-based short hairs and often minutely asperulous on the margins and midnerve beneath; venation dark beneath; petiole and rhachis together 4–16 cm long, densely covered with stiff bristly hairs and shorter pubescence and also often with short spinelets; petiolules 0.5–1 mm long; stipules

10–13 × 6–9 mm, broadly ovate, auriculate on one side only, velvety, eventually deciduous. Inflorescences axillary, 1–4-flowered; peduncle 1.3 cm long; rhachis 1–4.5 cm long; pedicels 12–16 mm long; bracts 6–7 × 3–5 mm; bracteoles asymmetrical, 10–16 × 5–10 mm, ovate or elliptic, velvety. Calyx densely pubescent and with longer sticky bristly hairs, 2-lipped; lips entire or 2–3-toothed at the apex, one 19–25 × 6–12 mm, ovate-lanceolate, the other, 15–20 × 6–11 mm, ovate-oblong. Standard yellow to orange, 3–4.7 cm long and wide, rounded, emarginate, puberulous above on outer surface; wings and keel yellow, the petals of the latter laciniate along their lower margins. Fruit spirally contorted, 10–14 cm long when unrolled, 6–17-jointed, densely covered with glandular bristly tubercular-based hairs with short pubescence as well; articles 6–8 × 7–9 mm, ± oblong or trapeziform, venulose beneath the dense indumentum. Seeds dark purplish-brown, 6 × 3.8 × 2 mm, reniform, with a slightly eccentric hilum, slightly beaked resembling the front of a canoe.

Zambia. B: Lake Lutende, c. 22.4 km east of Mongu, 18.xi.1959, *Drummond & Cookson* 6603 (K; LISC; PRE; SRGH). N: Mbala Distr., mouth of Lufubu River, fl. & fr. 15.viii.1958, *Lawton* 436 (K). C: 9.6 km below Kafue Bridge, fl. xi.1909, *Rogers* 8635 (K). **Malawi**. N: Khondowe (Kondowe) and Karonga, fl. & fr. vii.1896, *Whyte* (K). C: Nkhotakota (Nkota-Kota), fl. & fr. 2.v.1963, *Verboom* 892 (K; SRGH). S: River Shire, Lake Malombe (Pamalombe), fl. & fr. ix.1859, *Kirk* (K). **Mozambique**. N: Unango to Lake Shire, *W.P. Johnson* (K).

Throughout tropical Africa from Senegal, the Sudan and Ethiopia to Central Africa and Angola; also in Madagascar and cultivated in Egypt, Jawa and S America. Lake sides, pools and swamps bordering lakes and rivers, usually standing in 1–2 m of water; 470–1060 m.

13. **Aeschynomene americana** L., Sp. Pl.: 713 (1753). —Rudd in Reinwardtia **5**: 25 (1959). —Verdcourt in Man. New Guinea Leg.: 367 (1979). —Rudd in Rev. Fl. Ceylon **7**: 166 (1991). —Lock, Leg. Afr. Check-list: 101 (1989). Lectotype from Jamaica.

Erect or decumbent herb, 0.3–1.5(2) m tall. Stems glandular-hispid to subglabrous. Leaves 2–7 cm long, c. 20–60-foliolate, the petiole and rhachis hispidulous; leaflets glabrous, 2–several-costate, ciliate, 4–15 × 1–2 mm, subfalcate, apiculate, the base asymmetrically rounded; stipules peltate-appendiculate, glabrous or somewhat hispid at the point of attachment, striate, usually ciliate, (5)10–25 × 1–4 mm, the upper portion attenuate, 2–3 times longer than the lower acute or erose portion. Inflorescences axillary, racemose, few-flowered, the axes hispidulous; bracts c. 2–4 × 2–3 mm, cordate, acuminate or, sometimes, truncate-flabelliform, glabrous, ciliate; bracteoles 2–4 × 1–1.5 mm, linear to linear-ovate, acute to acuminate, serrate-ciliate. Flowers c. 5–10 mm long. Calyx glabrous to hispidulous, bilabiate, 3–6 mm long. Corolla yellowish to tan, usually with red or purplish lines, glabrous. Fruit 2–3 cm long, 3–9-articulate, the stipe c. 2 mm long, the articles semicircular, the upper margin essentially straight, the lower curved, 3–6 × 2.5–5 mm, glandular-hispidulous, or glabrous or nearly so, usually muricate, the margins thickened, reticulate-veined; seeds dark brown, 2–3 × 1.5–2 mm.

Malawi. S: Blantyre, Chilomoni Dam, fl. 15.iv.1970, *Brummitt & Williams* 9871 (K).

Widespread in tropical and subtropical America and now introduced and often naturalised in Sri Lanka, SE Asia, Jawa, Philippines and New Guinea. Well naturalised around the margin of Lake Malawi. Also cultivated at several Agricultural Stations, e.g. Matopos Research Station (S.S.D. 44) and at Mt. Makulu Agricultural Research Station (*van Rensburg* 3129); 1000 m.

In her revised treatment (1991) Rudd has dispensed with infraspecific categories and it seems unnecessary to recognize var. *americana* and var. *glandulosa* (Poir.) Rudd based on whether the fruit articles are ± glabrous or pubescent and glandular.

14. **Aeschynomene nodulosa** (Baker) Baker f. in J. Linn. Soc., Bot. **40**: 56 (1911); Legum. Trop. Africa: 291 (1929). —Verdcourt in Kirkia **9**: 394 (1974). —Lock, Leg. Afr. Check-list: 106 (1989). Type: Malawi, Mt. Chiradzulu, *Meller* (K, holotype).

 Smithia nodulosa Baker in F.T.A. **2**: 153 (1871).

 Aeschynomene shirensis Taub. in Engler, Pflanzenw. Ost-Afrikas C: 215 (1895). Type: Malawi, Shire Highlands, *Buchanan* 13 (B, holotype).

 Aeschynomene leptobotrya sensu E.G. Baker, Legum. Trop. Africa: 298 (1929) quoad *Teague* 128. —sensu Hutch., Botanist Southern Africa: 304, 344 (1946) non Harms ex Baker f.

Aromatic shrub 1.2–6 m tall; bark brown, lenticellate. Stems densely covered with short glandular tubercular-based hairs, or more rarely glabrous and glaucous; older stems appearing jointed at the nodes due to the scars formed by the deciduous stipules, glabrescent, often quite nodular due to remnants of petiole bases. Leaves 8–32-foliolate; leaflets 3–18 × (1.5)2–6 mm, narrowly oblong or linear-oblong, rounded or subtruncate and mucronulate at the apex, slightly subcordate at the base, glabrous or at the most slightly ciliolate when young; lateral nerves visible beneath; petiole and rhachis (1.5)4–8 cm long, with hairs like the stem or glabrous; petiolules c. 0.5 mm long; stipules conspicuous, 7–22 × 3–7 mm, ovate to oblong-lanceolate, acuminate, glabrous or pubescent, soon deciduous. Racemes axillary, in upper leaves, many-flowered, mostly overtopping the terminal bud and forming a large branched terminal inflorescence, (2)5–14 cm long, densely hairy like the stem or glabrous; peduncles (0.5)2–3 cm long, pedicels 9–16 mm long, both similarly hairy or glabrous; bracts 2–3-fid, 3–6.5 × 2–4 mm, oblong, very soon deciduous; bracteoles 4.5–7 × 2–4 mm, narrowly ovate, glabrous or minutely pubescent towards the apex, soon deciduous. Calyx reddish, glabrous or minutely pubescent towards the apex, 2-lipped; lips 10–13 × 6–8 mm, oblong or oblong-elliptic, one ± entire, the other 3-fid. Standard yellow or orange-yellow with red veins, 10–16 × 9–14 mm, broadly elliptic; wings orange-yellow, paler at the base, streaked red at the apex, free; keel pale yellow, sometimes veined. Fruit of 3–6(7) articles joined by narrow necks, each article semicircular, 5–8 × 4–6 mm, compressed, shortly pubescent with tubercular-based hairs, or glabrous. Seeds yellow-brown to deep reddish-brown, 3.6–4 × 2.5–2.8 × 1.5–2 mm, oblong-reniform; hilum eccentric.

Like other species in this group it varies in indumentum from glabrous to densely pubescent with glandular tubercular-based hairs. Since the glabrous variety has been given a name it has been retained here.

Var. **nodulosa** —Gillett in Kew Bull. **14**: 334 (1960). —Drummond in Kirkia **8**: 216 (1972). —Verdcourt in Kirkia **9**: 395 (1974). —Gonçalves in Garcia de Orta, Sér. Bot. **5**: 62 (1982).

Stems, petioles, leaf rhachis and inflorescence densely pubescent.

Zimbabwe. C: Makoni Distr., Forest Hill Kop, fl. & fr. vii.1917, *Eyles* 746 (K; SRGH). E: Chimanimani Distr., Farm Thornton, fl. & fr. 6.vii.1950, *Crook* M 52 (K; LISC; SRGH). S: Mberengwa Distr., Mt. Buhwa, fl. 10.xii.1953, *Wild* 4318 (K; SRGH). **Malawi**. S: Zomba Mt., Chingwe's Hole, fl. & fr. 22.vi.1961, *Chapman* 1393 (K; LISC; PRE; SRGH). **Mozambique**. Z: between Mocuba and Olinga (Maganja da Costa), 74.3 km from Mocuba, fl. & fr. 26.ix.1949, *Barbosa & Carvalho* in *Barbosa* 4182 (K; LISC; LMA). T: Serra da Zóbuè, fr. 3.x.1942, *Mendonça* 593 (LISC). MS: Catandica (Vila Gouveia), fl. & fr. 3.vii.1941, *Torre* 2986 (LISC).
Also in South Africa (Transvaal). Submontane grassland, bracken scrub, open woodland (e.g. *Uapaca*–bamboo) also on exposed rocky (mostly quartzite) slopes and summits, sometimes by streams; 750–1900 m.

Var. **glabrescens** J.B. Gillett in Kew Bull. **14**: 334 (1960). —Drummond in Kirkia **8**: 216 (1972). —Verdcourt in Kirkia **9**: 396 (1974). Type: Zimbabwe, Nyanga Distr., near Nyamingura River, *Phipps* 1232 (K, holotype; LISC; SRGH).

Stems, petioles, leaf rhachis and inflorescences (including articles of the fruit) glabrous or nearly so.

Zimbabwe. E: Mutasa Distr., near Nyamingura River, Rureche (Luleche) Tea Estate, st. 4.iv.1967, *Corby* 1597 (K; SRGH). **Mozambique**. MS: Sussundenga Distr., eastern foothills of Chimanimani Mts., Gossamer Falls, 26.iv.1974, *Pope & Müller* 1319 (K; SRGH).
Also in South Africa (Transvaal). Forest margins, open veld, scrub streamsides; 600–1000 m.

15. **Aeschynomene semilunaris** Hutch., Botanist Southern Africa: 532 (1946). —White, F.F.N.R.: 141 (1962). —Drummond in Kirkia **8**: 216 (1972). —Verdcourt in Kirkia **9**: 396 (1974). —Lock, Leg. Afr. Check-list: 108 (1989). Type: Zambia, Kaloswe, *Hutchinson & Gillett* 4074 (BM; K, holotype).

Shrub 1.5–4.5 m tall; bark rough, greyish-brown, some of the stems lenticellate, appearing jointed at the nodes due to the scars formed by the deciduous stipules;

young shoots densely covered with glandular tubercular-based hairs. Leaves 10–28-foliolate; leaflets 2.5–17 × (2)3.5–7 mm, oblong-elliptic to narrowly oblong, rounded to slightly emarginate and very shortly mucronulate at the apex, obliquely rounded to asymmetrically emarginate at the base, entire, with similar hairs to the young stems or glabrescent; venation visible beneath; petiole and rhachis 2–8.5 cm long, glabrescent or glandular-pubescent; petiolules 0.5 mm long; stipules conspicuous, 5–20(26) × 2.5–8 mm, oblong to lanceolate, deciduous or a few sometimes ± persistent, glabrous or pubescent. Racemes axillary, sometimes forming a "terminal" corymb, 3–7 cm long, hairy like the stems; peduncles 1.3–2 cm long; pedicels 5–6 mm long, both similarly pubescent; bracts sheathing, 3-fid or 2-fid at the apex, 4–9 × 3–5 mm, elliptic, obscurely ciliolate, very soon deciduous; bracteoles 4.5–6 × 2–2.5 mm, ovate-lanceolate, glabrous or ciliolate, deciduous. Calyx reddish, glabrous, 2-lipped; lips 8–9 × 4–5 mm, oblong-lanceolate or oblong-elliptic, one 2-fid, the other 3-fid. Standard pale yellow, sometimes veined lilac, or crimson, 10–11 × 7–8 mm, rounded oblong, broadly rounded; wings and keel yellow, the former free, the latter much exceeding the standard. Fruit of 1–2 articles, if 2 then joined by a narrow neck, each semicircular, 10–20 × 6–10 mm, compressed, glabrous or with one or two obscure hairs on the rounded margin. Seeds chestnut-coloured, 5 × 3.5–4 × 1.7 mm, oblong-reniform, rather sharply beaked to the outside of the small eccentric hilum.

Zambia. N: 12.8 km east of Kasama, fl. & fr. 18.viii.1960, *E.A. Robinson* 3783 (EA; K; SRGH). C: Kapiri Mposhi, fl. & fr. 6.viii.1957, *Fanshawe* 3448 (K; NDO). E: Nyika Plateau, by main road to c. 3.2 km SW of Rest House, fl. & fr. 21.x.1958, *Robson & Angus* 181 (BM; K; LISC; PRE; SRGH). **Malawi**. N: Chitipa Distr., Mafinga Hills, fl. 25.viii.1962, *Tyrer* 667 (SRGH).
    Confined to Zambia and Malawi. Rocky slopes and ridges in *Brachystegia–Protea* woodland and on margins of evergreen forest; 1200–2150 m.
    Greenway & Trapnell 5551 (Zambia N: Mukungwa Valley, bud, 1.viii.1938) is probably best considered to be a form of this but the leaflets attain 2 × 1.1 cm, the stipules 3 × 1.4 cm and the bracts 1.4 × 0.8 cm. Further material in flower and fruit is required. White (F.F.N.R.: 143 (1962)) has referred it to "*A.* sp. nr. *bella* Harms" but it does not agree well with *A. bella*.
    Torre & Correia 16434 (Mozambique N: Ribáuè, Serra Mepáluè, 9.xii.1967) may belong to this species but the material is sterile.

16. **Aeschynomene megalophylla** Harms in Repert. Spec. Nov. Regni Veg. **8**: 355 (1910). —E.G. Baker, Legum. Trop. Africa: 291 (1929). —Brenan in Mem. New York Bot. Gard. **8**: 253 (1953). —Drummond in Kirkia **8**: 216 (1972). —Verdcourt in Kirkia **9**: 397 (1974). — Lock, Leg. Afr. Check-list: 105 (1989). Syntypes: Malawi, Blantyre, *Buchanan* 210 (B) and Mulanje (Mlanje) Mt., Thuchila (Tuchila) Plateau, *Purves* in *Sharp* 33 (K)*.

Shrub or small tree 1.2–6 m tall; bark brown, smooth with conspicuous white lenticels. Stems densely covered with short glandular tubercular-based hairs or glabrous; older stems appearing jointed at the nodes due to the scars formed by the deciduous stipules. Leaves 12–18(20)-foliolate; leaflets 1–3.5 × 0.7–2.2 cm, oblong-elliptic, shortly emarginate at the apex, obscurely obliquely subcordate at the base, entire, glabrous, often glaucous; venation obscurely reticulate beneath; petiole and rhachis together 5–12 cm long, with similar hairs to the stem or glabrescent or glabrous; petiolules distinct, c. 1.5 mm long; stipules conspicuous forming a sheath over the young leaves but almost immediately deciduous, 17–28 × 5–8 mm, narrowly lanceolate, glabrous. Racemes axillary from upper leaves, overtopping the terminal bud and appearing as a large branched terminal inflorescence, 7–20 cm long, densely pubescent with hairs similar to those on the stems or glabrescent; peduncles 3–4 cm long; pedicels 5–15 mm long, both similarly pubescent; bracts 14 × 7 mm, oblong-elliptic, acuminate, cucullate and clasping, glabrous, very soon deciduous; bracteoles 7–10 × 5 mm, elliptic, cucullate and clasping, glabrous, soon deciduous. Calyx red, glabrous, 2-lipped; lips 13–18 × 5–6 mm, oblong, both entire or one with slight trace of 3-lobing. Standard bright yellow or orange, flushed crimson or veined brown outside, 1.7–2.2 × 1.4–1.8 cm, almost round or broadly elliptic, rounded or emarginate; wings and keel yellow, the former free. Fruit of (4)5–8 articles joined by

---

* Harms mentions Sharpe (Sharp) as the collector but the label on the syntype definitely gives Purves as the collector.

narrow necks, each article semicircular or elliptic, 6–7 × 5–6 mm, compressed, glabrous or sometimes with very scattered ciliae when young, rarely persistently bristly ciliate, smooth. Seeds yellow-brown, 4–5 × 3 × 2–2.2 mm, oblong-ellipsoid; hilum small, eccentric.

Zimbabwe. E: Chimanimani (Melsetter), fl. 26.vi.1937, *Obermeyer* in *Transvaal Mus.* 36549 (K; PRE). **Malawi**. S: Mulanje (Mlanje) Mt., slopes near Thuchila (Tuchila) River Valley, fl. & fr. 12.vii.1956, *G. Jackson* 1868 (BM; K; LISC). **Mozambique**. MS: Sussundenga Distr., Serra Zuira, planalto Tsetserra, c. 2 km from the cowshed on road to Chimoio (Vila Pery), fl. & fr. 6.xi.1965, *Torre & Pereira* 12723 (LISC).

Not known elsewhere. Secondary forest and edges of *Widdringtonia* relict patches, montane forest-woodland boundaries, scrubland on rocky grass slopes and ridges; 1400–2550 m.

Material from Zimbabwe has glabrous stems and leaves but, since similar variants are found on Mulanje, separation even as a variety seems unnecessary.

17. **Aeschynomene grandistipulata** Harms in Repert. Spec. Nov. Regni Veg. **8**: 355 (1910). — Eyles in Trans. Roy. Soc. South. Africa **5**: 377 (1916). —E.G. Baker, Legum. Trop. Africa: 291 (1929). —Drummond in Kirkia **8**: 216 (1972). —Verdcourt in Kirkia **9**: 398 (1974). — Lock, Leg. Afr. Check-list: 104 (1989). Type: Mozambique, Chimanimani Mts., Moribane, *W.H. Johnson* 232 (K, holotype).

Shrub 1.2–3 m tall. Young stems dark purplish-brown, often glaucous, glabrous or with dense to scattered tubercular-based hairs on some or all internodes, the older ones appearing jointed at the nodes due to the scars formed by the deciduous stipules, glabrous. Leaves rather fleshy, often glaucous, (8)14–22-foliolate; leaflets 1.2–3 × 0.6–2 cm, elliptic, oblong-elliptic or elliptic-ovate, rounded but mucronulate at the apex, obliquely rounded to cordate at the base, entire or sometimes ciliolate when young, glabrous; venation reticulate beneath; petiole and rhachis together (3)7–12 cm long, glabrous or rarely densely covered with tubercular-based hairs; petiolules distinct, c. 1 mm long; stipules large and conspicuous, 1.5–3.7 × 1.3–3 cm, ovate-elliptic, glabrous, reticulately veined, eventually deciduous. Inflorescences axillary, many-flowered, often branched, 7–15 cm long, the rhachis densely covered with tubercular-based hairs or glabrescent; peduncle 1.5–7 cm long; pedicels 3–7 mm long, all with similar hairs; bracts 7–13 × (3)4–8 mm, ovate or lanceolate, glabrous, deciduous; bracteoles 4–6 × 1.5–2.5 mm, lanceolate, glabrous, deciduous. Calyx glabrous, 2-lipped; lips 8–12 × (3.5)4–5.5 mm, oblong, one entire and the other obscurely 3-fid, practically entire. Standard yellow or orange-yellow with purple veins and a reddish basal patch inside, 11–13 × 7–8 mm, elliptic; wings and keel yellow or orange, the former free, the latter purplish at the tip. Fruit of 3 articles joined by very narrow necks, each article semicircular, 7–8 × 5 mm, compressed, glabrous or very sparsely ciliate, smooth. Seeds not seen.

Zimbabwe. E: Chimanimani Mts., Stonehenge, fl. 9.v.1958, *Chase* 6915 (K; LISC; PRE; SRGH). **Mozambique**. MS: Chimanimani Mts., Moribane, fl. 10.x.1907, *W.H. Johnson* 232 (K).

Confined to the Chimanimani Mts. Bracken scrub, sandstone crags, quartzite rocks; 1200–2100 m.

The holotype exhibits an interesting abnormality, in two cases opposite leaflets having become partially joined. Verboom has collected what appears to be a variant of this species on top of the Mafinga Mts. (Zambia N: fl. 20.x.1958, *Verboom* LK 98) but it differs in having up to 32 more oblong leaflets 2 × 1 cm, and very short internodes. More material is needed to assess its status.

18. **Aeschynomene chimanimaniensis** Verdc. in Kew Bull. **27**: 435 (1972); in Kirkia **9**: 399 (1974). —Lock, Leg. Afr. Check-list: 102 (1989). TAB. 3.6.**20**. Type: Mozambique, Chimanimani Mts., *Wild* 2888 (K, holotype; LISC; SRGH).

Small shrub c. 90 cm tall. Young stems densely covered with glandular tubercular-based hairs; older stems appearing jointed at the nodes due to the scars formed by the deciduous stipules. Leaves 12–22-foliolate; leaflets 6–15 × 3.1–6 mm, elliptic-oblong, acute or rounded and mucronulate at the apex, obliquely rounded or slightly emarginate at the base, entire (but sometimes appearing denticulate due to tubercles of hairs) with short tubercular-based hairs, the tubercles sometimes large enough to render the leaflet rough, particularly along the margins and on the

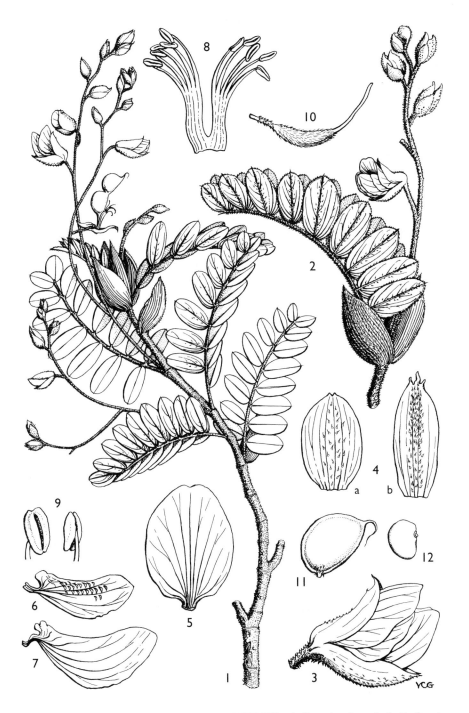

Tab. 3.6.**20**. AESCHYNOMENE CHIMANIMANIENSIS. 1, flowering branch (× 1); 2, twig with inflorescence (× 1⅓); 3, flower (× 3⅓); 4, calyx split in two, 4a, adaxial portion, 4b, abaxial portion (× 3⅓); 5, standard (× 3⅓); 6, wing (× 3⅓); 7, keel (× 3⅓); 8, androecium (× 3⅓); 9, anthers (× 6); 10, gynoecium (× 4), 1–10 from *Grosvenor* 396; 11, fruit (× 3⅓); 12, seed (× 3⅓), 11 & 12 from *Wild* 2888 in *GHS* 23888. Drawn by Victoria Friis. From Kew Bull.

midrib beneath; venation reticulate and sometimes raised beneath; petiole and rhachis 1.8–5 cm long, hairy like the stems; petiolules up to 1 mm long; stipules conspicuous, 6–19 × 5–7 mm, ovate to lanceolate, at length deciduous, densely shortly pubescent with tubercular-based hairs. Racemes axillary, often branched, 2.5–10 cm long, hairy like the stems; peduncles 1–2 cm long; pedicels 2–4 mm long, both similarly pubescent; bracts 2.5–6 × 1.5–4 mm, ovate or ovate-lanceolate, pubescent, acute, very soon deciduous; bracteoles 4.5 × 1 mm, oblong-lanceolate, ciliolate, deciduous. Calyx pubescent, 2-lipped; lips 10–11 × 5 mm, elliptic, the marginal areas not pubescent, one acute the other slightly 3-fid. Standard yellow with a 2-lobed purple spot at the base, 11 × 8 mm, almost round, slightly emarginate; wings and keel yellow, the former free, the latter not exceeding the standard. Fruit of 2 articles (1 sometimes undeveloped) joined by a narrow neck, each oblong-elliptic or roughly semicircular, 7 × 5 mm, rather sparsely covered with tubercular-based hairs particularly when young. Seeds chestnut in colour, 3.8 × 3 × 2 mm, ellipsoid-reniform, not beaked beyond the small eccentric hilum.

Zimbabwe. E: Chimanimani Mts., southern end, fl. & fr. viii.1965, *Morgan* CN 67 (K; SRGH). **Mozambique**. MS: Chimanimani Mts., near St. George's Cave, between The Saddle and Poacher's Cave, fl. 12.iv.1967, *Grosvenor* 396 (K; SRGH).
Apparently confined to the Chimanimani Mts. On rocky ground; 1500 m.

19. **Aeschynomene baumii** Harms in Warburg, Kunene-Samb.-Exped. Baum: 261 (1903). — E.G. Baker, Legum. Trop. Africa: 291 (1929). —Léonard in F.C.B. **5**: 271, fig. 19B–C (1954). —Hepper in F.W.T.A., ed. 2, **1**: 578 (1958). —White, F.F.N.R.: 141 (1962). —Torre in C.F.A. **3**: 197 (1966). —Verdcourt in F.T.E.A., Leguminosae, Pap.: 381 (1971); in Kirkia **9**: 400 (1974). —Lock, Leg. Afr. Check-list: 102 (1989). Type from Angola.

Subshrub or shrub, (0.6)1–2.4(6) m tall or, in fireswept situations, a woody herb or subshrub 0.2–1 m tall, with erect flowering stems from a thick woody rootstock, some leafy, others not. Stems glabrous to covered with tubercular-based hairs. Leaves 8–20(32)-foliolate; leaflets 4–25 × 2–13 mm, rounded, oblong or obovate-oblong, rounded or retuse at the apex, oblique at the base, sometimes ciliate and often with translucent marginal glands, entire or crenulate, glabrous; petiole and rhachis together 1–6 cm long, with indumentum like the stems; petiolules 0.8 mm long; stipules 3–8 × 1–6 mm, ovate to lanceolate, not appendaged, deciduous. Inflorescences axillary or terminal, ample, racemose or paniculate; peduncle 2 cm long; rhachis 2.5–5.5 cm long; pedicels 1–2(2.7) cm long; bracts 3–5 × 1.5–4 mm, ovate, 3-toothed to 3-lobed, deciduous; bracteoles 3.5–6 × 1–3 mm, elliptic to lanceolate, glabrous to pubescent, denticulate-ciliolate. Calyx glabrous to pubescent, 2-lipped; lips 5–9 × 2–4.5 mm, oblong, often ciliolate-serrulate, one 3-lobed, the other 2-lobed. Standard yellow or brownish, 10–15 × 6–10 mm, oblong, obovate or elliptic, emarginate, glabrous; wings yellow; keel petals yellow, longer than the standard, not laciniate but glandular-setose above. Fruit of 1, rarely 2 articles, 14–25 × 8–14 mm, each ± elliptic, flat and papery, glabrous, reticulate save on the wing-like margins which are 0.5–1 mm wide. Seeds pale brown, 6 × 4.5 × 1.5 mm, obliquely elliptic-reniform, very compressed, with a minute hilum in a groove and a small beak beyond it.

Var. **baumii** —Verdcourt in Kirkia **9**: 401 (1974).

Leaflets usually much smaller than in var. *kassneri*, typically c. 10 × 3–4 mm, scarcely, if at all crenulate.

Zambia. N: Mbala Distr., Kambole Escarpment, fl. & fr. 23.viii.1956, *Richards* 5932 (K; SRGH). W: Mwinilunga Distr., east of the Boma and 32 km west of River Kabompo, fl. 11.ix.1930, *Milne-Redhead* 1096 (K).
Also in Cameroon and Angola, and in western Tanzania. *Brachystegia* and "Chipya" woodland, short grassland, scrub, dambos, usually on sandy soil, sometimes on rocky slopes; 1500–1800 m.
The two sheets cited above represent extremes of the range of variation and at Kew the sheets from Zambia N had been suggested to belong to a new species but inspection of material from the Dem. Rep. Congo suggests that Léonard is correct to unite them as one species. Broadly speaking the northern and western material seems to fall into two geographical variants, the

plants from the north having densely pubescent stems, inflorescences, etc. and the leaves longer and mostly 20–28-foliolate, and those from the west often having glabrous or glabrescent stems, inflorescences, etc. and the leaves shorter, c. 12–16-foliolate. The possibility that this is linked to intensity of burning needs study in the field.

Var. **kassneri** (Harms) Verdc. in Kew Bull. **27**: 437 (1972); in Kirkia **9**: 401 (1974). Type from Dem. Rep. Congo.
    Aeschynomene kassneri Harms in Repert. Spec. Nov. Regni Veg. **8**: 357 (1910). —E.G. Baker, Legum. Trop. Africa: 292 (1929).

Leaflets larger and thicker, up to 25 × 13 mm, very distinctly crenulate.

Zambia. N: Mbala (Abercorn) to Mpulungu road, fl. & fr. 29.ix.1956, *Richards* 6322 (K).
Also in the Dem. Rep. Congo. Grassland with scattered bushes subject to seasonal burning; 1440–1620 m.
Léonard has sunk this into *A. baumii* completely but I agree with Gillett in adnot. that it should be maintained at some level.

20. **Aeschynomene inyangensis** Wild in Kew Bull. **8**: 93, fig. 2 (1953). —Drummond in Kirkia **8**: 216 (1972). —Verdcourt in Kirkia **9**: 402 (1974). —Lock, Leg. Afr. Check-list: 104 (1989). Type: Zimbabwe, Nyanga (Inyanga), Pungwe Source, *Wild* 1432 (K; LD; LISC; PRE; SRGH, holotype).

Branched shrub 0.3–1.2 m tall; young shoots densely covered with short bristly mostly tubercular-based glandular curved hairs; old shoots grey-brown or purplish, glabrescent, ridged. Leaves and petioles reduced to a linear vestige, 2–4.5 × 0.5–0.75 mm, cuspidate, minutely pellucid-punctate, 1–3-nerved, with margins hyaline and minutely setulose near the apex, mostly deciduous; or sometimes leaves 12–16-foliolate, c. 10 mm long with leaflets 3–4 × 0.5–0.7 mm, linear-oblong; stipules thick, congested, 2.5–5(7.5) × 1–1.5 mm, ovate or ovate-lanceolate, cuspidate, slightly convolute, 4–7-nerved, pellucid-punctate between the nerves and with minutely setulose hyaline margins, unappendaged. Inflorescences axillary, lax, 1–3-flowered, 10–15 mm long, pubescent with tubercular-based hairs; peduncle 3–5 mm long; pedicels 2–3 mm long, setulose; bracts mostly clearly 3-fid, 3 × 1.5 mm, ovate-elliptic, margins smooth or sparsely ciliolate, deciduous; bracteoles 2.5–3 × 1.5 mm, ovate to lanceolate, margins smooth or sparsely ciliolate, deciduous. Calyx glabrous, 2-lipped; lips 6 × 4–5 mm, oblong-ovate or ovate, upper acute, lower minutely 3-toothed. Standard yellow, lined with brown, 9–11 × 4.5–8 mm, obovate; wings and keel yellow, the petals of the latter not laciniate. Fruit of 1–2 articles joined by a narrow neck, each article elliptic, 5–7 × 3.5–4 mm, compressed, sparsely setulose, the margins thickened. Seeds purplish-brown, 3.5–4 × 2–3 mm, asymmetrically semicircular; hilum minute, asymmetrical.

Zimbabwe. E: Nyanga Distr., Matenderere (Mtendere) R. source, fl. 5.ix.1954, *Wild* 4592 (K; LISC; PRE; SRGH). **Mozambique**. MS: Báruè Distr., Serra de Chôa, Catandica (Vila Gouveia), fl. & fr. 7.ix.1943, *Torre* 5849 (K; LISC).
Restricted to Zimbabwe–Mozambique border in the Nyanga area. Montane grassland and scrub, often near streams; 1300–2520 m.

21. **Aeschynomene aphylla** Wild in Kew Bull. **8**: 95, fig. 3 (1953). —Drummond in Kirkia **8**: 216 (1972). —Verdcourt in Kirkia **9**: 402 (1974). —Lock, Leg. Afr. Check-list: 101 (1989). Type: Zimbabwe, Chimanimani Mts., *Wild* 2873 (K; SRGH, holotype).

Branched shrub 0.6–1 m tall; young stems densely covered with stout tubercular-based glandular hairs which become stiffer on the older shoots; older stems glabrescent, purplish and slightly ridged. Leaves and petioles reduced to a simple brown scale 0.3–0.5 mm long; stipules thick, congested, 6–16 × 1.5–3 mm, lanceolate, cuspidate, slightly convolute, 11–13-nerved, pellucid-punctate between the nerves and with densely setulose hyaline margins, unappendaged. Inflorescences axillary, lax, 1–4-flowered, 1–2 cm long, pubescent with tubercular-based hairs; peduncles 5 mm long; pedicels graceful, 3–4 mm long, setulose; bracts 2–3-fid, 3.5 × 2 mm, ovate, sparsely ciliolate, deciduous; bracteoles 3–3.5 × 1.5 mm, ovate to lanceolate, ciliolate, deciduous. Calyx glabrous, 2-lipped; lips 5 × 4 mm, oblong or ovate-oblong, the upper obtuse, the lower minutely 3-toothed. Standard yellow, 9 × 5–6 mm, oblong-obovate, obtuse; wings and keel yellow, the petals of latter not laciniate. Fruit of 1–2 articles joined by a narrow neck; each article 4 × 3 mm, oblong-semicircular, compressed, slightly reticulate, sparsely setulose, the margins thickened. Seeds not seen.

Zimbabwe. E: extreme southern end of Chimanimani Mts., fl. & fr. viii.1965, *Anderson* CN 3 (SRGH). **Mozambique**. MS: Chimanimani Mts., fl. 6.vi.1949, *Wild* 2889 (K; SRGH).
Apparently restricted to this mountain range. Grassland and mountain scrub, by streams; 1170–1800 m.

22. **Aeschynomene gazensis** Baker f. in J. Linn. Soc., Bot. **40**: 56 (1911); Legum. Trop. Africa: 299 (1929). —Drummond in Kirkia **8**: 216 (1972). —Verdcourt in Kirkia **9**: 403 (1974). —Lock, Leg. Afr. Check-list: 103 (1989). Type: Zimbabwe, Chimanimani Distr., *Swynnerton* 1457 (BM, holotype; K).

Shrub 0.9–1.8 m tall; young stems densely covered with short bristly tubercular-based hairs, older stems purplish, somewhat ridged and glabrescent. Leaves 14–24-foliolate; leaflets 1–4(5) × 0.6–1.2 mm, elliptic to oblong, mucronulate at the apex, obliquely rounded at the base, glabrous, the main nerve ± central; petiole and rhachis together 8–15 mm long, pubescent with bristly hairs; petiolules minute; stipules 3–9 × 0.5–2.5 mm, lanceolate, acuminate, unappendaged, glabrous, at length deciduous. Inflorescences 1.6–3 cm long, pubescent with tubercular-based hairs; peduncles 3–6 mm long; pedicels 2 mm long, similarly pubescent; bracts 2–3-fid, 1.5–2.5 × 0.5–1 mm, elliptic, deciduous; bracteoles 2 × 1 mm, ovate to lanceolate, slightly ciliolate at the apex, sometimes slightly 2–3-fid, deciduous. Calyx glabrous, 2-lipped; lips 6 × 3–4 mm, ovate or ovate-oblong, one scarcely emarginate, the other very slightly 3-toothed. Standard yellow or bronze-yellow, veined with brown, 10–12 × 6.5–7 mm, obovate; wings and keel yellow, the petals of the latter not laciniate. Fruit of 2 articles joined by a narrow neck, each article semicircular, 4–5 × 3.5–4 mm, glabrous or very sparsely setulose. Seeds not seen.

Zimbabwe. E: Chimanimani (Melsetter) Downlands, fl. 17.viii.1950, *Crook* M 79 (K; LISC; SRGH).
Restricted to the Chimanimani area; 1200–1900 m.
A specimen collected in the Chimanimani Mts. (Dead Cow Gulch, 4.vii.1965, *Corby* 1348 (K; SRGH)) is related to *A. gazensis* but has the leaves 3 cm long with 34 leaflets 5 × 1.2 mm, and pedicels 5 mm long. It thus appears very different but with only very few sheets of *A. gazensis* available it is not possible to decide if it represents a distinct entity or not. *West* 3627 from Chimanimani (Melsetter), Chimanimani Mts., 18.v.1958 (SRGH) has 45 leaflets.

23. **Aeschynomene solitariiflora** J. Léonard in Bull. Jard. Bot. État **24**: 70 (1954); in F.C.B. **5**: 273 (1954). —Verdcourt in F.T.E.A., Leguminosae, Pap.: 382 (1971); in Kirkia **9**: 404 (1974). —Lock, Leg. Afr. Check-list: 108 (1989). Type from the Dem. Rep. Congo.

Erect herb or small subshrub with many tufted stems, 10–30 cm long, from a woody rootstock, mostly coming up after burns. Stems glabrous or with very sparse tubercular-based hairs. Leaves 14–28-foliolate; leaflets 2.5–6 × 0.6–1.5 mm, linear to obovate-linear, rounded to truncate and mucronulate at the apex, narrowed at the base, glabrous; petiole and rhachis together 1.5–3.5 cm long, sparsely pubescent; petiolules 0.2 mm long; stipules leaf-like, 6–10 × 0.5–1.5 mm, linear-lanceolate, acuminate, truncate, subcordate or shortly unilaterally appendaged, nervose, persistent. Flowers mostly numerous, solitary, axillary; peduncle and pedicels together 5–17 mm long; bracts absent; bracteoles 2.5–4 × 0.5–0.7 mm, ovate-lanceolate, soon falling. Calyx glabrous, 2-lipped; lips 5–7 × 2–3.5 mm, ovate-oblong, one emarginate, the other slightly 3-lobed. Standard yellow with brown streaks, 7.5–10 × 4–5 mm, violin-shaped, emarginate; wings and keel petals yellow, the latter not laciniate. Fruit of 1–3 articles joined by a very narrow neck, each article semicircular, 5–6.5 × 3.5–4.5 mm, compressed, glabrous, reticulate. Seeds not examined.

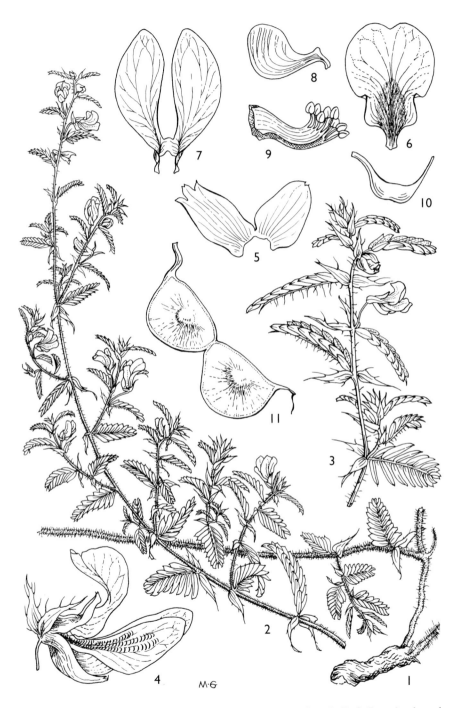

Tab. 3.6.**21**. AESCHYNOMENE LATERITICOLA. 1, base of plant (× 1); 2, flowering branch
(× 1); 3, apical portion of flowering branch (× 3); 4, flower (× 5); 5, calyx, opened out (× 5);
6, standard (× 5); 7, wings (× 5); 8, keel (× 5); 9, androecium (× 6); 10, gynoecium (× 6); 11,
fruit (× 5), 1–11 from *Milne-Redhead* 4362. Drawn by Mary Grierson. From Kew Bull.

Zambia. N: Mbala Distr., road down the Yendwe (Iyendwe) Valley, fl. 7.xii.1959, *Richards* 11891 (K). E: Nyika Plateau, by main road c. 3.2 km SW of the Rest House, fl. & fr. 21.x.1958, *Robson* 204 (BM; K; LISC; PRE; SRGH). **Malawi**. N: Nyika Plateau, Kasyaula Forest Reserve, fl. *Salubeni* 361 (K).

Also in the Dem. Rep. Congo and Tanzania (Mbeya District). Montane and escarpment grassland, often rough and subject to seasonal burning, also damp sandy pans; 1200–2150 m.

24. **Aeschynomene laterticola** Verdc. in Kew Bull. **24**: 5, fig. 1 (1970); in Kirkia **9**: 405 (1974) sphalm. "*latericola*". —Lock, Leg. Afr. Check-list: 104 (1989). TAB. 3.6.**21**. Type: Zambia, just north of Mwinilunga, *Milne-Redhead* 4362 (BM; BR; K; holotype; PRE).

Perennial prostrate herb with 2–4 stems 30–45 cm long from a woody rootstock. Stems much-branched, densely covered with setiform hairs 1–1.5 mm long. Leaves 12–24-foliolate; leaflets 1.5–3 × 0.5–0.8 mm, oblong or oblong-lanceolate, acute and aristulate at the apex, obliquely rounded at the base, glabrous; main nerve central or nearly so; petiole and rhachis together 8–10 mm long, setulose; petiolules minute; stipules 4–5 × 1–1.5 mm, lanceolate, acuminate, not appendaged at the base, leaf-like, nervose, ciliate, reflexed, persistent. Flowers axillary, solitary, pedicels 3–5.5 mm long, glabrous or sparsely setulose; bracteoles 3 × 0.5–0.7 mm, narrowly lanceolate, acute, ciliate on the margins. Calyx glabrous, 2-lipped; lips 3–5 × 2.5–3 mm, ovate, one rounded, the other shortly 3-fid. Standard orange-yellow, 7 × 4.5–5 mm, violin-shaped; wings and keel yellow, the petals of the latter not laciniate. Fruit of 1–2 articles joined by a narrow neck; articles almost round, 4.2 × 4 mm, compressed, glabrous. Seeds not seen.

Zambia. W: just north of Mwinilunga, fl. & fr. 26.i.1938, *Milne-Redhead* 4362 (BM; BR; K; LISC; PRE).

Known only from the type gathering. On shallow soil overlying laterite in open; 1350 m.

25. **Aeschynomene bracteosa** Welw. ex Baker in F.T.A. **2**: 150 (1871). —E.G. Baker, Legum. Trop. Africa: 294 (1929). —Léonard in F.C.B. **5**: 274 (1954). —Torre in C.F.A. **3**: 199 (1966). —Verdcourt in F.T.E.A., Leguminosae, Pap.: 383 (1971). —Drummond in Kirkia **8**: 216 (1972). —Verdcourt in Kirkia **9**: 405 (1974). —Lock, Leg. Afr. Check-list: 102 (1989). Type from Angola.

Erect, somewhat spreading or almost prostrate intricately branched sticky subshrub, 4–15 mm high, or tufted pyrophyte, 7–35 cm high, with a thick woody rootstock. Stems several–30, glabrous to densely pubescent with sticky tubercular-based short hairs, ± rugose. Leaves (6)12–56-foliolate; leaflets 1–7 × 0.3–1.5(3) mm, oblong or oblong-elliptic, acute to rounded at the apex, obliquely rounded at the base, glabrous, sometimes denticulate-ciliate; main nerve somewhat or very eccentric; petiole and rhachis together 0.5–3 cm long, pubescent or glabrous; petiolules 0.3 mm long; stipules 1.5–5(7.5) × 0.5–2(3) mm, ovate or ovate-lanceolate, not appendaged, sometimes ciliate. Inflorescences axillary or terminal, branched, several- to many-flowered; peduncle 0.2–3.5 cm long; rhachis 0.5–10(14) cm long, often zigzag; pedicels 2–8(17) mm long; bracts 2-lobed to the base, persistent, the lobes 1–3.5(6) × 0.5–1.5(2.5) mm, ovate to elliptic-lanceolate, entire or 2-fid, pubescent to glabrous; bracteoles 1.5–3(6) × 0.5–1(1.3) mm, lanceolate to ovate-lanceolate, often ciliate, persistent. Calyx glabrous, 2-lipped, one 3–5 × 2–3.5 mm, elliptic or ovate, emarginate or 2-toothed, the other 4–5(6.5) × 2–3.5 mm, oblong, 3-toothed, both ciliate above. Standard yellow or orange, rarely cream-coloured, 5–10 × 2.5–7 mm, violin-shaped, emarginate; wings and keel petals yellow, the latter not laciniate. Fruit of 1 article or sometimes 2, joined by a very narrow neck, each article rounded, 4–5(7) × 3.5–4.5(6) mm, compressed or slightly inflated, glabrous, densely and finely rugulose when adult. Seeds pale brown, compressed, 3.4 × 2.5 × c. 1 mm, almost semicircular in outline; hilum small, circular, eccentric.

A very variable species. Two taxa which I had thought might be species have been retained at varietal rank. *A. nambalensis* cited below is actually about intermediate between var. *bracteosa* and var. *delicatula*.

1. Leaves 12–56-foliolate; leaflets 0.3–1.5(2) mm wide; bracts up to 3.5 mm long · · · · · · · · ·
· i) var. *bracteosa*
– Leaves (6)8–16(20)-foliolate; leaflets c. 1.5–2.5(3) mm wide · · · · · · · · · · · · · · · · · · · 2
2. Bracts c. 2.5 mm long · ii) var. *delicatula*
– Bracts c. 6 mm long · iii) var. *major*

i) Var. **bracteosa** —Verdcourt in Kirkia **9**: 406 (1974).
   Aeschynomene nambalensis Harms in Warburg, Kunene-Samb.-Exped. Baum: 261 (1903).
   —E.G. Baker, Legum. Trop. Africa: 294 (1929) pro parte. Type from Angola.
   Aeschynomene elisabethvilleana De Wild. in Repert. Spec. Nov. Regni Veg. **11**: 503 (1913).
   —E.G. Baker, Legum. Trop. Africa: 294 (1929). Type from Dem. Rep. Congo.
   Aeschynomene zigzag De Wild. in Repert. Spec. Nov. Regni Veg. **11**: 506 (1913). —E.G.
   Baker, Legum. Trop. Africa: 294 (1929). Type from Dem. Rep. Congo.

Leaves 12–56-foliolate; leaflets 1–6.5 × 0.3–1.5(2) mm; stipules 1.5–5 × 0.5–2 mm.
Bracts 1–3.5 mm long.

   Zambia. B: Kaoma (Mankoya), fl. & fr. 17.x.1964, *Fanshawe* 8962 (K). N: Mbala Distr.,
Kawimbe, Nachalanga Hill, fl. 10.vii.1960, *Richards* 12856 (K; SRGH). W: Mufulira, fl. & fr.
4.v.1934, *Eyles* 8252 (K; SRGH). C: c. 8 km east of Lusaka, 17.ix.1955, *King* 140 (K). E: Makutu
Hills, fl. 17.viii.1979, *Chisumpa* 551 (K; NDO). S: Mumbwa, fl. & fr., *Macaulay* 997 (K).
Zimbabwe. N: Mazowe Distr., Mvurwi (Umvukwes) Range, 1.6 km north of Mtorashangu
(Mtoroshanga) Pass, fl. 19.xi.1961, *Leach* 11281 (K; LISC; PRE; SRGH). **Malawi**. N: Nkhata Bay
Distr., c. 33 km east of Mzuzu, fl. 4.iv.1971, *Pawek* 4570 (K; MAL). C: Angoniland, *Sharpe* 91 (K).
   Also in central Africa, Dem. Rep. Congo, Angola and southern Tanzania. Mainly in *Brachystegia*
woodland on sandy or rocky ground, also in seasonally burnt grassland; 900–1800 m.

ii) Var. **delicatula** (Baker f.) Verdc. in Kew Bull. **24**: 3 (1970); in Kirkia **9**: 407 (1974). Syntypes
   from Dem. Rep. Congo and Tanzania.
   Aeschynomene delicatula Baker f., Legum. Trop. Africa: 293 (1929).

Leaves (6)8–16(20)-foliolate; leaflets 2.5–5 × 1.5–2 mm; stipules 3–5 × 1.5–2 mm.
Bracts 2–2.5 mm long.

   Zambia. N: Mbala Distr., Lumi River flats, fl. 17.viii.1956, *Richards* 5855 (K).
   Also in Dem. Rep. Congo and southern Tanzania. Seasonally burnt grassland, woodland;
1680 m.

iii) Var. **major** Verdc. in Kew Bull. **24**: 4 (1970); in Kirkia **9**: 407 (1974). Type: Zambia,
   Kawambwa, *Fanshawe* 3518 (EA; K, holotype; SRGH).

Leaves 12–16(20)-foliolate; leaflets 5–7 × 1.5–2 mm; stipules 7.5 × 3 mm. Bracts 6
mm long.

   Zambia. N: Kawambwa, fl. & fr. 22.viii.1951, *Fanshawe* 3518 (EA; K; SRGH).
   Not known elsewhere. Bushland; 1415–1690 m.
   This variety is larger in most of its parts than the other two. *Richards* 3632 (Zambia N: pans
near Mbala (Abercorn), fl. 14.xii.1954 (K)) is intermediate between var. *delicatula* and var. *major*
in some respects, the leaflets attaining 6 × 3 mm but the stipules and bracts are small. Both
varieties come from an area well known for endemism. *Fanshawe* 3164 (Zambia W: Kitwe, fl. &
fr. 10.iv.1957 (K)) also has large leaflets and small bracts. But for some intermediates these
varieties could scarcely be classified with *A. bracteosa*.

26. **Aeschynomene minutiflora** Taub. in Engler, Pflanzenw. Ost-Afrikas **C**: 214 (1895). —E.G.
   Baker, Legum. Trop. Africa: 290 (1929). —Verdcourt in F.T.E.A., Leguminosae, Pap.: 384
   (1971). —Drummond in Kirkia **8**: 216 (1972). —Verdcourt in Kirkia **9**: 407 (1974). —
   Lock, Leg. Afr. Check-list: 105 (1989). Type from Tanzania.

Erect annual herb, 15–60 cm tall, rather sparsely branched with some of the lower
lateral branches spreading. Stems glabrous. Leaves 6–12-foliolate; leaflets 5–13 ×
2–2.7(4) mm, obcuneate to narrowly oblong-oblanceolate, rounded to emarginate
and mucronulate at the apex, cuneate, glabrous, with dense very minute translucent
dots; main nerve central save near base of leaflet; petiole and rhachis together 5–25
mm long, glabrous, produced as a bristle at apex; petiolules 0.5 mm long; stipules
leaf-like, 4–18 × 0.5–3(5) mm, lanceolate, apiculate, mostly with a slender subulate

appendage on one side at the base, persistent. Inflorescences terminating the lateral branches, 4–8-flowered; peduncle 0–25 mm long; rhachis 2.5–7 cm long; pedicels very slender, 5–15 mm long, with sparse tubercular-based hairs; bracts paired, 2–4.5 × 0.5 mm, linear-lanceolate, tapering-acute, persistent, sometimes in threes, 2 being stipules and the third probably representing the reduced leaf, thus indicating that the flowers are strictly solitary in the axils of reduced leaves; bracteoles 1.5–2 × 0.3–0.5 mm, linear-lanceolate, usually with 2–3 tubercular-based hairs. Calyx glabrous, 2-lipped, one 3.5 × 2 mm, ovate, 3-toothed, the other 3 × 1.4 mm, oblong, emarginate. Standard creamy-white, tinged dull orange-yellow inside and often veined green below, (3)5.5(9 in one variety) × 3 mm at apex, 1 mm wide below, spathulate; wings creamy-white with upper margins or tips orange-yellow; keel petals creamy-white, not laciniate. Fruit with a tenuous stipe 3–4 mm long, of 1–2 articles joined by a very narrow neck, each article kettle-drum-shaped in outline, straight above, very curved below, 2–2.5(3.5) × (1.5)2–2.2(3) mm, compressed or slightly inflated, at first glabrous, later densely minutely tuberculate, sometimes with a raised or even spinous ridge bisecting the article at right-angles to the upper suture. Seeds greenish, 2 × 1.5 × 0.9 mm, irregularly lentil-shaped, ± straight or emarginate near the minute round central hilum, margins compressed.

Subsp. **minutiflora** —Verdcourt in Kirkia **9**: 408 (1974). —Gonçalves in Garcia de Orta, Sér. Bot. **5**: 62 (1982). —Lock, Leg. Afr. Check-list: 105 (1989).

Mostly erect herb; corolla 3–4.5(5.5) mm long.

Zambia. C: 100–129 km east of Lusaka, Chakwenga Headwaters, fl. & fr. 27.iii.1965, *E.A. Robinson* 6499 (K; SRGH). E: Chipata Distr., Jumbe area, fl. & fr. 24.iii.1963, *Verboom* 456 (K). S: c. 22.5 km west of Pemba, Chepezami Dam, fl. & fr. 21.iv.1954, *E.A. Robinson* 697 (K). **Zimbabwe**. N: Hurungwe (Urungwe) Reserve, Msuku River, fl. & fr. iv.1956, *Davies* 1871 (K; SRGH). W: Matobo Distr., Besna Kobila Farm, fl. & fr. i.1961, *Miller* 7668 (K; SRGH). C: Harare Distr., Rumani Farm, near Munenga Stream, fl. & fr. 23.iii.1950, *Chase* 2113 (BM; COI; K; LISC; SRGH). **Malawi**. N: Mzimba Distr., 33.6 km south of Mzambazi, fl. & fr. 18.iv.1974, *Pawek* 8382 (K; MAL; MO; SRGH; UC). C: Chitala Escarpment, fl. & fr. 12.ii.1959, *Robson* 1560 (BM; K; LISC; SRGH). S: Machinga Distr., Chikala Hills, fl. 2.ii.1955, *G. Jackson* 1436 (K). **Mozambique**. N: 28 km from Ribáuè towards Malema, 21.iii.1964, *Correia* 222 (LISC). T: Macanga Distr., Monte Furancungo, fl. & fr. 15.iii.1966, *Pereira, Sarmento & Marques* 1685 (LMU).

Also in Tanzania. *Brachystegia* and *Colophospermum* woodland, often in thin sandy soil over rocks, sandy river beds, etc.; 500–1830 m.

Subsp. **grandiflora** Verdc. in Kew Bull. **27**: 437 (1972); in Kirkia **9**: 409 (1974). —Lock, Leg. Afr. Check-list: 105 (1989). Type: Mozambique, Namapa, *Balsinhas & Marrime* 325 (BM; COI; K, holotype; LISC; LMA; PRE).

Suberect herb with side branches to 15 cm long; corolla 6.5–9 mm long.

Mozambique. N: Namapa, near the C.I.C.A. Experimental Station, track to R. Lúrio, fl. & fr. 29.iii.1961, *Balsinhas & Marrime* 325 (BM; COI; K; LISC; LMA; PRE). Z: Ile Distr., Errego, at c. 3 km, Monte Ile, fl. & fr. 3.iii.1966, *Torre & Correia* 15004 (LISC).

Not known elsewhere. Open *Brachystegia* woodland, rocky soil; 200–450 m.

Some isotypes have a distinctly upwardly directed spine on the articles but they are much less evident in the holotype. The collectors remark on a very pleasant scent.

27. **Aeschynomene pawekiae** Verdc. sp. nov.* TAB. 3.6.**22**. Type: Malawi, 32 km WSW of Karonga, near Kayelekera, 09°59'S, 33°40'E, 9.vi.1989, *Brummitt* 18452 (K, holotype; MAL). *Aeschynomene sp.* E of Verdcourt in Kirkia **9**: 377, 414 (1974).

Diffuse ± glaucous herb with horizontal stems or prostrate, up to 1 m long, with sparse white ± stiff hairs 0.5–1 mm long, from thickened bases but otherwise glabrous; older stems minutely tuberculate where hairs have broken off. Leaves

---

* **Aeschynomene pawekiae** Verdc. sp. nov. in subgenere *Rueppellia* (A. Rich.) J. Léonard posita; affinis *A. bracteosae* Welw. ex Baker sed ramulis sparse albo-pilosis vel glabris (haud viscoso-hirsutis) foliolis glaucescentibus latioribus usque 7 mm latis (haud 2.5(3) mm) basi 5–7-nervibus differt. Typus: Malawi, Karonga Distr., 9.vi.1989, *Brummitt* 18452 (K, holotype; MAL).

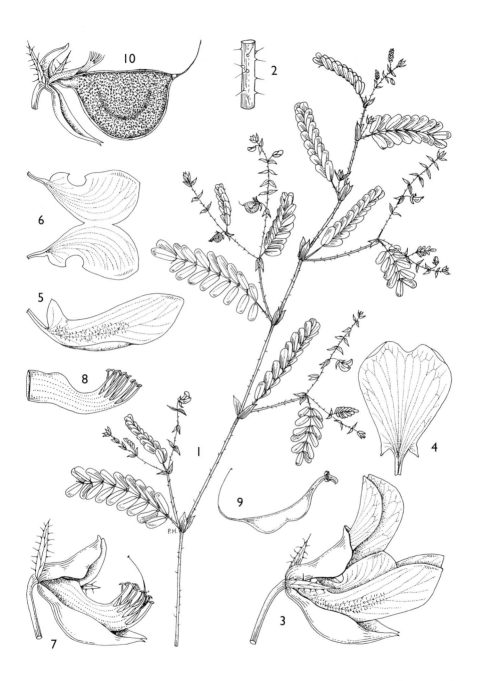

Tab. 3.6.**22**. AESCHYNOMENE PAWEKIAE. 1, flowering and fruiting branch ($\times \frac{2}{3}$); 2, detail of portion of stem, much enlarged, 1 & 2 from *Pawek* 4195; 3, flower ($\times$ 6); 4, standard ($\times$ 6); 5, wing ($\times$ 6); 6, keel, opened out ($\times$ 6); 7, calyx and androecium ($\times$ 6); 8, androecium ($\times$ 6); 9, gynoecium ($\times$ 6), 3–9 from *Brummitt* 18452; 10, fruit ($\times$ 4), from *Pawek* 11093. Drawn by Pat Halliday.

(6) 10–14-foliolate; leaflets glaucous, (5) 7–14 × 2–7 mm, oblong, rounded to truncate at the apex and when young with fine white hair terminating the mucro soon breaking off, bleakly rounded at the base, glabrous, 5–7-nerved from the base including the distinctly eccentric main nerve; petiole and rhachis together 1.5–5.3 cm long, with sparse bristly hairs; stipules 3–7 × 0.8–2.5 mm, ovate-lanceolate to lanceolate, 7–13-nerved, not appendaged. Inflorescences axillary, 1.5–2.5(4) cm long; peduncles up 1.5 cm long; pedicels 2.5–4 mm long; bracts 3–5 × 1.1–1.5(2) mm long, lanceolate, persistent, 7-nerved, with marginal bristle-like hairs; bracteoles 3 × 0.5–0.8 mm, narrowly lanceolate, sparsely white bristly ciliate. Calyx lobed, the upper 3–4 × 2 mm, entire, the lower 4.2–5 × 2 mm, 3-lobed. Standard yellow, obovate; blade 6 × 4–4.8 mm, emarginate at apex with triangular acute auricles at base and claw 1.1 mm long; wings narrowly oblong, blade 7 × 2.2 mm with triangular auricle and claw 1.3 mm long; keel yellow, 5.8 mm long including claw, the petals joined for c. 1.5 mm at the base and with a sinus beyond the basal auricle, not laciniate. Fruit with 1 deleoped circular article 5–5.5 × 4.5–5.5 mm, glabrous, beaked with a second undeveloped article and sometimes also persistent style or, occasionally terminal article is developed and undeveloped one below it; stipe 1.2 mm long. Seeds brown, 3.5 × 3 mm, semi-circular reniform in outline, compressed.

Malawi. N: Chitipa Distr., 16 km east of Chitipa, Kaseye Mission, fl. & fr. 18.iv.1976, *Pawek* 11093 (K; MAL; MO; SRGH; UC). **Mozambique**. N: Lago Distr., Maniamba, c. 7 km on new track from Metangula towards Maniamba, fl. & fr. 29.ii.1964, *Torre & Paiva* 10916 (LISC).

Known only from the Flora Zambesiaca area. *Brachystegia allenii* woodland with *Julbernardia globiflora, Parinari curatellifolia*, etc. on gravelly soil; 700–1260 m.

28. **Aeschynomene mossambicensis** Verdc. in Kew Bull. **27**: 437 (1972); in Kirkia **9**: 409 (1974).
—Lock, Leg. Afr. Check-list: 105 (1989). TAB. 3.6.**23**. Type: Mozambique, Zambézia Prov., Namagoa Estate, *Faulkner* 206 (COI; K, holotype; PRE; SRGH).

Spreading ± prostrate annual or sometimes perennial herb 60–90 cm long, usually distinctly branched. Stems glabrous or with sparse short glandular somewhat tubercular-based hairs. Leaves 8–12(16)-foliolate; leaflets 3.5–12 × 1–5 mm, obcuneate to oblanceolate-oblong, broadly rounded and mucronulate at the apex, narrowed to an unequally rounded base, glabrous, obscurely minutely punctate, midvein central save at base of leaflets; petiole and rhachis together 0.4–4.5 cm long, glabrous, produced as a seta at the apex; petiolules 0.5 mm long; stipules 10–23 × 2–3.5 mm, lanceolate, mostly obliquely appendaged on one side for 0–1 mm, closely veined, ± persistent. Inflorescences axillary, 2.5–1.6 cm long, mostly few-flowered but with numerous groups of spaced bracts, with sparse (rarely dense) tubercular-based hairs; peduncle 3–15 mm long; pedicels 1.5–6(11 in fruit) mm long with very sparse (rarely dense) tubercular-based hairs; bracts 3-partite, persistent, each lobe 3–7 × 0.5–1.2 mm, lanceolate; bracteoles 2–3 × 0.3–0.6 mm, lanceolate, ± persistent. Calyx glabrous, 2-lipped; lips 2.5–3.5 × c. 2 mm, elliptic-oblong, one emarginate, the other shortly 3-fid. Standard orange or yellow, (4)5.5–6.5 × 3.2 mm, ± violin-shaped, the upper part rounded. Fruit of 1–2(4) articles joined by a narrow neck, each article three-quarters-circular, one margin straight, the other strongly curved, (3)4–5 × (2.5)4–4.5 mm, compressed, with a few scattered small tubercles and usually a median vertical row of tubercles or a marked row of spines on the faces and a very few scattered tubercular-based hairs; stipe long and slender, 7–11(15) mm long. Seeds chestnut-brown, 3 × 2.2–2.8 × 1.5 mm, rounded-reniform, fairly compressed, the margin ± angled; hilum elliptic, ± central.

Subsp. **mossambicensis** —Verdcourt in Kirkia **9**: 410 (1974). —Lock, Leg. Afr. Check-list: 106 (1989).

Fruit stipe mostly under 1 cm long; articles of fruit larger.

Mozambique. N: Nampula, fl. & fr. 12.iii.1937, *Torre* 1190 (COI; LISC). Z: near Olinga (Maganja da Costa), fl. & fr. *Torre* 5208 (LISC).

Not known elsewhere. *Brachystegia–Julbernardia* woodland, dry rocky places, usually on sandy soil; 10–430 m.

A. mossambicensis var. *longistipitata* (Verdc.) Vollesen occurs in southern Tanzania; its true relationship to *A. nematopoda* must await proper observations on the fruiting of the latter.

Tab. 3.6.**23**. AESCHYNOMENE MOSSAMBICENSIS. 1, flowering and fruiting branch (× 1);
2, part of inflorescence (× 6); 3, calyx split in two, a, adaxial portion, b, abaxial portion (× 8);
4, standard (× 8); 5, wing (× 8); 6, keel (× 8); 7, androecium (× 8); 8, gynoecium (× 8);
9, fruit (× 4); 10, seed (× 4), 1–10 from *Faulkner* 206. Drawn by Victoria Friis. From Kew Bull.

29. Aeschynomene sp. B
Aeschynomene sp. C of Verdcourt in Kirkia **9**: 410 (1974).

Perennial herb, presumably with a woody rootstock and prostrate stems to 70 cm. Young shoots glabrous. Leaves 8–10-foliolate; leaflets 3–7.5 × 1–2 mm, oblanceolate or elliptic-oblanceolate, rounded and shortly mucronulate at the apex, narrowed to an obliquely rounded base, glabrous, minutely pellucid-punctate; main nerve ± central; petiole and rhachis together 4–14 mm long; petiolules very short; stipules 2–5 × 0.5–1.2 mm, lanceolate, appendaged obliquely on one side for up to 1 mm, leaf-like, veined, ± persistent. Inflorescences axillary, relatively short and condensed, 1–4.5 cm long, often bearing a 4-foliolate leaf at the base subtending one or several side branches, few- to several-flowered but often with 10 or so sterile bracts, zigzag, glabrous, peduncle 5–10 mm long; pedicels 2–3 mm long, bracts 3-partite, persistent, each lobe 1–1.5 × 0.5 mm, narrowly lanceolate; bracteoles 1 × 0.5 mm, ovate-lanceolate, glabrous, persistent. Calyx glabrous, 2-lipped; lips 1.5–3 × 2 mm, elliptic-oblong, one emarginate, the other shortly 3-fid. Standard orange-yellow, consisting of an apical elliptic portion 2.5 × 4 mm and a narrow part 2–3 × 1.3 mm, emarginate; wings and keel orange-yellow, the petals of the latter not laciniate. Fruit of 1–2 articles joined by a narrow neck, each semicircular, one margin straight, the other strongly curved, 3–3.5 × 2.5–3 mm, compressed, covered with small raised tubercles and a raised median line but no spine; stipe 3.5–5 mm long. Seeds dark purplish-brown, 2.3 × 2 × 0.8 mm, semicircular-reniform, compressed; hilum round, very small, rather eccentric.

Mozambique. MS: Gondola Distr., Garuso, fl. & fr. 2.iii.1948, *Pedro* 3616 (LMA). Not known elsewhere. Rocky places; c. 700 m.

30. Aeschynomene nematopoda Harms in Notizbl. Bot. Gart. Berlin-Dahlem **11**: 813 (1933); in op. cit. **13**: 419 (1936). —Verdcourt in F.T.E.A., Leguminosae, Pap.: 386 (1971); in Kirkia **9**: 411 (1974). —Lock, Leg. Afr. Check-list: 106 (1989). Type from Tanzania.

Diffuse annual trailing herb c. 30 cm long. Stems slender, glabrous or almost so. Leaves 4–16-foliolate; leaflets 2.5–11 × 1–2.7 mm, oblong-oblanceolate, oblanceolate, narrowly oblong or obovate-obcuneate, rounded to slightly emarginate at the apex, narrowed to an obliquely rounded base, glabrous; main nerve ± central; petiole and rhachis together 5–15 mm long, produced as a bristle; petiolules 0.3 mm long; stipules 4–8 × c. 1 mm, ovate-lanceolate to lanceolate, acuminate, shortly appendaged on one side at the base, persistent. Inflorescences axillary, few- to several-flowered; peduncle obsolete; rhachis 1–4 cm long; pedicels 6–13 mm long, filiform, glabrous; bracts in threes, 2–3 × 0.5 mm, persistent; bracteoles 1–2 × 0.5 mm, ovate-lanceolate to lanceolate, glabrous, persistent. Calyx glabrous, 2-lipped; lips 3–4.5 × 1–1.5(2) mm, oblong or elliptic, one slightly emarginate, the other minutely 3-toothed. Standard dark orange-yellow, 6–7 mm long, 4 mm wide above, 1 mm wide below, spathulate; wings and keel petals yellow, the latter not laciniate. Fruit unknown but developing underground, the stipe accrescent, 1.5–8 cm long at least.

Mozambique. N: 10 km from Lalaua towards Ribáuè, fl. 22.i.1964, *Torre & Paiva* 10122 (LISC). Also in southern Tanzania. Open *Brachystegia* woodland on sandy soil; 400 m.
Fruiting material of this species is needed; since the fruits develop beneath ground they become detached as the plant is pulled up.

31. Aeschynomene rhodesiaca Harms in Repert. Spec. Nov. Regni Veg. **8**: 356 (1910). —E.G. Baker, Legum. Trop. Africa: 290 (1929). —Drummond in Kirkia **8**: 216 (1972). —Verdcourt in Kirkia **9**: 412 (1974). —Lock, Leg. Afr. Check-list: 107 (1989). Type: Zimbabwe, Bulawayo, *Gardner* 17 (K, holotype).
Aeschynomene sp. (*Gardner* 17) of Eyles in Trans. Roy. Soc. South. Africa **5**: 378 (1916).

Perennial herb with prostrate or straggling stems to 15–60 cm long from a woody rootstock; young shoots glabrous or with sparse tubercular-based sticky hairs, soon glabrous. Leaves 8–14-foliolate; leaflets 4–15 × 1–3.5 mm, oblanceolate or oblong-oblanceolate, rounded and mucronulate at the apex, cuneate at the base, glabrous;

main nerve ± central; petiole and rhachis together 10–25 mm long; petiolules very short; stipules 8–28 × 1.2–8 mm, ovate, oblong-lanceolate or lanceolate, appendaged obliquely on one side for 1–3 mm, leaf-like, closely veined, ± persistent. Inflorescences axillary, often apparently branched, 3–12 cm long, few–several-flowered, pubescent; peduncle 1–2 cm long, pedicels 3–7 mm long, pubescent with tubercular-based hairs; bracts 3-partite, persistent, each lobe 2–6 × 0.5–1 mm, lanceolate; bracteoles 2–3 × 0.5–1 mm, lanceolate, ciliolate, sometimes deeply 2-partite, persistent. Calyx with a few short tubercular-based hairs save near margins, 2-lipped; lips 4–5 × 2–3 mm, elliptic-oblong, one emarginate, the other 3-fid. Standard orange or yellow, often veined red (and in one instance flowers stated to be pink), 6–10 × 4–6.5 mm, violin-shaped, emarginate, wings and keel yellow or orange, the petals of the latter not laciniate. Fruit of 1–3(4) articles joined by very narrow necks, each article semicircular, one margin straight the other strongly curved, 2.7–3.5 × 2.5–3 mm, compressed, with a few tubercular-based hairs on the faces and margins; stipe c. 3–6 mm long. Seeds chestnut-brown, 2 × 1.5 × 0.7 mm, semicircular-reniform, compressed; hilum round, very small, slightly eccentric.

Zimbabwe. W: Bulawayo, fl. i.1905, *Gardner* 17 (K). C: Marondera Distr., Dombo-Dombo Kopje fort, fl. & fr. 24.xii.1966, *Corby* 1702 (K; SRGH). E: Nyanga Distr., near top of Mutarazi (Mtirazi) Falls, fl. 12.ii.1961, *Goodier* 1038 (K; LISC; SRGH). S: Bikita Distr., Old Bikita, fl. & fr. 16.xii.1953, *Wild* 4405 (K; PRE; SRGH). **Mozambique**. MS: Bárué Distr., Serra de Chôa, ao km 24 de Catandica (Vila Gouveia), picada com rumo à fronteira, fl. 26.iii.1966, *Torre & Correia* 15414 (LISC).
Not known elsewhere. Grassland, often in damp rocky areas; 1050–2250 m.
The inflorescences often bear a leaf near the base supporting a side inflorescence. Actually all the flowers are axillary, the leaves being suppressed and the bracts derived from stipules.

32. **Aeschynomene sparsiflora** Baker in Bull. Misc. Inform., Kew **1897**: 259 (1897). —E.G. Baker, Legum. Trop. Africa: 290 (1929). —Verdcourt in F.T.E.A., Leguminosae, Pap.: 390 (1971); in Kirkia **9**: 412 (1974). —Lock, Leg. Afr. Check-list: 108 (1989). Syntypes: Malawi, Nyika Plateau, *Whyte* 256 (K); between Mpata and the commencement of the Nyasa–Tanganyika Plateau, *Whyte* (K).

Probably a straggling subshrub, 10–30 cm tall, with many suberect or spreading slender stems from a woody rootstock. Stems mostly glabrous. Leaves 6–10-foliolate; leaflets 5–10 × 2.5–5.5 mm, elliptic-oblong or obovate-oblong, rounded and mucronulate at the apex, cuneate to obliquely rounded at the base, glabrous; main nerve eccentric and curved, there being 2–3 basal nerves in the wider side of the leaflet; petiole and rhachis together 10–25 mm long; petiolules 0.5 mm long; stipules 8–12 × 3–5 mm, elliptic, acute, subcordate and mostly unilaterally appendaged at the base, glabrous or pubescent, at length deciduous. Inflorescences axillary, few- to 10-flowered, often branched; rhachis 7 cm long, densely pubescent with tubercular-based hairs; peduncle 5 mm long; pedicels 5–6 mm long; bracts 4–6 × 1.5–2.8 mm, ovate-elliptic or elliptic, acute, pubescent with tubercular-based hairs and appearing glandular-serrulate, persistent; bracteoles similar, 4–5 × 1.2–1.6 mm, lanceolate. Calyx pubescent with glandular-based hairs, 2-lipped; lips c. 6–7 × 3–5 mm, oblong or ovate, one emarginate, the other slightly 3-fid. Standard golden-yellow, 10–13 × 7–8 mm, violin-shaped, keel petals not laciniate. Fruit of 1 article, 6.5 × 6 mm, almost round, compressed, sparsely covered with tubercular-based hairs. Seeds not seen.

Zambia. N: Mafinga Mts., fl. 24.v.1973, *Fanshawe* 11970 (K). **Malawi**. N: Nyika Plateau, fl. vi.1896, *Whyte* 256 (K).
Also known from one old gathering from southern Tanzania. There are only five specimens known, *Brachystegia* woodland and montane grassland; 600–2100 m.

33. **Aeschynomene sp. C**
   Aeschynomene sp. D of Verdcourt in Kirkia **9**: 413 (1974).

Tufted perennial herb c. 10 cm tall, with numerous erect or procumbent stems from a woody rootstock, mostly flowering after burns when few or no leaves are apparent. Stems densely pubescent with glandular hairs. Leaves 6–8-foliolate; leaflets 3–9 × 1.5–4 mm, elliptic or oblong-obovate, obtuse but mucronulate, narrowed at the base, pubescent; main nerve very eccentric; petiole and rhachis together 8–12 mm long; petiolules minute; stipules 4–11 × 1.5–5 mm, elliptic, elliptic-oblong or narrowly ovate acute, laterally expanded at the base to the side of the oblique attachment but scarcely appendaged, venose, pubescent. Inflorescences axillary but sometimes accounting for most of the plant in the early stages; peduncle 5–15 mm long; rhachis 5–9 cm long, pubescent; pedicels 3.5–10 mm long; bracts divided to the base or only half-way into 2 elliptic lobes, 4–8 × 1–2.5 mm, acute, pubescent, persistent; bracteoles 3–4.5 × 0.7 mm, elliptic-lanceolate, pubescent, persistent. Calyx puberulous, 2-lipped; lips 5–6 × 2–2.8 mm, oblong, one shortly 2-fid the other shortly 3-fid, the lobes all blunt. Standard orange-yellow or orange, 11.5 × 8 mm, obovate-spathulate, the lower narrow portion truncate, not narrowed into the claw; keel petals not laciniate. Fruit of 2 articles joined by a narrow neck, each article (immature) elliptic-semicircular, 8–9 × 5 mm, rather shortly glandular pubescent. Seeds not seen.

Malawi. N: 4.8 km west of Chisenga, fl. & fr. 26.viii.1962, *Tyrer* 614 (BM).
Also in Tanzania. In open *Vellozia* community on exposed mountain ridge; c. 2100 m.

34. **Aeschynomene oligophylla** Harms in Repert. Spec. Nov. Regni Veg. **8**: 356 (1910). — E.G. Baker, Legum. Trop. Africa: 290 (1929). —Brenan, Check-list For. Trees Shrubs Tang. Terr.: 407 (1949). —Léonard in F.C.B. **5**: 275, pl. 21 (1954). —Verdcourt in F.T.E.A., Leguminosae, Pap.: 390 (1971); in Kirkia **9**: 415 (1974). —Lock, Leg. Afr. Check-list: 106 (1989). TAB. 3.6.**24**. Syntypes: Malawi, Mt. Masisi, *McClounie* 72 (K); and from Tanzania.

Erect subshrub 6–45 cm tall, with many stems from a woody rootstock, flowering after burns. Stems glabrous or pubescent. Leaves erect, pressed to the stem, 2–4-foliolate; leaflets 4–13 × 2–6 mm, obovate to oblanceolate-oblong, somewhat falcate, mostly rounded at the apex, narrowed to the base; rather thick, glabrous, 3–6-nerved from the base; venation reticulate; petiole and rhachis together 2–4(10) mm long, with a line of hairs beneath; petiolules obsolete; stipules 6–18 × 3–6 mm, obovate-oblong or oblong-elliptic, leaf-like, obliquely appendaged at the base on one side, persistent. Inflorescences axillary, several-flowered; peduncle 1–1.5 cm long; rhachis 2–6 cm long, sparsely pubescent or glabrous; pedicels 2–4(10) mm long; bracts 2-lobed, persistent, the lobes 2–6.5 × 0.7–1.5 mm, lanceolate, some of the lowest 3-partite, corresponding to 2 stipules and a reduced leaf; bracteoles 3–5.5 × 1–2 mm, ovate-lanceolate or lanceolate; both bracts and bracteoles appendaged. Calyx 2-lipped; lips 5.5–7 × 3.54 mm, ovate, the one 2-toothed, the other 3-toothed. Standard yellow or orange, 10–13 × 7–8 mm, violin-shaped; keel petals not laciniate. Fruit reflexed, of 1 or 2 articles joined by a very narrow neck, each article ± semicircular, lower margin rounded, upper margin straight, both somewhat thickened, 6–8 × 5–7 mm, compressed, glabrous, reticulate. Seeds chestnut-brown, 4.5 × 3 × 2 mm, ellipsoid-reniform; hilum elliptic, with a distinct flange-like rim-aril present.

Zambia. E: Nyika Plateau, fl. & fr. 24.xi.1955, *H.M.N. Lees* 81 (K). **Malawi.** N: Nyika Plateau, by main road, c. 3.2 km SW of Rest House, fl. & fr. 21.x.1958, *Robson & Angus* 226, 226A (BM; K; LISC; PRE; SRGH).
Also in southern Dem. Rep. Congo and southern Tanzania. Seasonally burnt montane grassland; 1755–2400 m.
Coxe 28 (Zambia, near Nyika Rest House, fl. 19.ix.1960 (SRGH)) represents a state of this species flowering before its leaves appear, whilst still only 3–5 cm tall.

Tab. 3.6.**24**. AESCHYNOMENE OLIGOPHYLLA. 1, habit (× ¹/₂); 2, detail of branch showing stipule with leaf (× 2); 3, bipartite bract (× 3); 4, flower, longitudinal section (× 5); 5, bracteoles and lips of calyx (× 3); 6, standard, internal face (× 3); 7, wings joined by the appendices, external face (× 3); 8, keel, opened out, internal face (× 3); 9, androecium (× 3); 10, fruit (× 2), 1–10 from *van Meel* in *G. de Witte* 7457. Drawn by J.M. Lerinckx. From Fl. Congo Belge. Reproduced with permission of Jardin Botanique National de Belgique.

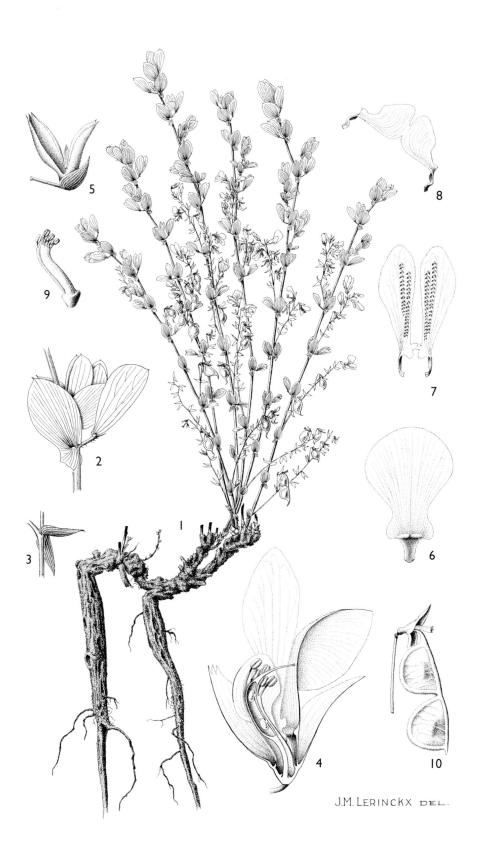

J.M. LERINCKX DEL.

35. **Aeschynomene glauca** R.E. Fr., Wiss. Ergebn. Schwed. Rhod.-Kongo-Exped.: 84 (1914). —
E.G. Baker, Legum. Trop. Africa: 297 (1929). —Brenan in Mem. New York Bot. Gard. **8**:
254 (1953). —Léonard in F.C.B. **5**: 280 (1954). —Verdcourt in F.T.E.A., Leguminosae,
Pap.: 389 (1971); in Kirkia **9**: 416 (1974). —Gonçalves in Garcia de Orta, Sér. Bot. **5**: 61
(1982). —Lock, Leg. Afr. Check-list: 103 (1989).   Type: Zambia, north end of Lake
Bangweulu, Nsombo (Msombo), *Fries* 1076 (UPS, holotype).

Erect subshrub with tufts of usually many stems, 10–40 cm tall, from a very
woody rootstock, usually flowering after burns.   Stems at first with short
tubercular-based hairs, later ± glabrous.   Leaves 12–24-foliolate; leaflets 3–9 ×
1–2.5 mm, linear-oblong to oblong, truncate or rounded and mucronulate at the
apex, obliquely rounded at the base, glabrous or slightly ciliolate when young,
glaucous; petiole and rhachis together 6–25 mm long, at first pubescent, later
glabrescent; petiolules 0.5 mm long; stipules 4.5–10 × (1)1.5–2 mm, ovate-
lanceolate to lanceolate, acuminate, unilaterally appendaged, leaf-like, glabrous,
persistent.   Inflorescences axillary, several- to 10-flowered, sometimes branched;
peduncle 5–15 mm long; rhachis 2.5–8 cm long; pedicels 2–7 mm long,
pubescent; bracts 2–3-fid, 2–3 mm long, persistent; bracteoles 2–3 × 0.5–1 mm,
lanceolate to ovate-lanceolate, slightly ciliolate.   Calyx glabrous, 2-lipped; lips
4–4.5 × 3–3.5 mm, ovate, one 2-toothed, the other 3-toothed.   Standard yellow, 7–8
× 5 mm, violin-shaped; wings and keel petals yellow, the latter not laciniate.   Fruit
of 1 article, or 2 joined by a very narrow neck, rounded or ± semicircular, 4–7 ×
3.5–5 mm, glabrous, reticulate.   Seeds dark chestnut-brown, 3.5 × 3 × 1 mm,
elliptic-reniform; hilum small, circular, slightly eccentric.

Zambia. N: Mpika to Shiwa Ngandu, fl. & fr. 17.ix.1938, *Greenway* 5697 (EA; K). W: Solwezi,
fl. & fr. 10.ix.1952, *White* 3205 (FHO; K). C: Great North Road, 19.2 km north of Kabwe
(Broken Hill), Nasuleka River Bridge, fl. 23.ix.1947, *Brenan & Greenway* 7911 (K). **Malawi**. C:
Dedza, fl. 13.ix.1946, *Brass* 17628 (K; NY; SRGH). S: Mt. Mulanje Masif, between Karama
Village, Fort Lister Gap and the Phalombe Gorge, fl. & fr. 8.x.1987, *J.D. & E.G. Chapman* 8900
(K; MO). **Mozambique**. N: Lichinga (Vila Cabral), fl. 30.x.1934, *Torre* 244 (BM; COI; K;
LISC). T: Macanga Distr., between Furancungo and Angónia, fl. & fr. 28.ix.1942, *Mendonça*
533 (LISC).
Also in Dem. Rep. Congo and southern Tanzania.   *Brachystegia, Marquesia* and 'Chipya'
woodland, also open dambos; 600–1500? m.
Richards 22352 (Zambia N: Mbala Distr., new road to Chinakila from Kambole Road, fl. & fr.
4.x.1967) is atypical in having the young leaves hairy and the articles pubescent.   *Mutimushi*
2633 (Zambia W: Kitwe, 13.viii.1968) has the young leaves pubescent but the articles practically
glabrous.
Torre 246 (Mozambique N: Litunde to Lichinga (Vila Cabral), fl. & fr. 15.xi.1934 (COI;
LISC)) fits neither into *A. glauca* nor *A. heurckeana* yet is close to both.   It is described as a
caespitose herb and the shoots are very branched from the base.   The leaves are c. 3 cm long
with the rhachis bearing a rather dense row of curved bristly hairs, leaflets glabrous with rather
prominent venation.   The inflorescences are short and few-flowered; articles 5 × 4 mm,
glabrous.   Some of the developing ovaries are galled, resulting in fusiform bodies closely
covered with tubercles.   Better flowering material is needed of this but it seems to be an extreme
variant of *A. glauca*.   The bracts are not persistent in the fruiting state.

36. **Aeschynomene sp. D**
Aeschynomene sp. F of Verdcourt in Kirkia **9**: 417 (1974).

Subshrubby herb 20–60 cm long with several to many procumbent wiry basal
stems from a woody rootstock; young shoots ascending?, covered with short
tubercular-based hairs, later glabrous, but often some of the older stems
roughened with small tubercles.   Leaves 10–40-foliolate; leaflets 2.5–7.5 × 0.8–2
mm, linear-oblong with slight tendency to be obovate, rounded but mucronulate
at the apex, obliquely rounded at the base, glabrous, midvein ± central; petiole and
rhachis together 1.5–3.5 cm long, with a row of tubercular-based bristly hook-like
hairs; petiolules short but distinct; stipules 6 × 1 mm, lanceolate, conspicuously
appendaged, glabrous or ciliolate, at length deciduous.   Inflorescences axillary or
terminal, 2–8 cm long, pubescent with tubercular-based hairs, several-flowered;
peduncles 5–20 mm long; pedicels 3–4 mm long; bracts 2-lobed, persistent, the
lobes 2 × 1 mm, narrowly ovate, pubescent, sometimes with an extra smaller
median lobe; bracteoles 2.5–3 × 0.7–1 mm, lanceolate, pubescent, ± persistent.

Calyx with sparse tubercular-based hairs, 2-lipped; lips 3–5 × 3 mm, oblong or ovate, one emarginate, the other 3-fid. Standard yellow with dark veining above, 6–6.5 × 4 mm, ± oblong but widest at the apex and slightly constricted in the middle; keel yellow, not laciniate. Fruit of 1–2 elliptic or almost semicircular articles 6–7 × 4.5 mm, compressed, glabrous or with a few very short hairs on lower margin. Seeds not seen.

Zambia. W: Kitwe, fl. 3.iii.1968, *Mutimushi* 2503 (K).
Brachystegia woodland; 1230 m.
Known from three specimens from Kitwe. It has been confused with *A. bracteosa* but differs in its appendaged stipules and is presumably closely related to *A. glauca* and *A. heurckeana* and may merely be a variant of the former. The procumbent habit and more numerous leaflets give it a different appearance. More information on the exact habit is needed and whether or not these might be plants which have lived for several seasons unburnt.
Richards 10023 (Kambole Escarpment, 1500 m, 5.vi.1957 (K)) is a very similar semi-prostrate plant with the articles having sparse very short hairs on the faces.

37. **Aeschynomene stolzii** Harms in Bot. Jahrb. Syst. **54**: 384 (1917). —E.G. Baker, Legum. Trop. Africa: 290 (1929). —Brenan, Check-list For. Trees Shrubs Tang. Terr.: 407 (1949); in Mem. New York Bot. Gard. **8**: 253 (1953). —Verdcourt in F.T.E.A., Leguminosae, Pap.: 388 (1971); in Kirkia **9**: 418 (1974). —Lock, Leg. Afr. Check-list: 108 (1989). Type from Tanzania.

A much branched prostrate cushion-forming subshrub, or forming mats or sometimes an erect bushy herb, 0.2–1.2(1.5) m long or tall. Stems glabrous or almost so, or pubescent with tubercular-based hairs in Malawi variants. Leaves 10–24-foliolate; leaflets 1–7 × 0.5–2 mm, oblong, obovate-oblong or almost linear, acute to obtuse and mucronulate at the apex, obliquely rounded at the base, glabrous; main nerve somewhat eccentric; rhachis and petiole together 5–22 mm long with a row of spaced bristly hairs along the back; petiolules minute; stipules 3–9 × 0.6–2 mm, lanceolate, acuminate, conspicuously appendaged on one side at the base, striate, glabrous. Inflorescences axillary, few-flowered; rhachis 1–2 cm long, ± glabrous; peduncle 5–20 mm long; pedicels 5–14 mm long, laxly covered with very distinctly tuberculate hairs; bracts mostly 2-fid, 2–4 × 1.5–2.2 mm, ovate or ovate-lanceolate, auriculate at the base, nerved, glabrous, persistent; bracteoles 2.5–3.5 × 1–1.5 mm, ovate, acute or slightly acuminate, nerved, persistent. Calyx with rather sparse tubercular-based hairs, 2-lipped; lips 3–6 × 2–2.5 mm, ovate or oblong, one subentire, the other 3-toothed. Standard yellow with dark veining, 8–10 × 4–5.5 mm, obovate-oblong; keel petals not laciniate. Fruit of 1 or mostly of 2 articles joined by a narrow neck, each article almost round, one margin slightly curved, the other strongly curved, 3.5 × 3 mm, compressed, at first smooth save for venation, later minutely rugulose and usually with very sparse tubercular-based hairs. Seeds not seen.

Malawi. N: Nyika Plateau, fl. 11.viii.1946, *Brass* 17170 (BM; K; NY; SRGH).
Also in southern Tanzania. Forest fringe; 2140–2440 m.

38. **Aeschynomene heurckeana** Baker in J. Linn. Soc., Bot. **20**: 130 (1883). —Brenan in Mem. New York Bot. Gard. **8**: 253 (1953). —Léonard in F.C.B. **5**: 279 (1954). —Verdcourt in F.T.E.A., Leguminosae, Pap.: 388 (1971); in Kirkia **9**: 418 (1974). —Lock, Leg. Afr. Check-list: 104 (1989). Syntypes from Madagascar.
      Aeschynomene dissitiflora Baker in Bull. Misc. Inform., Kew **1897**: 259 (1897). —E.G. Baker, Legum. Trop. Africa: 293 (1929). Type: Malawi, Nyasa–Tanganyika Plateau, Chitipa (Fort Hill), *Whyte* (K, holotype).

Erect, spreading or trailing subshrub 0.6–4.5 m long or tall. Stems glabrescent to densely covered with short tubercular-based hairs. Leaves 12–32-foliolate; leaflets 2–10(15) × 1–3.5 mm, oblong, obovate-oblong or linear-oblong, rounded and mucronulate at the apex, obliquely rounded at the base, glabrous or slightly pubescent beneath, often ciliolate when young; petiole and rhachis together 1–5(6) cm long, pubescent; petiolules 0.5 mm long; stipules 6–25 × 1.5–3.5 mm, ovate to ovate-lanceolate, acuminate, obliquely appendaged on one side, leaf-like, glabrous or pubescent, persistent. Inflorescences axillary, several- to 10-flowered, often branched; peduncle 1–2(5) cm long; rhachis 1–9.5 cm long, sometimes zigzag; pedicels 3–13 mm long; bracts 1.5–4 mm long, 2–3-lobed, ± persistent; bracteoles 2–4 × 0.5–1 mm, ovate-lanceolate, usually ciliolate. Calyx glabrous or slightly pubescent, 2-lipped; lips 4–6 × 3–4 mm, ovate, one 2-toothed, the other 3-toothed. Standard orange-yellow with reddish veins, 7–11 × 4–7 mm, violin-shaped; wings orange-yellow, paler towards the base; keel petals pale greenish-yellow, not laciniate. Fruit of 2 (rarely 1) articles joined by a very narrow neck, each article rounded-ellipsoid or semicircular, 4–6.5 × 3–4 mm, compressed or slightly inflated, glabrous, reticulate, sometimes with a slightly granular texture, the margins slightly thickened. Seeds reddish-brown, 3.5 × 2.3 × 1.5 mm, oblong-reniform; hilum small, elliptic.

Zambia. N: Lake Bangweulu, near Samfya Mission, fl. & fr. 30.viii.1952, *White* 3179 (FHO; K). W: Kitwe, fl. & fr. 20.iv.1957, *Fanshawe* 3190 (K). **Malawi**. N: Mugesse Forest Reserve, ix.1953, *Chapman* 130 (FHO; K). C: Ntcheu Distr., west foot of Dzonze Hill, fl. vii.1983, *Seyani & Tawakali* 1176 (K; MAL). **Mozambique**. N: Lichinga (Vila Cabral), fl. 8.vi.1954, *Torre* 261 (BM; COI; LISC).
Also in Dem. Rep. Congo, southern Tanzania and Madagascar. Fringing woodland, fringing evergreen forest, also *Mitragyna–Syzygium* swamp forest edges; 1050–2200 m.
Verboom 632 (Zambia E: Nyika Plateau, 7.vi.1962 (K; LISC; SRGH)) is probably a glabrous-leaved variant.

39. **Aeschynomene venulosa** Verdc. in Kew Bull. **24**: 8, fig. 2/1, 3–11 (1970); in F.T.E.A., Leguminosae, Pap.: 387 (1971); in Kirkia **9**: 419 (1974). —Lock, Leg. Afr. Check-list: 109 (1989). Type from Tanzania.

Subshrub or perennial herb c. 40 cm tall with suberect or decumbent glabrous stems with peeling bark. Leaves 4-foliolate; leaflets 5–12 × 3–9 mm, obovate, obovate-falcate or almost semicircular, rounded and obliquely mucronulate at the apex, subcordate or in upper leaflets very oblique, glabrous, 5–6-nerved from the base, the venation prominent on both sides; rhachis and petiole together 1–1.8 cm long, rhachis with scattered tubercular-based hairs; petiolules very short; stipules 4–18 × 2.5–12 mm, broadly elliptic or rounded, rounded at the apex, conspicuously appendaged on one side at the base, leaf-like, glabrous, conspicuously nervose, at length deciduous. Inflorescences axillary, lax, 1.5–6 cm long, 10–15-flowered; peduncles 5–15 mm long; pedicels 1.7–4.8 mm long; bracts 2–6.5 × 1.8–4 mm, oblong, nervose, glabrous, deeply divided into 2 obtuse lobes, soon deciduous; bracteoles 3–3.5 × 0.7 mm, lanceolate, acuminate. Calyx glabrous or very sparsely setulose, 2-lipped; lips 3.5–5.5 × 2–3 mm, elliptic or oblong, one obtuse, the other divided into 3 acute lanceolate lobes. Standard yellow, 6–7.8 × 3–6 mm, violin-shaped; wings and keel yellow, the petals of the latter not laciniate. Fruit of 1–2 articles joined by a narrow neck, ± semicircular, 4.5–5.5 × 4–4.5 mm, compressed, glabrous or with some hairs on the margins, the faces reticulate. Seeds not seen.

Var. **venulosa** —Verdcourt in Kirkia **9**: 420 (1974). TAB. 3.6.**25**, fig. A.

Leaflets 5–8 × 3–5 mm; stipules 4–12 × 2.5–9 mm with appendage 1–2 mm long.

Zambia. N: Mbala Distr., at the Kalambo Falls, 15.v.1936, *Burtt* 6366 (EA; K).
Also in Tanzania. *Brachystegia* woodland on stony ground; 1040–1350 m.

Var. **grandis** Verdc. in Kew Bull. **24**: 8, fig. 2/2 (1970); in Kirkia **9**: 420 (1974). TAB. 3.6.**25**, fig. B. Type: Zambia, Mbala Distr., Kambole, *Richards* 8249 (EA; K, holotype; SRGH).

Leaflets 7–14 × 5–9 mm; stipules 1.8 × 1.2 cm with appendage 3–6 mm long.

Zambia. N: Mbala Distr., Kambole Escarpment, above the Ngozye (Ngoshi) River, 22.iv.1969, *Richards* 24514 (K).
Only known from the Mbala District. Woodland on stony steep banks and short grassland; 1650–1828 m.

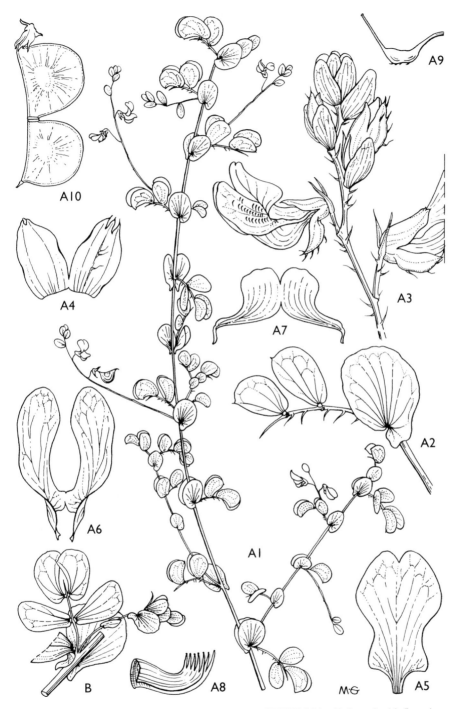

Tab. 3.6.**25**. A. —AESCHYNOMENE VENULOSA var. VENULOSA. A1, branch with flowering shoots (× 1); A2, leaf with one leaflet of each pair removed, showing stipule and tubercular-based hairs (× 3); A3, inflorescence (× 3); A4, calyx (× 6); A5, standard (× 6); A6, wings (× 6); A7, keel (× 6); A8, androecium (× 6); A9, gynoecium (× 6); A10, fruit (× 3), A1–A10 from *Richards* 12804. B. —AESCHYNOMENE VENULOSA var. GRANDIS, leaf and stipules with axillary shoot (× 1), from *Richards* 8249. Drawn by Mary Grierson. From Kew Bull.

40. **Aeschynomene fulgida** Welw. ex Baker in F.T.A. **2**: 149 (1871). —E.G. Baker, Legum. Trop. Africa: 298 (1929). —Léonard in F.C.B. **5**: 283 (1954). —White, F.F.N.R.: 142 (1962). — Torre in C.F.A. **3**: 200 (1966). —Verdcourt in F.T.E.A., Leguminosae, Pap.: 392 (1971); in Kirkia **9**: 420 (1974). —Lock, Leg. Afr. Check-list: 103 (1989). Type from Angola.

An erect branched shrub or undershrub 0.6–3.6 m tall. Stems covered with dense viscid tubercular-based hairs, later glabrescent or practically glabrous and ridged. Leaves 16–70-foliolate; leaflets 2–12 × 0.5–3 mm, linear or linear-oblong to obovate-oblong, retuse to rounded and mucronulate at the apex, obliquely rounded at the base, glabrous (rarely ciliate in some Angolan specimens), densely minutely pellucid-dotted; main nerve central; petiole and rhachis together 1–6.5 cm long, glabrous or pubescent; petiolules 0.5 mm long; stipules 4–15 × 1–3.5 mm, linear to ovate, acuminate, not appendaged or with slight traces of an appendage, deciduous. Inflorescences axillary and/or terminal, branched or simple, sometimes aggregated to form leafy panicles, 3- to usually many-flowered; peduncle 5–20 mm long, densely pubescent; rhachis 3–13(24) cm long; pedicels 2–7 mm long, densely pubescent; bracts 2–3-fid, 4–7 × 3–4.5 mm, elliptic, often ciliate, soon falling; bracteoles 5–7 × 2–3.5 mm, ovate, glabrous, sometimes ciliolate. Calyx glabrous or almost so, 2-lipped; lips 8–10 × 4–5 mm, ovate-oblong or elliptic, one emarginate, the other slightly 3-lobed. Standard orange, 16–19 × 9–12 mm, violin-shaped, emarginate; wings orange; keel petals orange, not laciniate. Fruit of 1 or 2 articles joined by a very narrow neck, each article elliptic, one margin slightly curved or straight, the other strongly curved, 8–12 × 5–8 mm, compressed at first, glabrous, smooth and reticulate, but sometimes slightly rugulose when adult. Seeds chestnut-brown, 4.5–5.5 × 3–3.5 × 1.5 mm, elliptic-reniform, compressed; hilum small, broadly elliptic.

Zambia. N: Mbala Distr., Mululwe River, between Kitumbula and Senga Hill, fl. & fr. 6.vi.1936, *B.D. Burtt* 6156 (BM; K). W: north of Mwinilunga, Luakela (Luakera) Falls, fl. & fr. 25.i.1938, *Milne-Redhead* 4346 (K; PRE). C: Serenje Distr., Kundalila Falls, fl. & fr. 13.x.1963, *E.A. Robinson* 5705 (K; SRGH).
Also in southern Tanzania and Angola. Riverine marshes and banks, dambos, swamp forest and evergreen forest margins sometimes in very swampy places and even in standing water; 1200–1680 m.

41. **Aeschynomene pygmaea** Welw. ex Baker in F.T.A. **2**: 148 (1871). —E.G. Baker, Legum. Trop. Africa: 295 (1929). —Torre in C.F.A. **3**: 201 (1966). —Verdcourt in Kirkia **9**: 421 (1974). —Lock, Leg. Afr. Check-list: 107 (1989). Type from Angola.

Herb or subshrub with several caespitose stems from a tough woody rhizome-like rootstock. Stems 0.07–1.5 m tall, in burnt areas mostly unbranched and often leafless when flowering but in the absence of fire forming a branched leafy shrub; young shoots densely covered with short and rather long bristly tubercular-based yellowish hairs, older stems glabrescent, roughened by the hair bases or bark peeling to reveal a slightly fissured surface. Leaves alternate or subfasciculate, 14–38-foliolate; leaflets 1.5–7 × 0.3–1.8 mm, linear-oblong or linear, often slightly to distinctly falcate, acute or rounded and mucronulate at the apex, obliquely rounded at base, somewhat coriaceous, glabrous or ciliate; main nerve subcentral or submarginal; petiole and rhachis together 4–35 mm long, hairy like the young stems; petiolules 0.3 mm long; stipules 2.5–13 × 0.5–3 mm, linear-lanceolate to ovate-lanceolate, not appendaged or at the most slightly subcordate, ciliolate, scarious, conspicuously parallel-veined, sometimes subpersistent. Racemes or panicles terminal or axillary, 2–16 cm long hairy like the young stems; peduncles 0.3–4 cm long; pedicels 1.5–6(7) mm long, pubescent; bracts entire or 2–3-fid, 1.5–5 × 0.5–1.5 mm, ovate or ovate-lanceolate, ciliolate, soon deciduous; bracteoles 2–4.5 × 0.5–1 mm, ovate or ovate-lanceolate, ciliolate, deciduous. Calyx flushed crimson, glabrous or pubescent, 2-lipped; lips 3–6 × 2–3.5 mm, elliptic, one entire or narrowly emarginate, the other slightly 3-fid. Standard orange-yellow, ± rectangular, or wider at the base of the lamina, 4–11.5 × 2.5–4.5 mm, somewhat constricted in the middle, emarginate at the apex, truncate or subauriculate at the base; wings and keel orange-yellow, the petals of the latter not laciniate. Fruit of 1–2 articles joined by a narrow neck, each article ± semicircular, 5–8 × 4.5–6 mm, pubescent on margins or all over with fine tubercular-based hairs, reticulate. Seeds dark reddish-brown, 3.5 × 2.5 × 1 mm, rounded-reniform, slightly beaked beyond the small circular eccentric hilum.

Var. **hebecarpa** J. Léonard in Bull. Jard. Bot. État **24**: 75 (1954); in F.C.B. **5**: 284 (1954). — Verdcourt in Kirkia **9**: 422 (1974). Type from Dem. Rep. Congo.

    Aeschynomene hockii De Wild. in Repert. Spec. Nov. Regni Veg. **11**: 504 (1913). —E.G. Baker, Legum. Trop. Africa: 297 (1929). Type from Dem. Rep. Congo.

    Aeschynomene homblei De Wild. in Repert. Spec. Nov. Regni Veg. **13**: 107 (1914). —E.G. Baker, Legum. Trop. Africa: 295 (1929). Type from Dem. Rep. Congo.

    Aeschynomene humilis N.E. Br. in Bull. Misc. Inform., Kew **1921**: 292 (1921). —E.G. Baker, Legum. Trop. Africa: 295 (1929). Type from Dem. Rep. Congo.

    Aeschynomene recta N.E. Br. in Bull. Misc. Inform., Kew **1921**: 293 (1921). —E.G. Baker, Legum. Trop. Africa: 296 (1929). Type from Dem. Rep. Congo.

    Aeschyomene youngii Baker f. in J. Bot. **73**: 295 (1935). Type from Dem. Rep. Congo.

Sepals mostly ciliolate. Ovary pubescent all over. Articles of the fruit pubescent on the faces.

Zambia. N: Mbala Distr., Kambole Escarpment, 5.vi.1957, *Richards* 10022 (K). W: Mwinilunga Distr., 27.2 km south of Mwinilunga on Kabompo road, 6.vi.1963, *Loveridge* 834 (K; LISC; SRGH).

Also in the Dem. Rep. Congo. Open *Brachystegia* woodland and bushland, grassland, dambo margins; 1200–1600 m.

Var. *pygmaea*, which occurs in Angola, is distinguished by its sepals not being ciliolate and the ovary and fruit being pubescent on the margins only.

42. **Aeschynomene mimosifolia** Vatke in Oesterr. Bot. Z. **29**: 224 (1879). —E.G. Baker, Legum. Trop. Africa: 296 (1929). —Brenan, Check-list For. Trees Shrubs Tang. Terr.: 406 (1949). —Léonard in F.C.B. **5**: 285 (1954). —White, F.F.N.R.: 142 (1962). —Verdcourt in F.T.E.A., Leguminosae, Pap.: 392, fig. 56 (1971). —Drummond in Kirkia **8**: 216 (1972). —Verdcourt in Kirkia **9**: 423 (1974). —Lock, Leg. Afr. Check-list: 105 (1989). TAB. **3.6.26**. Type from Kenya.

    Aeschynomene walteri Harms in Bot. Jahrb. Syst. **54**: 385 (1917). —E.G. Baker, Legum. Trop. Africa: 300 (1929). —Brenan, Check-list For. Trees Shrubs Tang. Terr.: 406 (1949). Type from Tanzania.

    Aeschynomene nyikensis var. *gracilis* Suess. in Proc. & Trans. Rhodesia Sci. Assoc. **43**: 17 (1951). Type: Zimbabwe, Marondera (Marandellas), *Dehn* 169 (M, holotype; SRGH).

Subshrub with slender somewhat woody erect or prostrate stems, 0.3–2.7 m tall, from a woody rootstock. Stems pubescent with glandular tubercular-based hairs. Leaves often fasciculate, 8–34-foliolate; leaflets 1–7 × 0.5–1.5 mm (rarely 2 mm in Mozambique), linear to oblong, sometimes subfalcate, acute to rounded and mucronulate at the apex, obliquely rounded at the base, glabrescent or glabrous but often conspicuously glandular-ciliate, somewhat thick; main nerve ± marginal; petiole and rhachis together mostly very short, 0.5–3 cm long, bristly-pubescent; petiolules 0.2 mm long; stipules 3–6 × 0.5–1 mm, linear- or ovate-lanceolate, acuminate, not appendaged, scarious, nervose, ± persistent. Inflorescences axillary, often aggregated towards the apices of shoots to form leafy panicles; peduncle 5–15 mm long; rhachis 1–10 cm long; pedicels 2–5 mm long; bracts 2–3-fid, 2.5–3 × 2 mm, ovate, ciliolate, soon falling; bracteoles 2–3 × 0.5–1 mm, ovate-lanceolate or lanceolate, ciliolate. Calyx puberulous, 2-lipped; lips 3.5–6 × 3–4 mm, oblong, one 2-lobed, the other 3-lobed and often ciliolate. Standard orange-yellow with purple-red veins, 7–8 × 4–5 mm, violin-shaped, emarginate; wings pale yellow, longer than the standard; keel petals greenish with reddish-purple nerves, not laciniate. Fruit of 1–2 articles joined by a very narrow neck, each article ± semicircular, 4.5–9 × 4–6.5 mm, glabrous or the margins puberulous, reticulate at first, later very minutely roughened. Seeds chestnut-brown, 5 × 3.5 × 1.5 mm, elliptic-reniform, compressed, slightly obtusely beaked beyond the small eccentric round hilum.

Zambia. N: Mbala Distr., Pans, fl. & fr. 1.x.1956, *Richards* 6331 (K). E: Petauke Distr., Great East Road, 62.4 km east of Beitbridge near Luangwa River, 18.iv.1952, *White* 2407 (FHO; K; PRE). **Zimbabwe**. N: Gokwe, fl. & fr. 19.v.1962, *Bingham* 270 (K; LISC; SRGH). C: 11.2 km SE from Gweru (Gwelo), fl. & fr. 14.xi.1966, *Biegel* 1438 (K; LISC; PRE; SRGH). E: Mutare (Umtali), fl. & fr. 21.ix.1911, *Rogers* 4056 (BM; K). S: Masvingo (Victoria), *Monro* 783. **Malawi**. N: Chitipa Distr., NE slopes of Mafinga foothills, fl. & fr. 19.xii.1964, *E.A. Robinson* 6312 (K). **Mozambique**. N: Unango, fl. & fr. 1.xi.1934, *Torre* 249 (BM; COI; K; LISC). MS: Moribane, flats of Rio Messambuze, fl. 10.ix.1942, *Salbany* 14 (LISC) (exceptionally long leaves and large leaflets).

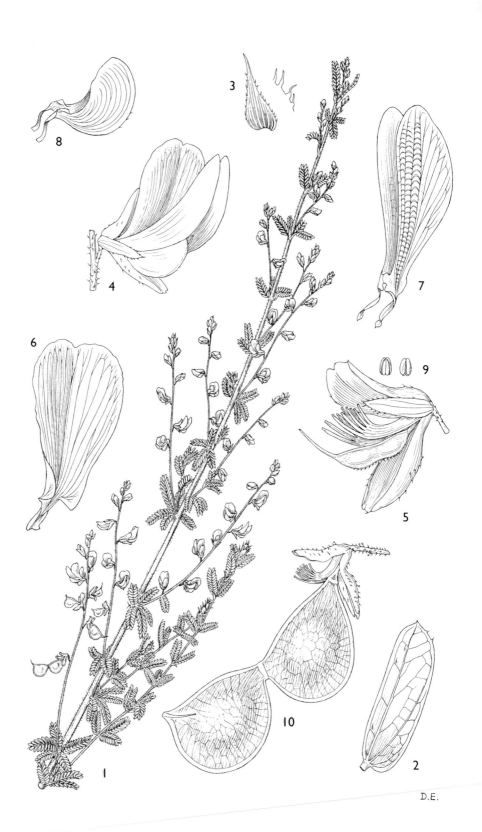

D.E.

Also in Kenya, Tanzania and Dem. Rep. Congo. Open woodland on sandy soil, *Brachystegia* woodland, vlei edges, grassland by water; 1175–1680 m.

43. **Aeschynomene nyikensis** Baker in Bull. Misc. Inform., Kew **1897**: 259 (1897). —E.G. Baker, Legum. Trop. Africa: 297 (1929). —Brenan, Check-list For. Trees Shrubs Tang. Terr.: 406 (1949); in Mem. New York Bot. Gard. **8**: 254 (1953). —Verdcourt in F.T.E.A., Leguminosae, Pap.: 394 (1971); in Kirkia **9**: 424 (1974). —Lock, Leg. Afr. Check-list: 106 (1989). Type: Malawi, Nyika Plateau, *Whyte* (K, holotype).

Erect branched shrub 1.2–3(?4.5) m tall. Stems at first pubescent with viscid tubercular-based hairs, sometimes peeling to leave a roughened surface. Leaves fasciculate, 18–26(40)-foliolate; leaflets 1.5–3.5 × 0.5–0.7 mm, linear-oblong, acute and mucronulate at the apex, obliquely rounded at the base, rather thick, imbricated, pubescent or glabrous, ciliolate; main nerve eccentric but not marginal, other venation prominent; petiole and rhachis together 4–8 mm long; petiolules minute; stipules 3 × 0.5 mm, lanceolate, acute, not appendaged, striate, at length falling. Inflorescences axillary, usually aggregated and forming a terminal panicle; peduncle 3–12 mm long; rhachis 1.5–5 cm long; pedicels 3 mm long; bracts 2-fid, 3.5 × 2 mm, ovate, striate, pubescent, soon deciduous; bracteoles 5.5 × 1.5 mm, lanceolate, ciliolate, at length deciduous. Calyx slightly pubescent, 2-lipped; lips 7–10 × 2.2–5 mm, ovate or oblong, one slightly to deeply emarginate, the other slightly 3-fid. Standard yellow or orange, sometimes lined with purple or red, 12–15 × 5–11 mm, violin-shaped; wings yellow or orange; keel petals green or yellow, not laciniate. Fruit of 1 or 2 articles joined by a narrow neck; articles 7.5–10 × 5.6–7 mm, elliptic, with upper edge almost straight and lower edge strongly curved, compressed, margined, glabrous, reticulately venose, later minutely verruculose. Seeds brown, 5 × 3.5 × 1.5 mm, ellipsoid-reniform, compressed; hilum small, round, eccentric.

Malawi. N: Viphya (Vipya), Mtangatanga, fl. & fr. 9.vii.1952, *G. Jackson* 957 (K). C: Dedza, fl. & fr. 23.vii.1946, *Brass* 16883 (K; NY; SRGH). S: Mulanje Distr., Chongoni–Lilongwe Road, fl. & fr. 20.vi.1962, *Richards* 16776 (K; LISC; SRGH).
   Also in southern Tanzania. *Brachystegia* woodland, edges of woodland, marshland margins; 1200–2255 m.

44. **Aeschynomene schliebenii** Harms in Notizbl. Bot. Gart. Berlin-Dahlem **13**: 420 (1936). — Brenan, Check-list For. Trees Shrubs Tang. Terr.: 407 (1949). —Verdcourt in Kew Bull. **24**: 9 (1970); in F.T.E.A., Leguminosae, Pap.: 395 (1971); in Kirkia **9**: 425 (1974). —Lock, Leg. Afr. Check-list: 107 (1989). Type from southern Tanzania.

Erect irregularly branched shrub or small tree 1–4.5(5) m tall. Stems at first pubescent, glabrescent or setulose-hairy, but bark soon breaking down to form a distinctive reddish-brown powdery coating. Leaves mostly fasciculate, 20–40-foliolate; leaflets 3–4 × 0.5–1 mm, lanceolate or linear-oblong, acute or obtuse and mucronulate at the apex, obliquely rounded at the base, rather thick, imbricated, glabrous or pubescent but mostly glandular-ciliate; main nerve eccentric but not marginal, other venation obscure or prominent; petiole and rhachis together 5–20 mm long; petiolules minute; stipules 4.5 × 1 mm, lanceolate, acute, shortly appendaged, striate, soon falling or persistent. Inflorescences axillary, short, often occurring over considerable lengths of shoots and giving a characteristic appearance to the plant; peduncle 0–5 mm long; rhachis 1–5 cm long; pedicels 3–5 mm long; bracts 2-fid, 4–4.5 × 1.2–2 mm, ovate-lanceolate or ovate, acute, striate, pubescent, soon falling; bracteoles 3–5 × 1.2–1.5 mm, ovate to lanceolate, pubescent, soon falling. Calyx pubescent or glabrescent, 2-lipped; lips 6.5–9(10) × 3.5–4 mm, oblong,

Tab. 3.6.**26**. AESCHYNOMENE MIMOSIFOLIA. 1, stem (× ²/₃), from *Graham in Agric. Dept.* 2406; 2, leaflet (× 10); 3, bract with enlarged detail of margin (× 4); 4, flower (× 4); 5, flower with petals removed (× 4); 6, standard (× 6); 7, wings (× 6); 8, keel petals (× 6); 9, anthers (× 6), 2–9 from *Peter* 44199; 10, fruit (× 4), from *Graham in Agric. Dept.* 2406. Drawn by Derek Erasmus. From F.T.E.A.

one slightly emarginate, the other slightly 3-fid. Standard yellow with purple lines outside or all yellow, 9–12 × 5–6 mm, obovate-oblong; wings yellow; keel greenish-yellow, not laciniate. Fruit of 1 or 2 articles joined by a narrow neck; articles 6.5–9.5 × 5–6 mm, semicircular, the upper edge almost straight, lower edge strongly curved, margined, compressed, entirely glabrous or pubescent or puberulous, at least on the margins, somewhat venose, sometimes minutely verrucose. Seeds reddish-brown, 4 × 3 × 1.5 mm, reniform, compressed; hilum small, round, eccentric.

Var. **mossambicensis** (Baker f.) Verdc. in Kew Bull. **24**: 9 (1970); in F.T.E.A., Leguminosae, Pap.: 395 (1971). —Drummond in Kirkia **8**: 216 (1972). —Verdcourt in Kirkia **9**: 426 (1974). —Gonçalves in Garcia de Orta, Sér. Bot. **5**: 63 (1982). Type: Mozambique, Niassa, Mepoche, Maniamba, *Torre* 269 (BM, holotype; COI; K; LISC).
    Aeschynomene nyikensis var. *mossambicensis* Baker f. in J. Bot. **76**: 21 (1938).

Branchlets rather stouter. Leaflets in 10–12 pairs. Articles of the fruit usually puberulous.

Zambia. E: Chama-fork on road to Nyika Plateau, km 12.8, fl. 20.x.1958, *Robson & Angus* 175 (BM; K; LISC; PRE; SRGH). **Zimbabwe**. N: Mazowe Distr., Iron Mask Hill, i.1907, *Eyles* 523 (BM; SRGH). C: Goromonzi Distr., Ruwa, Farm Tanglewood, vii.1957, *Miller* 4443 (K; LISC; PRE; SRGH). E: Nyanga Distr., Britannia, fl. 31.xii.1964, *Corby* 1207 (K; SRGH). **Malawi**. N: Chitipa Distr., Great North Road passing foothills of Mafinga (Mfingi) Mts., between Chitipa (Fort Hill) and Mzimba, fl. & fr. 24.vi.1936, *Burtt* 6155 (BM; K). C: Dedza Distr., Chongoni Mt., fl. 3.ii.1959, *Robson* 1432 (BM; K; LISC; SRGH). S: Machinga Distr., Chikala Hills, fl. & fr. 17.ii.1975, *Brummitt et al.* 14362 (K). **Mozambique**. N: 65.6 km north of Mandimba, Massangulo Mt., fl. 26.v.1961, *Leach & Rutherford-Smith* 11035 (K; LISC; SRGH). T: between Furancungo and Angónia, fl. & fr. 25.viii.1941, *Torre* 3335 (LISC).
Also in southern Tanzania. *Brachystegia* bushland and woodland, often on rocky slopes; 480–2200 m.
Var. *schliebenii* occurs in southern Tanzania and is distinguished by its more slender branchlets, its glabous articles and its leaflets in up to 20 pairs.

45. **Aeschynomene abyssinica** (A. Rich.) Vatke in Oesterr. Bot. Z. **29**: 224 (1879). —E.G. Baker, Legum. Trop. Africa: 299 (1929). —Brenan, Check-list For. Trees Shrubs Tang. Terr.: 405 (1949). —Léonard in F.C.B. **5**: 286, fig. 19G (1954). —Hepper in F.W.T.A., ed. 2, **1**: 578 (1958). —White, F.F.N.R.: 142 (1962). —Verdcourt in F.T.E.A., Leguminosae, Pap.: 396, fig. 54/6 (1971). —Drummond in Kirkia **8**: 216 (1972). —Verdcourt in Kirkia **9**: 426 (1974). —Gonçalves in Garcia de Orta, Sér. Bot. **5**: 61 (1982). —Lock, Leg. Afr. Check-list: 101 (1989). TAB. **3.6.16**, fig. 6. Type from Ethiopia.
    Rueppellia abyssinica A. Rich., Tent. Fl. Abyss. **1**: 203, t. 37 (1847).
    Aeschynomene ruppellii Baker in F.T.A. **2**: 149 (1871). Type as for *A. abyssinica*.
    Aeschynomene kilimandscharica Taub. ex Engl. in Abh. Königl. Akad. Wiss. Berlin. **2**: 262 (1892). Type from Tanzania.
    Aeschynomene glutinosa Taub. in Bot. Jahrb. Syst. **23**: 190 (1896). —E.G. Baker, Legum. Trop. Africa: 299 (1929) (non Schinz). —M.A. Exell in Bol. Soc. Brot., sér. 2, **12**: 9 (1937). Syntypes: Malawi, without locality, *Buchanan* 256, 521 (B, syntypes; BM) and Shire Highlands, *Buchanan* (B, syntype).

Erect, branched shrub or herb 0.8–4.5 m tall. Stems puberulous with sometimes viscid tubercular-based hairs when young, later glabrous; epidermis at length peeling in thin pieces. Leaves often fasciculate, 10–30(48)-foliolate; leaflets 2–12 × 0.5–3 mm, linear to linear-obovate, truncate to rounded and mucronate at the apex, obliquely rounded at the base, glabrous; main nerve ± central; petiole and rhachis together 0.6–7 cm long, somewhat pubescent; petiolules 0.3 mm long; stipules 3–9 × 1–2 mm, ovate-lanceolate, acuminate, not appendaged but ± subcordate, nervose, mostly deciduous. Inflorescences axillary or terminal, numerous, often lax, several-flowered; rhachis 2–15 cm long, peduncle 2–10 mm long; pedicels 1–6 mm long, glabrous to sparsely puberulous; bracts 2–3 × 2–2.5 mm, 2-lobed, ovate, soon falling; bracteoles 2.5–3 × 1–1.5 mm, ovate. Calyx 2-lipped; lips 3–7 × 2–4.5 mm, ovate or oblong, one emarginate, the other slightly 3-lobed. Standard yellow with brown or purple veins or pinkish, tinged purple, 5–13 × 3.5–6 mm, violin-shaped; wings pale yellow or orange; keel petals orange to very pale yellow, often veined with reddish-purple, not laciniate. Fruit of 1–2 articles joined by a very narrow neck, each article elliptic, rounded or semicircular, 5–9 × 4–6 mm, compressed or somewhat inflated, glabrous, reticulate and later finely to quite strongly rugulose, the tubercles sometimes arranged reticulately. Seeds cream to reddish-brown, 3.6–5 × 2.6–3.5 × 1.8 mm, ovoid-reniform, compressed; hilum small round, eccentric.

Zambia. N: Mbala Distr., Kalambo Falls road, fl. & fr. 15.ii.1964, *Richards* 19020 (K; SRGH). W: Kitwe, fl. & fr. 12.iii.1957, *Fanshawe* 3035 (K). C: Chakwenga Headwaters, 100–129 km east of Lusaka, fl. & fr. 27.iii.1965, *E.A. Robinson* 6512 (K; SRGH). S: Mumbwa area, *Macaulay* 351 (K). **Zimbabwe**. N: Hurungwe Distr., Hurungwe Reserve (Urungwe), fl. & fr. iv 1956, *Davies* 1894 (K; SRGH). C: Goromonzi, fl. & fr. 17.iv.1927, *Eyles* 4881 (K; SRGH). E: Mutare (Umtali), Christmas Pass, fl. & fr. 9.i.1967, *Corby* 1724 (K; SRGH). **Malawi**. N: Kande Forest, c. 67 km south of Nkhata Bay junction, c. 650 m, fl. & fr. 25.vi.1977, *Pawek* 12821 (K; MAL; MO). C: Dedza Mt., fl. & fr. 19.i.1959, *Robson & Jackson* 1250 (BM; K; LISC; SRGH). S: Blantyre, fl. & fr. 3.iii.1948, *Faulkner* Kew 217 (COI; K; SRGH). **Mozambique**. N: Massangulo, fl. iii.1933, *Gomes e Sousa* 1335 (BM; K). Z: between Milange and Mocuba, 3.iii.1943, *Torre* 4874 (LISC). T: between Fíngoè and Chiputu (Vila Vasco da Gama), 12.5 km from Fíngoè, fl. & fr. 27.vi.1949, *Barbosa & Carvalho* in *Barbosa* 3321 (K; LMA; SRGH). MS: between Manica (Macequece) and the frontier, fl. & fr. 4.iv.1948, *Mendonça* 3867 (LISC; LMU).

Also in northern Nigeria, Cameroon, Ethiopia, Dem. Rep. Congo, and east Africa. Grassland, *Brachystegia* and other woodland and bushland, often in rocky places; 580–1900 m.

An attempt was made to split this species (known from several hundred sheets) into two subspecies, the southern one having shorter leaf rhachis and fewer leaflets, but in my opinion the variation is not sufficiently correlated and I have followed Léonard in maintaining one taxon. If kept distinct, the southern form corresponds to *A. glutinosa*. *Norlindlh & Weimarck* 4339 (K; LD) (Nyanga (Inyanga), near Cheshire, 15.i.1931) has galled fruits which are densely covered with fleshy tubercles. If normal fruits were lacking such specimens could be deceptive.

46. **Aeschynomene multicaulis** Harms in Bot. Jahrb. Syst. **51**: 226 (1914). —E.G. Baker, Legum. Trop. Africa: 300 (1929). —Brenan, Check-list For. Trees Shrubs Tang. Terr.: 407 (1949). —Léonard in F.C.B. **5**: 287 (1954). —Verdcourt in F.T.E.A., Leguminosae, Pap.: 398 (1971); in Kirkia **9**: 428 (1974). —Lock, Leg. Afr. Check-list: 106 (1989). Syntypes from NE Burundi.

Erect subshrub, 7–40 cm tall, with many stems from a woody rootstock, flowering after burns. Stems glabrous or sparsely pubescent with tubercular-based hairs. Leaves 8–42-foliolate; leaflets 3.5–17 × 1.5–5 mm, oblong or obovate-oblong, obtuse to emarginate and mucronulate at the apex, oblique at the base, glabrous; main nerve ± central; venation very prominent beneath; petiole and rhachis together 1.5–11 cm long, glabrous or slightly pubescent; petiolules 0.3 mm long; stipules 5–26 × 1.5–4 mm, lanceolate to oblong-lanceolate, acuminate, mostly appendaged on one side or subcordate, leaf-like, glabrous, persistent. Inflorescences axillary or terminal, sometimes branched, few–several-flowered; peduncle 1.2–2.5 cm long, glabrous to puberulous; rhachis 2.5–8 cm long; pedicels 3–5.5(12 fide Léonard) mm long, glabrous to puberulous; bracts, 4–5 × 2–2.5 mm, ovate to ovate-elliptic entire or with a few small teeth at the apex, soon falling; bracteoles 2.5–4 × 0.7–1 mm, ovate-lanceolate, soon falling. Calyx glabrous, 2-lipped; lips 5–7 × 3–4.5 mm, ovate, one 2-lobed the other 3-lobed. Standard bright yellow, 9–12 × 4–5 mm, violin-shaped, emarginate; wings bright yellow; keel petals greenish, not laciniate. Fruit of 1–3 articles joined by very narrow necks, each article semicircular, 5.5–7 × 4.5–5.5 mm, glabrous, reticulate, the margins thickened. Seeds not seen.

Zambia. N: Mporokoso, 6.i.1944, *Bredo* 5907 (BR).
Also in western Tanzania and Dem. Rep. Congo. In Tanzania it occurs in seasonally burnt grassland and light *Brachystegia* woodland on stony soil.

47. **Aeschynomene mossoensis** J. Léonard in Bull. Jard. Bot. État **24**: 79 (1954); in F.C.B. **5**: 289 (1954). —Verdcourt in F.T.E.A., Leguminosae, Pap.: 398 (1971); in Kirkia **9**: 429 (1974). —Lock, Leg. Afr. Check-list: 106 (1989). Type from Burundi.

Erect subshrub 0.45–1(1.5) m tall. Stems densely pubescent, mostly sticky, glabrescent below. Leaves 10–46-foliolate; leaflets 4–17 × 1.5–7 mm, oblong to obovate-oblong, rounded and mucronulate at the apex, oblique at the base, glabrous or slightly pubescent beneath when young, ciliolate or almost denticulate; main nerve ± central, venation slightly prominent; petiole and rhachis together 2–9.5(12) cm long, pubescent; petiolules 0.5 mm long; stipules 5–14 × 1–2.5(6) mm, linear to ovate-lanceolate, acuminate, unappendaged, subcordate or appendaged on one side, leaf-like, at first pubescent, later glabrescent, ciliolate, persistent. Inflorescences axillary, several-flowered; peduncle 2 cm long; rhachis 4 cm long, pubescent; pedicels 4–8 mm long, pubescent; bracts 3–4 mm wide, 3-toothed, ± oblong, glabrous but ciliolate, very soon falling; bracteoles 2–3 × 0.5–1 mm, ovate-lanceolate or lanceolate, glabrous, soon falling. Calyx glabrous, 2-lipped; lips 5–6 × 2–4 mm, ovate, sometimes ciliolate, one 2-toothed, the other 3-toothed at the apex. Standard bright yellow, 10–11 × 3–4.5 mm, violin-shaped; wings bright yellow; keel petals pale yellow, not laciniate. Fruit of 1 article or 2 joined by a very narrow neck, each article semicircular, 5.5–10 × 4.5–7.5 mm, reticulate, either puberulous on the margins only or glandular pubescent all over. Seeds dark reddish-brown, 4.8 × 3.8 × 1 mm, semicircular-reniform; hilum small, round, eccentric, the seed slightly beaked beyond it.

Var. **pubescens** J. Léonard in Bull. Jard. Bot. État **24**: 81 (1954). —Verdcourt in Kirkia **9**: 429 (1974). Type: Zambia, Mbala Distr., Kalambo Falls, *B.D. Burtt* 6164 (BM; BR, holotype; EA; K).

Ovary and articles pubescent all over with tubercular-based hairs.

Zambia. N: Mbala Distr., hill on left hand side of road to Nkali (Kali) Dambo, fl. & fr. 4.iv.1955, *Richards* 5301 (K).
Also in southwestern Tanzania. *Brachystegia* woodland and grassland, mostly in rocky places; 1350–1620 m.
The typical variety, distinguished by ovary and articles ± glabrescent, occurs in western Tanzania and Burundi but is not yet recorded for the Flora Zambesiaca area.

48. **Aeschynomene pseudoglabrescens** Verdc. in Kew Bull. **24**: 11 (1970); in Kirkia **9**: 429 (1974). —Lock, Leg. Afr. Check-list: 107 (1989). Type: Zambia, Kitwe, *Fanshawe* 7093 (K, holotype; SRGH).

Subshrubby herb with 1–5 erect stems 15–35 cm tall from a woody rootstock bearing some long tuberous roots; young stems glabrous or very sparsely setulose. Leaves 8–46-foliolate; leaflets 4–13 × 2–4.5 mm, oblong or narrowly oblong, truncate or rounded and mucronulate at the apex, subtruncate or obliquely rounded at the base, glabrous, often glaucous, obscurely minutely punctulate; main nerve central, rest of venation prominent; rhachis and petiole together 1.5–10 cm long, glabrous or with a row of bristly hairs; petiolules 0.5 mm long; stipules 6–2 × 3–6.5 mm, ovate-lanceolate, oblong-lanceolate or lanceolate, acute or acuminate, not appendaged or conspicuously unequally appendaged on both sides at the base, leaf-like, glabrous, nervose, persistent. Inflorescences terminal and axillary, 3–12.5 cm long, 6–16-flowered, mostly glabrous; peduncles 10–25 mm long; pedicels 4–12 mm long, glabrous or very sparsely setulose; bracts 3–4 × 1.5–2.5 mm, ovate-oblong, shortly divided into 2 acute triangular lobes, soon deciduous or rarely subpersistent; bracteoles 2–4 × 0.8–1.5 mm, lanceolate, acute, glabrous, at length deciduous. Calyx glabrous, 2-lipped; lips 6–9 × 2–4.5 mm, oblong, one emarginate, the other 3-toothed. Standard yellow, 11 × 5 mm, oblong, slightly widened above; keel petals not laciniate. Fruit of 1 (perhaps sometimes 2 since 2-ovuled ovaries have been seen) almost semicircular article, 8.5–9.5 × 6–6.5 mm, compressed, glabrous, reticulately nerved. Seeds not seen.

Zambia. N: Kasama Distr., by River Mifundu, 12.xii.1964, *Richards* 19350 (K). W: Kitwe, 16.xi.1965, *Fanshawe* 9420 (K; SRGH).
Restricted to Zambia. *Brachystegia* woodland, 1200 m.

49. **Aeschynomene stipulosa** Verdc. in Kew Bull. **24**: 13, fig. 3 (1970); in Kirkia **9**: 430 (1974). —Lock, Leg. Afr. Check-list: 108 (1989). TAB. **3.6.27**. Type: Zambia, Mwinilunga, *Milne-Redhead* 3342 (BM; BR; K, holotype; LISC; PRE).

Perennial prostrate herb with 5–20 branches 8–20 cm long from a tuberous rootstock which bears narrow roots 11–15 cm long; young shoots sparsely setulose. Leaves 8–36-foliolate; leaflets 4–12 × 1.5–4 mm, oblong, rounded and mucronulate at the apex, truncate or obliquely rounded at the base, glabrous, minutely obscurely

Tab. 3.6.**27**. AESCHYNOMENE STIPULOSA. 1, habit (× 1); 2, inflorescence (× 2); 3, calyx (× 5); 4, standard (× 5); 5, wings (× 5); 6, keel petal (× 5); 7, androecium (× 5); 8, gynoecium (× 5); 9, style and stigma (× 18), 1–9 from *Milne-Redhead* 3342. Drawn by Mary Grierson. From Kew Bull.

punctulate; main nerve central; rhachis and petiole together 1–4.5 cm long, ± setulose; petiolules short; stipules 8–15 × 2.5–4 mm, ovate or ovate-lanceolate, acute or acuminate at the apex, subcordate or appendaged on one side at the base, leaf-like, glabrous, nervose, persistent. Inflorescences axillary, shorter than the leaves, 5–6 cm long, 5–6-flowered, glabrous or sparsely setulose; peduncles 1.5–2 cm long; pedicels 2.5–6.5 mm long; bracts deciduous or subpersistent, 2-fid or 3-fid, the lobes 2.5–5 × 0.5–0.8 mm, lanceolate; bracteoles 3.5 × 0.5–1 mm, lanceolate, acute. Calyx glabrous, 2-lipped; lips 6–7.5 × 3–4 mm, ovate-oblong, one subacute, the other minutely 3-fid. Standard apricot-coloured, 7.5–9.5 × 3.5–4.5 mm, somewhat violin-shaped, slightly widened above; wings free; keel petals not laciniate. Fruit of 1–2 articles joined by a narrow neck; articles ± semicircular, 6 × 3.5 mm, glabrous. Seeds not seen.

Zambia. W: Mwinilunga, slopes east of River Lunga, fl. & fr. 22.xi.1937, *Milne-Redhead* 3342 (BM; BR; K; LISC; PRE).
Not known elsewhere. *Brachystegia* woodland; 1310 m.

50. **Aeschynomene trigonocarpa** Taub. ex Baker f., Legum. Trop. Africa: 298 (1929). —Brenan, Check-list For. Trees Shrubs Tang. Terr.: 407 (1949). —White, F.F.N.R.: 142 (1962). —Verdcourt in F.T.E.A., Leguminosae, Pap.: 399, fig. 54/8 (1971). —Drummond in Kirkia **8**: 216 (1971). —Verdcourt in Kirkia **9**: 431 (1974). —Lock, Leg. Afr. Check-list: 108 (1989). TAB. 3.6.**16**, fig. 7. Type from Tanzania.
Aeschynomene trigonocarpa Taub. in Engler and Prantl, Nat. Pflanzenfam. **3** (3): 320, fig. 124E (1894) nomen and figure of fruit only. —Harms in Engler, Pflanzenw. Afrikas [Veg. Erde 9] **3** (1): fig. 298 (1915).
?*Aeschynomene goetzei* Harms in Bot. Jahrb. Syst. **30**: 328 (1901). —E.G. Baker, Legum. Trop. Africa: 300 (1929). —Brenan, Check-list For. Trees Shrubs Tang. Terr.: 406 (1949). Type from Tanzania (see note).

Small shrub 0.8–3 m tall or sometimes an ascending woody herb. Stems slender, at first glandular pubescent with tubercular-based hairs, later glabrous; bark on older stems at length peeling to reveal a powdery rusty-red surface. Leaves 16–40-foliolate; leaflets 3–8 × 1–2.5 mm, linear to linear-oblong or oblanceolate, rounded and mucronulate at the apex, obliquely truncate at the base, glabrous; main nerve central; other venation rather obscure; petiole and rhachis together 1.5–3.5 cm long, with a line of bristly hairs beneath; petiolules 0.3 mm long; stipules 7–12 × 1–2.3 mm, narrowly lanceolate, acuminate, conspicuously appendaged on one side at the base, soon falling or sometimes ± persistent. Inflorescences axillary, lax, several-flowered; peduncle 1–2 cm long; rhachis 3–13 cm long; pedicels 3–5.5 mm long, with tubercular-based hairs; bracts 2–3-fid, 4 × 1.6 mm, elliptic-lanceolate, glabrous, soon falling; bracteoles 2.5 × 0.6 mm, narrowly oblong-lanceolate, soon falling. Calyx glabrous, 2-lipped; lips 5–7 × 2–3 mm, elliptic, one slightly emarginate, the other shortly 3-toothed or almost entire. Standard dark yellow, tinged brown outside, 7.5–10 × 3.5–4.5 mm, oblong but slightly narrowed at the middle, ± truncate; wings yellow; keel petals pale yellow, not laciniate. Fruit of 1 or less often 2 articles joined by a narrow neck; articles 7–11 × 6–8 mm, triangular with acute angles or rarely rounded, compressed, glabrous or puberulous on the thickened margins, reticulate but smooth or at length very finely rugulose. Seeds chestnut-brown, 5 × 4 × 1.5 mm, irregularly compressed-reniform, slightly beaked beyond the small round eccentric hilum.

Zambia. N: Mbala Distr., Kawimbe, Nachalanga Hill, fl. & fr. 15.xii.1959, *Richards* 11989 (K; SRGH). C: Chilanga, Mt. Makulu Agriculture Research Station, fl. & fr. 14.iv.1956, *E.A. Robinson* 1468 (K; SRGH). S: Choma Dstr., Siamambo Forest Reserve, 14.i.1952, *White* 1910 (FHO; K; PRE). **Zimbabwe**. N: Mazowe Distr., Wengi River, Mutorashanga (Mtorashanga)–Concession road, fl. & fr. 2.iii.1965, *Corby* 1252 (K; SRGH). C: Harare Distr., Twentydales Road, fl. & fr. 3.ii.1961, *Rutherford-Smith* 495 (K; LISC; PRE; SRGH). **Malawi**. N: Mzimba–Lundazi, km 22.4, fl. 29.iv.1952, *White* 2508 (FHO; K; SRGH) (needs confirmation).
Also in southern Tanzania. Mostly in *Brachystegia* or *Julbernardia* woodland, also in *Acacia –Commiphora* thicket-woodland, often on rocky hills, less often in grassland or old cultivations; 1200–1740 m.
Unfortunately Taubert's original mention of this name is inadequate for validation. It seems likely that *A. goetzei* Harms in Bot. Jahrb. Syst. **30**: 328 (1901) (type from Tanzania) is the correct name for this species, but no fruit was present on the type specimen. I have always felt unwilling

to upset so apt a name as *A. trigonocarpa*, particularly as the name *goetzei* has long been misapplied to an undescribed species in some herbaria, and now conservation of specific names is possible the name *A. trigonocarpa* will be proposed for conservation.

51. **Aeschynomene sp. E**

> *Aeschynomene sp. F* of Verdcourt in F.T.E.A., Leguminosae, Pap.: 401 (1971).
> *Aeschynomene sp. G* of Verdcourt in Kirkia **9**: 432 (1974).

Erect subshrubby herb or small shrub, 0.2–0.9(1.2) m tall. Stems glabrous to densely covered with short tubercular-based glandular hairs and sometimes shorter eglandular pubescence as well; bark apparently never powdery; underground parts not preserved, presumably woody. Leaves 10–44-foliolate; leaflets 1–5 × 0.4–1.5 mm, narrowly oblong, oblong-oblanceolate or oblong-elliptic, acuminate or abruptly mucronate at the apex, obliquely rounded at the base, glabrous, rarely (in Malawi) ciliolate; main nerve ± central; lateral nerves somewhat prominent beneath; petiole and rhachis together 0.9–3.5 cm long, with a row of setiform hairs beneath; petiolules 0.2 mm long; stipules 8–16 × 1.3–3 mm, lanceolate or oblong-lanceolate, tapering-acute, appendaged at the base on one side, the lobes usually divergent, glabrous or rarely ciliolate, persistent. Inflorescences axillary, 3–16-flowered; peduncle 0.7–4 cm long; rhachis 3.5–10.5 cm long; pedicels 0.2–1.2 cm long, all the axes glabrous or pubescent like the stems; bracts deeply 3-fid, 3.5 × 2.5 mm, obcuneiform in outline, very soon falling or rarely subpersistent on some shoots; bracteoles 3.5–4.5 × 0.7–1.5 mm, ovate or lanceolate, often ciliolate, deciduous. Calyx glabrous, 2-lipped; lips 5.5–8.5 × c. 1.5–3 mm, oblong, one slightly 2-fid the other very shortly 3-fid. Standard yellow, 7.5–12 × 3–6 mm, violin-shaped, much broader at the apex, or almost rectangular; keel petals not laciniate. Fruit of 1 or 2 articles joined by a very narrow neck, each article semicircular, straight above, very curved below, 7–9 × 5–5.5 mm, compressed, glabrous, nervose. Seeds not seen.

Malawi. N: Nyika Plateau, bridge over River Chelinda, 10.i.1967, *Hilliard & B.L. Burtt* 4388 (K; LISC). **Mozambique**. N: Mts. east of Lake Malawi (Nyasa), *W.P. Johnson* (K).
Also in Tanzania.
This is exceedingly similar to *A. trigonocarpa* Taub. ex Baker f. in the flowering state, but is usually less shrubby, the bark never powders and the leaflets are more numerous. Much more fruiting material is needed from the areas concerned before this group can be elucidated, particularly as two sheets of undoubted *A. trigonocarpa* show rounded not angular articles. This species has been frequently identified with *A. goetzei* Harms but the description of that leads one to believe it is probably the same as *A. trigonocarpa*. It has also been misidentified with *A. katangensis* De Wild., but the standard is differently shaped, the habit different and the articles of different shape and texture.

52. **Aeschynomene tenuirama** Welw. ex Baker in F.T.A. **2**: 150 (1871). —Fries, Wiss. Ergebn. Schwed. Rhod.-Kongo-Exped.: 85 (1914). —E.G. Baker, Legum. Trop. Africa: 296 (1929). —Léonard in F.C.B. **5**: 290 (1954). —White, F.F.N.R.: 142 (1962). —Torre in C.F.A. **3**: 201 (1966). —Verdcourt in Kirkia **9**: 433 (1974). —Lock, Leg. Afr. Check-list: 108 (1989). Type from Angola.

> *Aeschynomene newtonii* Schinz in Bull. Herb. Boissier, sér. 2, **2**: 948 (1902). —E.G. Baker, Legum. Trop. Africa: 297 (1929). Type from Angola.
> *Aeschynomene subaphylla* De Wild. in Repert. Spec. Nov. Regni Veg. **11**: 505 (1913). —E.G. Baker, Legum. Trop. Africa: 298 (1929). Type from Dem. Rep. Congo.

Subshrub with a woody rhizome-like rootstock and several branched or unbranched, sometimes almost leafless erect stems 0.3–2 m tall, pubescent with short tubercular-based glandular hairs when young or glabrescent, finally glabrous. Leaves 10–100-foliolate; leaflets 1.5–8 × 0.7–2.5 mm, obovate-oblong to oblong, acute, rounded or retuse and mucronulate at the apex, rounded at the base, glabrous; main nerve central, the lateral nerves sometimes very raised and distinct below and anastomosing near the margin; the leaflets are larger and more numerous on the non-flowering stems; petiole and rhachis together 0.7–5.5 cm long, puberulous or glabrous; petiolules very short; stipules on flowering branches 5–8 × 1–1.7 mm, lanceolate, acuminate, unilaterally appendaged or not sometimes on the same branch, glabrous or ciliolate, deciduous or persistent; stipules on sterile branches 1–1.7 cm long with appendage c. 1–3 mm long, very deciduous. Racemes or panicles axillary or axillary and terminal, 4–12 cm long, puberulous or glabrous with many well-spaced flowers; peduncles 1–4 cm long; pedicels 3–8 mm long, puberulous or glabrous; bracts 2.5 × 1–1.5 mm, elliptic or oblong, 2–3-toothed or -partite, very deciduous; bracteoles 1.5–3 × 0.5–1 mm, lanceolate, deciduous.   Calyx brownish-purple, glabrous, 2-lipped; lips 3.5–6 × 2–4 mm, elliptic-oblong, ciliolate or not, one 2-lobed, the other 3-lobed.   Standard yellow to orange, sometimes tinged red outside, 5.5–7 × 2–4.5 mm, violin-shaped or ± rectangular, emarginate, wings and keel petals yellow or green, the latter not laciniate.   Fruit of 1–2 articles joined by a very narrow neck, each article semicircular, 7–9 × 5–7 mm (but see note), glabrous or rarely pubescent, reticulate, or finely rugulose when adult.   Seeds not seen ripe, pale brown, 3.2 × 2.2 mm, semicircular-reniform; hilum minute, eccentric.

Var. **tenuirama** —Verdcourt in Kirkia **9**: 434 (1974).

Articles 7–9 × 5–6 mm, glabrous or with only few hairs on upper margin.

Zambia. N: Mbala Distr., Kambole Escarpment, fl. 19.ii.1957, *Richards* 8248 (K). W: Mwinilunga Distr., near River Musangila, fl. & fr. 25.i.1938, *Milne-Redhead* 4324 (K). C: Mkushi Distr., Chiwefwe, fl. 20.ix.1964, *Mutimushi* 1004 (K; SRGH). **Malawi**. N: Rumphi Distr., Nyika Plateau, c. 1.6 km in entrance road, fl. & fr. 9.ix.1976, *Pawek* 11767 (K). S: Mulanje Mt., Thuchila (Tuchila) Skyline Path, fl. & fr. 10.iv.1970, *Brummitt* 9784 (K). **Mozambique**. N: near Marrupa, fl. 31.i.1981, *Nuvunga* 450 (K; LISC; LMU).
Also in Dem. Rep. Congo and Angola. *Brachystegia* and *Uapaca–Monotes* woodland, bushland; (800) 1180–1740 m.
Hiern divided the species into three varieties based on Welwitsch's own ideas. Léonard and Torre have dismissed these. Var. *sculpta* Welw. ex Hiern has the lateral nerves prominent and is frequently found in the Flora Zambesiaca area. The variation in this character undoubtedly renders it useless. Three sheets from Angola, (Benguela, Bié and Huíla) have fruit articles quite unlike those of Zambian material — they are thin, 13 × 10 mm and these specimens are surely distinct at some level. Since the material of var. *huillensis* Welw. ex Hiern is not in ripe fruit it is not possible to decide if this name applies to the variant. Far too little material in fruit has been collected to satisfactorily answer the question, but attached to an isolectotype of var. *tenuirama* is a packet containing almost ripe fruits and these measure 7 × 5.5 mm, so the problem does not concern the Flora Zambesiaca area as far as is known.

Var. **hebecarpa** Verdc. in Kew Bull. **24**: 14 (1970); in Kirkia **9**: 435 (1974). Type: Malawi, Nyika Plateau, Mt. Mwenembwe (Mwanemba), *McClounie* 134 (K, holotype).

Articles 7–9 × 6–7 mm, pubescent all over.

Zambia. N: Isoka Distr., Mafinga Hills, fl. & fr. 24.viii.1979, *Chisumpa* 593 (K; NDO). **Malawi**. N: Nyika Plateau, Mt. Mwenembwe (Mwanemba), ix.1902, *McClounie* 134 (K).
Not known elsewhere. Submontane grassland; 2000–2300 m.
A sheet from Malawi (Rumphi Distr., Nyika Plateau, above Nchenachena, fr. 4.v.1952, *White* 2592) may belong to this species but the leaflets and stipules are glandular-ciliate, the thickened hair bases making them appear denticulate. The inflorescences are entirely axillary and there is a certain difference in facies. Without further material I would hesitate to consider it more than a variant of *A. tenuirama*.

53. **Aeschynomene katangensis** De Wild. in Ann. Mus. Congo, Sér. IV, Bot. [Études Fl. Katanga] **1**: 188 (1903). —E.G. Baker, Legum. Trop. Africa: 295 (1929). —Léonard in F.C.B. **5**: 291 (1954). —Verdcourt in Kirkia **9**: 435 (1974). —Lock, Leg. Afr. Check-list: 104 (1989). Type from Dem. Rep. Congo.

Subshrub with several stems 9–80 cm tall from a woody rhizome-like rootstock. Stems unbranched or sparsely branched, erect, the flowering ones often almost leafless, all pubescent with tubercular-based hairs or glabrescent when young, later glabrous and with a peeling bark. Leaves 10–24-foliolate; leaflets 2–5 × 0.5–1 mm, linear, obtuse to acute and mucronulate at the apex, asymmetric at the base, glabrous, usually not ciliate; main nerve submarginal to central, lateral nerves indistinct; petiole and rhachis together 5–18 mm long, puberulous or glabrous; petiolules very short; stipules 5–15 × 0.5–3 mm, lanceolate or ovate-lanceolate, acuminate, unilaterally obliquely appendaged or sometimes subcordate or quite unappendaged, glabrous or ciliolate, entire or 2–3-partite, mostly persistent. Inflorescences terminal and also in upper axils, often forming a terminal panicle, 3–24 cm long, glabrous or sparsely puberulous; peduncles 5–20 mm long; pedicels 3–13 mm long, glabrous or sparsely puberulous; bracts 4–5 × 1–1.5 mm, elliptic or elliptic-lanceolate, entire, denticulate or 2-fid, ± ciliolate, very deciduous; bracteoles 3–4 × 1 mm, ovate-lanceolate, deciduous. Calyx glabrous, 2-lipped; lips 5–8 × 3.5–4.5 mm, oblong, not or scarcely ciliolate, one 2-lobed the other 3-lobed. Standard yellow or orange-yellow, 7–10 × 3–4.5 mm, violin-shaped or ± rectangular, slightly emarginate; wings and keel yellow, the petals of the latter not laciniate. Fruit of 1–2 articles joined by a very narrow neck; articles semicircular, 5–8 × 4.5–6.5 mm, glabrous, reticulate and finely covered with raised dots when adult. Seeds dark purplish-brown, 4.7 × 3 × 1.2 mm, semicircular-ellipsoid; hilum small, almost circular, eccentric.

Subsp. **sublignosa** (De Wild.) J. Léonard in F.C.B. **5**: 292 (1954). —Verdcourt in Kirkia **9**: 436 (1974). —Lock, Leg. Afr. Check-list: 104 (1989). Type from Dem. Rep. Congo.
   Aeschynomene sublignosa De Wild. in Repert. Spec. Nov. Regni Veg. **11**: 505 (1913). —E.G. Baker, Legum. Trop. Africa: 298 (1929).
   Aeschynomene racemosa De Wild. in Repert. Spec. Nov. Regni Veg. **11**: 504 (1913). —E.G. Baker, Legum. Trop. Africa: 296 (1929) non Vogel. Type from Dem. Rep. Congo.
   Aeschynomene rogersii N.E. Br. in Bull. Misc. Inform., Kew **1921**: 293 (1921). Syntypes from Dem. Rep. Congo.
   Aeschynomene tenuirama sensu E.G. Baker, Legum. Trop. Africa: 297 (1929) pro parte non Welw. ex Baker.
   Aeschynomene bracteosa sensu E.G. Baker, Legum. Trop. Africa: 294 (1929) pro parte non Welw. ex Baker.
   Aeschynomene subaphylla sensu E.G. Baker, Legum. Trop. Africa: 298 (1929) pro parte non De Wild.

Stems 15–80 cm tall. Stipules often 2–3-partite. Pedicels 3–13 mm long.

Zambia. W: Chingola, fl. & fr. 18.x.1955, *Fanshawe* 2539 (K; LISC).
Also in Dem. Rep. Congo. *Brachystegia* woodland; 1500 m.
   Subsp. *katangensis*, from Katanga, is a small pyrophyte with very contracted inflorescences (only the type collection was known when Léonard wrote it up).

54. **Aeschynomene leptophylla** Harms in Repert. Spec. Nov. Regni Veg. **8**: 356 (1910). —E.G. Baker, Legum. Trop. Africa: 300 (1929). —Brenan, Check-list For. Trees Shrubs Tang. Terr.: 406 (1949). —Léonard in F.C.B. **5**: 294, fig. 22 (1954). —White, F.F.N.R.: 142 (1962). —Torre in C.F.A. **3**: 202 (1966). —Verdcourt in F.T.E.A., Leguminosae, Pap.: 402 (1971). —Drummond in Kirkia **8**: 216 (1972). —Verdcourt in Kirkia **9**: 436 (1974). —Lock, Leg. Afr. Check-list: 104 (1989). TAB. 3.6.**28**. Syntypes: Zambia, Stevenson Road, *Scott-Elliot* 8388 (BM; K, syntype?) and from Tanzania.

Erect subshrub or shrub 0.4–3 m tall, with several shoots from a woody rootstock with some roots tuberous, sometimes leafless when flowering; mostly appearing after burns. Stems glabrous to densely viscid scabrid-pubescent or rarely stiffly spinulose with persistent hair-bases. Leaves 28–104-foliolate; leaflets 3–22(55) × 0.7–7(17) mm, linear to oblong, obtuse or emarginate and mucronulate at the apex, obliquely rounded at the base, glabrous; main nerve ± central; petiole and rhachis 3–21(30) cm long, glabrous to hispidulous; petiolules 1 mm long; stipules 6–16 × 1–5 mm, ovate to linear-lanceolate, acuminate, slightly subcordate but not appendaged at the base, sometimes ciliolate. Inflorescences terminal or from the upper axils, slightly to much branched, lax and many-flowered; peduncle 2.5–6 cm long; rhachis 3–45 cm long; pedicels 3–11 mm long, glabrous or pubescent; bracts 2.5–4 × 1.5–2.5 mm, oblong-elliptic, entire or 2–3-toothed, soon falling; bracteoles 2.5–3.5 × 1–1.5 mm, lanceolate to obovate, obtuse at the apex, glabrous, soon falling. Calyx glabrous, 2-lipped; lips 5.5–10 × 3–5 mm, ovate or oblong, often ciliolate, one emarginate, the other 3-lobed at the apex. Standard yellow with purple veins, 8–12 × 4.5–7 mm, ovate or ovate-elliptic, emarginate; wings yellow with purple veins, about equalling the keel; keel petals yellow with purple veins, 8–15 mm long, not laciniate; filament-sheath usually as long as or longer than the 7–13

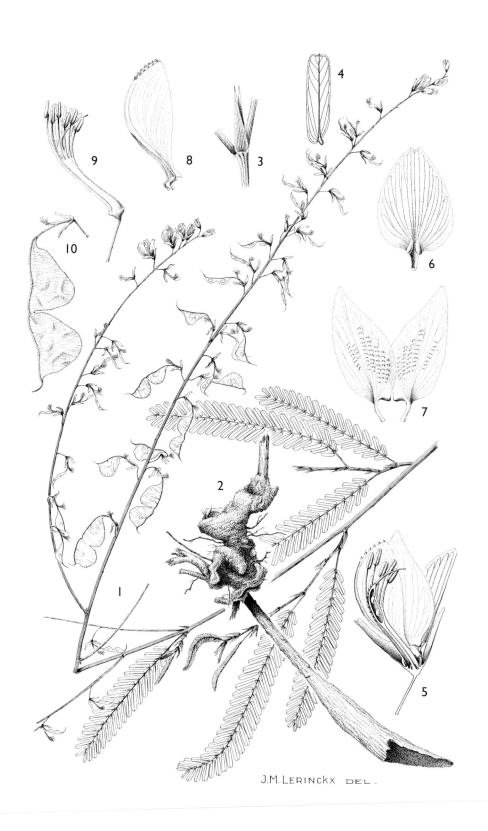

J.M. LERINCKX DEL.

mm long fruit stipe. Fruit of 1–3 articles joined by a very narrow neck consisting of practically nothing but the suture; articles 12–21 × 8–15 mm, semicircular, papery, glabrous or very sparsely pubescent, reticulate. Seeds yellow-brown, 5.5 × 4.5 × 1.1 mm, irregularly compressed-reniform; hilum small, elliptic, eccentric.

Var. **leptophylla** —Verdcourt in Kirkia **9**: 437 (1974).

Stems glabrous to densely pubescent, the tubercles of the hairs not very thick. Leaflets under 2 cm long.

Zambia. N: Mbala Distr., Dhul'miti (D'hulmiti), fl. 6.v.1955, *Richards* 5575 (K; SRGH). C: Mkushi–Lunsemfwa River junction, Bell Point, fl. & fr. 4.v.1957, *Fanshawe* 3265 (K). E: Chipata, fl. ii.1962, *Verboom* 460 (K). **Zimbabwe**. N: Gokwe Distr., Charama Plateau, fl. 18.iii.1962, *Bingham* 182 (K; LISC; SRGH). W: Gwampa Forest Reserve, fr. v.1956, *Goldsmith* 94/56 (K; SRGH; PRE in fruit only). **Malawi**. N: Chitipa Distr., Songa Stream, 14.4 km east of crossroads towards Karonga, fl. & fr. 19.iv.1969, *Pawek* 2250 (K).

Also in Burundi, Dem. Rep. Congo and Angola. Mostly in *Brachystegia* woodland and bushland or sometimes grassland, often on rocky hillsides; 900–2280 m.

White (loc. cit.) has doubted the distinctness of this species, but the wing character coupled with the longer stipe render the naming of any specimen easy and I see no reason to doubt its distinctness at some level. The first two specimens cited by White under *A. nyassana* are in fact *A. leptophylla*.

Var. *crassituberculata* Verdc. from Tanzania, has stiffly spinulose stems. Subsp. *magnifoliolata* J. Léonard, from upper Katanga, is described by Léonard as having leaflets up to 55 × 17 mm, however this needs further study.

55. **Aeschynomene nyassana** Taub. in Bot. Jahrb. Syst. **23**: 190 (1896). —E.G. Baker, Legum. Trop. Africa: 301 (1929). —Brenan, Check-list For. Trees Shrubs Tang. Terr.: 406 (1949). —Léonard in F.C.B. **5**: 292, fig. 19A (1954). —White, F.F.N.R.: 142 (1962) pro parte. — Verdcourt in F.T.E.A., Leguminosae, Pap.: 403, fig. 54/5 (1971). —Drummond in Kirkia **8**: 216 (1972). —Verdcourt in Kirkia **9**: 438 (1974). —Gonçalves in Garcia de Orta, Sér. Bot. **5**: 62 (1982). —Lock, Leg. Afr. Check-list: 106 (1989). TAB. 3.6.**16**, fig. 8. Syntypes: Malawi, without locality, *Buchanan* 28 (B, syntype; BM), 1366 (B, syntype; K) and Mt. Mulanje (Mlanje), *Whyte* (B, syntype; BM; K).

Subshrub or shrub mostly erect with several stems 0.15–3 m tall, from a woody rootstock, often coming up after burns and then frequently flowering when leafless or when the leaves are only slightly developed; secondary roots long and tuberous. Stems glabrous to densely viscid-pubescent or scabrid. Leaves 14–80-foliolate; leaflets (3)6–24 × (1.5)2–8 mm, oblong or obovate-oblong, obtuse to emarginate and mucronulate at the apex, obliquely rounded at the base, glabrous; main nerve mostly central; petiole and rhachis together 3–25 cm long, often pubescent; petiolules 0.5–1 mm long; stipules 7–24 × 2–4 mm, ovate- to linear-lanceolate, acuminate, somewhat auriculate but not appendaged. Inflorescences terminal, mostly well-branched and many-flowered with a rhachis 10–35 cm long, also some shorter less branched inflorescences in the axils of the upper leaves; peduncle 2.5–5 cm long, glabrous to densely pubescent; pedicels 4–11 mm long; bracts 3–10 × 2–6 mm, elliptic to ovate, entire or denticulate, soon falling; bracteoles 3.5 × 2 mm, ovate, obtuse or acute at the apex, glabrous but often ciliolate, soon falling. Calyx glabrous, 2-lipped; lips 4.5–8 × 3–7 mm, ovate, one emarginate, the other 3-lobed at the apex. Standard yellow with purple veins, 6–13 × 2–5 mm, ovate-elliptic, emarginate; wings pale yellow with purple veins, 8–17 mm long; exceeding the keel; keel petals pale yellow with purple veins, not laciniate; filament sheath mostly shorter than the fruit stipe, which is mostly 4–5(10) mm long. Fruit of 1–2 articles joined by a neck 2–4 mm wide; articles 12–27 × 9–13 mm, obovate, elliptic or

Tab. 3.6.**28**. AESCHYNOMENE LEPTOPHYLLA. 1, habit (× ¹⁄₂), from *Becquet* 117; 2, rhizome (× ¹⁄₂), from *Germain* 6530; 3, stipules (× 2); 4, leaflet (× 2); 5, flower, longitudinal section (× 3); 6, standard, internal face (× 3); 7, wings, external face (× 3); 8, keel (× 3); 9, androecium (× 3), 3–9 from *Becquet* 117; 10, fruit (× 1), from *Becquet* 2132. Drawn by J.M. Lerinckx. From Fl. Congo Belge. Reproduced with permission of Jardin Botanique National de Belgique.

semicircular, flat and papery, glabrous save for the slightly pubescent margins, reticulate. Seeds reddish-brown, 5 × 4 × 2 mm, elliptic-reniform, beaked beyond the small round eccentric hilum.

Zambia. N: Kasama Distr., Mungwi, fl. & fr. 21.ix.1960, *E.A. Robinson* 3844 (K). E: Nyika Plateau, fl. 24.ix.1956, *Benson* 152 (BM). W: Ndola Distr., Lake Ishiku, fl. & fr. 18.x.1953, *Fanshawe* 427 (K; SRGH). C: Serenje Distr., Kundalila Falls, fl. & fr. 13.x.1963, *E.A. Robinson* 5692 (K; SRGH). E: Nyika Plateau, fl. & fr. 21.x.1958, *Verboom* LK 104 (K). **Zimbabwe**. W: Matobo, fl. & fr. i.1954, *Miller* 2031 (K; LISC; PRE; SRGH). C: Marondera Distr., Grasslands Research Station, railway fireguard, fl. & fr. 22.x.1966, *Corby* 1660 (K; SRGH). E: Nyanga Distr., c. 9.6 km north of Troutbeck, Gairesi Ranch on Mozambique border, fl. & fr. 20.xi.1956, *E.A. Robinson* 1959 (K; LISC; SRGH). **Malawi**. N: Nkhata Bay Distr., Viphya (Vipya), 46.4 km south of Mzuzu, fl. l.x.1967, *Pawek* 1456 (SRGH). C: Dedza Distr., Ciwawo (Chiwao) Hill, fl. & fr. 21.xi.1966, *Jeke* 33 (SRGH). S: Zomba Plateau, fl. & fr. 23.x.1941, *Greenway* 6356 (EA; K). **Mozambique**. N: Ribáuè Distr., c. 2.5 km from Ribáuè towards Lalaua, fl. & fr. 22.i.1964, *Torre & Paiva* 10088 (LISC). Z: Lugela Distr., Namagoa Estate, fl. & fr. xi.1946, *Faulkner* PRE 197 and 272 (BM; COI; K; PRE; SRGH). T: between Zóbuè and Moatize, fl. & fr. 21.x.1941, *Torre* 3703 (LISC). MS: Rotanda, between Rio Mussapa and frontier, Tandara, km 3 from residence of Sr. V. Carvalho, fl. 18.xi.1965, *Torre & Correia* 13121 (LISC).

Also in Dem. Rep. Congo, southern Tanzania and South Africa (Transvaal). Woodland (mostly *Brachystegia*) grassland and bushland, often by swamps and on dambo edges, sometimes in rocky places; 60–2300 m.

There has been some discussion as to whether *A. nyassana* is truly distinct from the Angolan plant *A. siifolia* Baker. I am not convinced that they are conspecific since the venation is a little different and the flower-structure is not clear in the Angolan plant. Until more material has been collected I am content to follow Léonard and leave them distinct.

56. **Aeschynomene glabrescens** Welw. ex Baker in F.T.A. **2**: 148 (1871). —E.G. Baker, Legum. Trop. Africa: 295 (1929). —Léonard in F.C.B. **5**: 296, fig. 19E (1954). —Torre in C.F.A. **3**: 203 (1966). —Verdcourt in Kirkia **9**: 439 (1974). —Lock, Leg. Afr. Check-list: 103 (1989). Type from Angola.

Subshrub 20–45 cm tall with long fusiform secondary roots; young stems glabrous or nearly so, rarely pubescent. Leaves 10–54-foliolate; leaflets 5–20 × 1.5–6 mm, elliptic or elliptic-oblong, obtuse, rounded or mucronulate at the apex, obliquely rounded at the base, glabrous; main nerve central, lateral nerves prominent, all reaching the margin and joining to form a distinct marginal nerve; petiole and rhachis together 2–9.5 cm long, at first with a few sparse hairs but soon glabrous; petiolules up to 1 mm long; stipules 5–10 × 1–3 mm, linear-lanceolate to ovate-lanceolate, acuminate, not appendaged at the base, glabrous, deciduous. Inflorescences terminal and axillary, lax and many-flowered, 8–25 cm long; peduncles 4–7 cm long; pedicels 5–9 mm long; bracts 3–4 × 2–2.5 mm, elliptic, entire, very deciduous; bracteoles 2.5–3 × 1–1.5 mm, elliptic, mostly emarginate and mucronulate, at length deciduous. Calyx glabrous, 2-lipped; one lip 4.5–6 × 5–5.5 mm, rounded-ovate, emarginate, the other 5–6.5 × 4–5 mm, ovate-oblong, shortly 3-fid. Standard yellow or orange, 9–10 × 3–4 mm, elliptic-oblong, emarginate; wings exceeding the keel which is not laciniate; filament sheath very much shorter than the fruit stipe which is 8–11 mm long. Fruit of 1 article, 14–18 × 10–12 mm, broadly elliptic or semicircular, flat and papery, glabrous, reticulate. Seeds not seen.

Var. **glabrescens** —Verdcourt in Kirkia **9**: 440 (1974).

Stems glabrous or nearly so.

Zambia. W: Mwinilunga Distr., 25.6 km west of River Kabompo and east of the boma, fl. & fr. 11.ix.1930, *Milne-Redhead* 1109 (K).

Also in Dem. Rep. Congo and Angola. Sandy plains subject to burning; 1350 m.

Torre has synonymised *A. curtisiae* Johnston (type from Angola) with this but both Mr. Milne-Redhead and I have borrowed and examined this type on different occasions and concluded that it differed in venation, with the marginal nerve not uniform and strong around the entire margin and the basal nerves extending to the leaflet apex.

Var. *pubescens* J. Léonard, from Kasai in Dem. Rep. Congo, has pubescent stems.

57. **Aeschynomene pararubrofarinacea** J. Léonard in Bull. Jard. Bot. État **24**: 73 (1954); in F.C.B. **5**: 278 (1954). —White, F.F.N.R.: 141 (1962). —Verdcourt in Kew Bull. **24**: 14 (1970); in Kirkia **9**: 440 (1974). —Lock, Leg. Afr. Check-list: 106 (1989). Type from Dem. Rep. Congo.
   Humularia bianoensis P.A. Duvign. in Bull. Soc. Roy. Bot. Belgique **86**: 178 (1954); in F.C.B. **5**: 306, fig. 20 (1954). Type from Dem. Rep. Congo.

Shrub or small tree 1.5–4.5 m tall; all but youngest branchlets covered with thick brownish-red powdery bark; young stems covered with short glandular tubercular-based hairs. Leaves (12)18–26-foliolate; leaflets 6–23 × 2–10 mm, oblong, lanceolate, ovate-lanceolate or elliptic-oblong, rounded to truncate and mucronulate at the apex, obliquely rounded at the base, glabrous, midrib ± central with c. 3 basal nerves below it; petiole and rhachis 5.5–10 cm long, glabrous or at first slightly puberulous; petiolules 1–1.5 mm long; stipules 5–13 × 2.5–4.5 mm, ovate to ovate-lanceolate, cordate at the base or ± unilaterally appendaged, glabrous or at most ciliate, deciduous. Inflorescences axillary, zigzag, several-flowered, 2.5–10 cm long, pubescent like the young stems; true peduncles very short; pedicels 5–8 mm long, glandular-pubescent; bracts green, leaf-like, overlapping but not hiding the flowers, persistent, 4–10 × 3–12 mm, deeply bilobed, each lobe elliptic, scarious, nervose, glabrous inside, pubescent outside, denticulate-ciliate with glandular hairs; about half the bracts are sterile; bracteoles 4–7 × 1–3 mm, elliptic-lanceolate, ciliate, persistent. Calyx 2-lipped; lips 7–9 × 4–6 mm, ovate to oblong-elliptic, one 2-toothed, the other shortly 3-lobed. Standard yellow, 8–10 × 5–6 mm, violin-shaped, rounded at the apex, wings yellow; keel petals yellow, not laciniate. Fruit of 1–2 articles joined by a narrow neck, each article semicircular, 5–7 × 4–5.5 mm, compressed, glabrous, smooth, veined. Seeds reddish-brown, 5 × 4 × 2 mm, reniform; hilum small, circular, slightly eccentric; a trace of a rim-aril present.

Zambia. W: Solwezi, fl. & fr. 30.vii.1964, *Fanshawe* 8868 (K).
Also in southern Dem. Rep. Congo. *Brachystegia* woodland, often in rocky places; 1500–2000 m.

58. **Aeschynomene rubrofarinacea** (Taub.) F. White, F.F.N.R.: 455 (1962). —Verdcourt in Kew Bull. **24**: 15 (1970); in F.T.E.A., Leguminosae, Pap.: 404 (1971); in Kirkia **9**: 441 (1974). —Lock, Leg. Afr. Check-list: 107 (1989). Type from Tanzania.
   Smithia rubrofarinacea Taub. in Engler, Pflanzenw. Ost-Afrikas **C**: 216 (1895).
   Geissaspis rubrofarinacea (Taub.) Baker f. in J. Bot. **46**: 114 (1908). —De Wildeman in Bull. Jard. Bot. État **4**: 121 (1914). —E.G. Baker, Legum. Trop. Africa: 318 (1929). — Brenan, Check-list For. Trees Shrubs Tang. Terr.: 429 (1949).
   Geissaspis chiruiensis R.E. Fr., Wiss. Ergebn. Schwed. Rhod.-Kongo-Exped.: 88 (1914). — E.G. Baker, Legum. Trop. Africa: 318 (1929). Type: Zambia, Lake Bangweulu, Chilubi (Chirui) Island, *R.E. Fries* 1046 (S, holotype).
   Geissaspis clevei De Wild., in Bull. Jard. Bot. État **4**: 109 (1914). —E.G. Baker, Legum. Trop. Africa: 314 (1929). —Brenan, Check-list For. Trees Shrubs Tang. Terr.: 429 (1949). Type from Tanzania.
   Geissaspis maclouniei De Wild., in Bull. Jard. Bot. État **4**: 117 (1914). —E.G. Baker, Legum. Trop. Africa: 318 (1929). Type: Malawi, Nyika Plateau, Nacheri Hill, Sept. 1902, *McClounie* 159 (K, holotype).
   Geissaspis scott-elliotii De Wild., in Bull. Jard. Bot. État **4**: 122 (1914). —E.G. Baker, Legum. Trop. Africa: 318 (1929). Type: Zambia, Stevenson Road, *Scott-Elliot* 8284 (BM; K, holotype).
   Humularia rubrofarinacea (Taub.) P.A. Duvign. in Bull. Soc. Roy. Bot. Belgique **86**: 177 (1954).
   Humularia maclouniei (De Wild.) P.A. Duvign., loc. cit.

Lax erect aromatic shrub up to 0.6–4.5 m tall. Branchlets covered with thick brownish-red powdery bark; young stems densely covered with sticky tubercular-based hairs. Leaves (6)8–16-foliolate; leaflets 5–17 × 2.5–8 mm, obovate to oblong or elliptic, emarginate to acute at the apex, obliquely rounded at the base, glabrous on both surfaces or pubescent beneath with tubercular-based hairs, denticulate-ciliate or ciliate or in some areas glabrous on the margins, several-nerved from the base, the main nerve basally eccentric; petiole and rhachis together (1)2–4 cm long; petiolules 0.5 mm long; stipules 4.5–9 × 2–4 mm, ovate-lanceolate or ovate, acute, rounded or slightly auriculate or appendaged at the base, veined. Inflorescences axillary, sometimes together with several leaves on short shoots or even fasciculate; rhachis 3–5 cm long; pedicels 2–3 mm long; bracts purple or green, crowded, overlapping but not hiding the flowers, large, 8–17 × 10–35 mm, deeply to shallowly bilobed, each lobe obovate, rhombic or rounded, scarious, nervose, glabrous or sparsely pubescent, mostly finely denticulate-ciliate or in some areas glabrous on the margins; bracteoles 6–9 × 1–3 mm, elliptic-lanceolate. Calyx 2-lipped, yellow, lips thin, 10–11 × 4.5 mm, oblong-elliptic, one 2-toothed the other shortly 3-lobed. Standard yellow or orange, 10–14 × 5–7.5 mm, very slightly narrowed in the middle, obovate-oblong or oblong, rounded at the apex; wings yellow or green; keel petals white, green or yellow, not laciniate. Fruit of 1–2 articles, each article semicircular, 6–7 × 5–7 mm, smooth or slightly rugulose. Seeds dark reddish-brown, 3.5 × 2.8 × 1 mm, elliptic-reniform; hilum small, elliptic, slightly eccentric, with a trace of a rim-aril present.

Zambia. N: Kasama Distr., between Lua Lua and the Kasama Boma, fl. 18.x.1947, *Brenan & Greenway* 8140 (EA; FHO; K). W: Kitwe, fl. 18.vi.1955, *Fanshawe* 2336 (K; LISC; SRGH). C?: Kanona–Mpika road, fl. 5.vii.1960, *Richards* 12828 (SRGH). E: Nyika Plateau, 8.8 km SW of Rest House, fl. & fr. 25.x.1958, *Robson & Angus* 352 (BM; K; LISC; PRE; SRGH). **Malawi**. N: Chisenga, foothills of Mafinga Mts., fl. & fr. 13.xi.1958, *Robson & Fanshawe* 599 (BM; K; LISC; PRE; SRGH). C: Dzalanyama Forest, Kafi Village, fl. 13.viii.1954, *Adlard* 170 (FHO; K; PRE).

Also in southern Tanzania. *Brachystegia* and *Combretum* woodland, wooded grassland and grassland, *Protea* scrub, etc. often in rocky places; 1170–2250 m.

This species varies in the indumentum of the leaflets and bracts. Material from the Mbala area has glabrous leaves and bracts whereas typically the bracts at least are ciliate. The glabrous form is equivalent to *Geissaspis scott-elliotii* De Wild. which could be used at varietal level if thought necessary. Duvigneaud and White have both considered it indistinguishable from *rubrofarinacea*.

Duvigneaud (Bull. Soc. Roy. Bot. Belgique **86**: 176 (1954)) erected a section of *Humularia*, sect. *Rubrofarinaceae* for *H. rubrofarinacea*, *H. maclouniei* and *H. bianoensis* P.A. Duvign. Although he kept the former two distinct he admitted that intermediates due to introgressive hybridization are frequent; I agree with White that it is best to consider them to be a single species of *Aeschynomene* although admittedly forming a link with *Humularia*. The third species does not appear to be distinguishable from *A. pararubrofarinacea* J. Léonard.

59. **Aeschynomene micrantha** (Poir.) DC., Prodr. **2**: 321 (1825). —Harvey in F.C. **2**: 226 (1862). —Mogg in Macnae & Kalk, Nat. Hist. Inhaca Isl., Moçamb.: 146 (1958). —Verdcourt in Kirkia **9**: 443 (1974). —Lock, Leg. Afr. Check-list: 105 (1989). Type from Madagascar.

Hedysarum micranthos Poir., Encycl. Méth. Bot. **6**: 446 (1805).

Patagonium racemosum E. Mey., Comment. Pl. Afr. Austr. **1**, 1: 123 (1836). Type from South Africa.

Short-lived perennial (?) herb with prostrate stems 10–60 cm long, rather densely pubescent with short grey usually spreading hairs and also with some longer bristly tubercular-based hairs on the young shoots. Leaves with an equal or unequal number of leaflets, 5–11-foliolate; leaflets 2.5–11.5 × 1.5–6.5 mm, obovate-oblong to rounded-obovate, rounded but mucronulate at the apex, obliquely rounded or slightly subcordate at the base, glabrous above, pilose on margins and beneath, particularly on the central main nerve; venation openly and prominently reticulate beneath; rhachis and petiole together 0.5–2 cm long; petiolules up to 0.5 mm long; stipules 3–5 × 0.8–1.8 mm, ovate-lanceolate to lanceolate, acuminate or acute, not appendaged, pubescent and with longer tubercular-based glandular marginal hairs, conspicuously nervose, persistent. Inflorescences axillary, 1–2(several?)-flowered, 0.5–5 cm long, peduncles 6–20 mm long; pedicels 3–6 mm long; bracts 2 × 1 mm, asymmetrically ovate, acuminate, nervose, ciliate, persistent; bracteoles 2 × 1.5 mm, elliptic, nervose, ciliate, persistent. Calyx pubescent, 3 mm long, obscurely 2-lipped, 5-lobed, lobes c. 1.5 mm long, 2 just over 1 mm wide, 3 just under 1 mm wide. Standard yellow or orange, 7 mm long and wide, almost round, sparsely pubescent outside. Wings and keel yellow, the latter not laciniate. Ovary densely silky. Fruit of 1–4 articles joined by narrow necks but contiguous, each article semicircular, 3.5–4.5 × 3–3.5 mm, appressed puberulous, slightly reticulate, strongly margined; stipe 2–5 mm long. Seeds pale olive-coloured, 2.3 × 1.8 × 1 mm, ellipsoidal-reniform, with an eccentric minute raised dark hilum.

Mozambique. GI: Chongoéne, 1 km from hotel, fl. 18.i.1965, *Pereira, Marques & Balsinhas* 254 (LMU). M: Inhaca Island, fl. & fr. 30.i.1962, *Mogg* 29863 (K; PRE; SRGH).

Also in South Africa, Madagascar and Mascarene Islands. Sandy places, dunes, grassland and woodland; 0–150 m.

This species is extremely close to several American ones and a sheet from Zambia, Chipata, Katapola Farm Institute, fl. & fr. 9.v.1963, *van Rensburg* 2109 (K; SRGH) and another from Zimbabwe S: Choma Conservation area test plots, fl. & fr. 23.xii.1957, *Verboom* 119 (PRE) are deceptively similar to *A. micrantha* but differ in the appressed stem indumentum, appressed pubescence on the upper surface of the leaflets, up to 7-articled fruit and slightly different leaflet shape. These specimens appear to be *A. falcata* (Poir.) DC. and occur as an impurity in seed from South America.

76. KOTSCHYA Endl.

Kotschya Endl., Nov. Stirp. Dec. **1**: 4 (1839). —Dewit & Duvigneaud in Bull. Soc. Roy. Bot. Belgique **86**: 207–214, fig. 1–3 (1954). —Verdcourt in Kew Bull. **24**: 17 (1970); in Kirkia **9**: 447 (1974).

Herbs, subshrubs or shrubs, rarely small trees, erect or sometimes ± spreading, the stems generally covered with tubercular-based often glandular hairs. Leaves pinnately 4–many-foliolate; leaflets alternate, asymmetric at the base, with several basal nerves, the main nerve usually submarginal; stipules not spurred below the base, persistent or soon falling; stipels absent. Inflorescences mostly dense, less often 1–few-flowered, axillary or falsely terminal, ± scorpioid, often strobiliform, the flowers reflexed, arranged distichously, often with sterile bracts at the base; bracts entire, shorter than the flowers, scarious, striate, brownish and persistent; bracteoles free or connate, persistent, also scarious and striate. Calyx scarious, 5-lobed; tube short, the lower 3 lobes partly joined to form a lip, the upper pair usually almost completely joined to form a 2-fid lip. Corolla small to medium-sized, white, yellow or blue; standard obovate or rounded, often emarginate, narrowed into a claw; wings elliptic or obovate, usually shortly spurred and with a series of small pockets, free; keel petals sometimes shortly spurred, shortly joined just below the apex. Stamens monadelphous; anthers uniform. Ovary stipitate, linear, 2–9-ovuled; style inflexed, glabrous; stigma terminal, small. Fruit stipitate, of 1–9 joints, if segments more than 1 then fruit folded like a concertina, enclosed in the calyx, the segments rounded, smooth, indehiscent, deciduous. Seeds reniform or depressed semi-globose; hilum minute, eccentric.

A genus of c. 31 species restricted to tropical Africa and Madagascar.
The name *Sarcobotrya* Viguier, proposed as a genus, corresponds to the group with fasciculate leaves (species 14–17) and could perhaps be recognized as a section, but the relationships are too reticulate for this to be advisable.

1. Annual herb with flowers in ± globose clusters; ovary glabrous · · · · · · · · · 19. *capitulifera*
- Perennial herbs, shrubs or small trees · 2
2. Small tree or shrub with branches up to 25 mm in diameter bearing a thick corky bark; leaves congested at the apices of the branchlets; standard 18 mm long; ovary 1-ovuled, densely pilose · 10. *suberifera*
- Herbs or shrubs or rarely small tree *(K. bullockii)* without such corky branchlets · · · · · · 3
3. Calyx lobes ovate, obovate or broadly elliptic, 4 large, 10–14 × 5.5–9.5 mm but central lobe of lower lip shorter and narrower, elliptic, 6–8 × 2–2.5 mm; lips deeply divided, almost to the base; leaflets up to 14 × 6 mm · 18. *eurycalyx*
- Calyx lobes less conspicuous, the lower lip with central lobe about equalling or longer than the lateral lobes; lobes oblong or elliptic or if ovate then not so large · · · · · · · · · · · · 4
4. Leaves distinctly fasciculate, at least on most shoots*, flowers in distinct, mostly sessile strobilate clusters · 5
- Leaves not distinctly fasciculate · 9
5. Bracteoles joined for three-quarters of their length; standard 10–12 mm long · · 17. *coalescens*
- Bracteoles free or nearly so · 6

---

* It must be admitted that this is not always easy to assess but is too useful a character to be avoided. In certain species where the internodes are short on young shoots mistakes are easily made and in these cases it is advisable to run through both sections of the key.

6. Leaves 20–50-foliolate; bracts acute · 14. *imbricata*
   – Leaves 10–24 foliolate; bracts long cuspidate or acuminate · · · · · · · · · · · · · · · · · · · 7
7. Standard 6–9 mm long; stems covered with appressed ascending brisiitly hairs; ovary hairy
   · 15. *strigosa*
   – Standard exceeding 10 mm in length · 8
8. Corolla not or scarcely exserted from the calyx; standard 1–12(13) mm long · · · · · · · · ·
   · 15. *strigosa* var. *grandiflora*
   – Corolla exserted from the calyx; standard 14 mm long · · · · · · · · · · · · · · · · 16. *speciosa*
9. Ovary glabrous; claw of standard conspicuous, over half the length of the lamina; calyx 4–8
   mm long · 11. *aeschynomenoides*
   – Ovary pubescent; claw of standard not so conspicuously well-developed, less than half the
   length of the lamina · 10
10. Ovary 3–9-ovuled · 11
    – Ovary 2-ovuled · 15
11. Lower calyx lip divided only to about the middle, or sometimes to about two-thirds of its
    length · 12
    – Lower calyx lip divided to near the base · 22
12. Leaflets very distinctly curved, 4–9-jugate; leaves ± glabrous (if distinctly hairy see possible
    hybrids with *K. thymodora* subsp. *septentrionalis*) · · · · · · · · · · · · · · · · · · · 4. *recurvifolia*
    – Leaflets not curved or apices only slightly falcate, 8–22-jugate · · · · · · · · · · · · · · · · · 13
13. Calyx thicker, 10–22 mm long; lobes of lower lip diverging, the central one usually longest;
    inflorescence not strobilate · 1. *africana*
    – Calyx thinner, 5.5–16.5 mm long; lobes parallel, subequal, obtuse; inflorescence often
    strobilate · 14
14. Calyx 12–16.5 mm long · 2. *uguenensis*
    – Calyx 5.5–10 mm long · 3. *thymodora* subsp. *septentrionalis*
15. Flowers large, standard 2.4 cm long; shoots and leaf rhachis covered with short stiff closely
    appressed, upwardly directed very bulbous-based hairs; leaflets 12–20-jugate, 0.8–2 mm
    wide · 9. *bullockii*
    – Flowers smaller; shoots and leaf rhachis with ± spreading less thickened hairs · · · · · · · 16
16. Flowers in dense usually subsessile, strobiliform clusters; pedicels mostly not visible · · · 17
    – Flowers in more lax inflorescences, the pedicels visible · 22
17. Leaflets 3–9-jugate, often curved and most distinctly so if over 6-jugate · · · · · · · · · · · 18
    – Leaflets 7–22-jugate, some 10-jugate leaves always present; leaflets not distinctly curved or,
    if falcate and somewhat curved, then stems and leaf rhachis densely covered with spreading
    yellow bristles · 20
18. Leaflets several-nerved from base but main one distinctly submarginal; inflorescences with
    long yellow bristly hairs on the bracts and rhachis; leaflets 4–9-jugate, 3–9(11) mm long,
    distinctly curved · 4. *recurvifolia*
    – Leaflets several-nerved from the base, all nerves equally developed or main one subcentral;
    inflorescences often with shorter whitish hairs; leaflets 3–5(6)-jugate; stipules well
    developed · 19
19. Calyx lobes distinctly lanceolate; standard yellow, 10–15 mm long; leaflets 4–5(6)-jugate
    · 13. *prittwitzii*
    – Calyx lobes rounded or ovate, very obtuse; standard mostly blue or pinkish, rarely white or
    yellow, c. 6 mm long; leaflets mostly 3–4-jugate · · · · · · · · · · · · · · · · · · · 12. *strobilantha*
20. Stems and leaf rhachis scabrid with very short bulbous-based hairs well under 0.5 mm long;
    midribs of leaflets with similar hairs but inflorescences with long yellow bristly hairs · · · ·
    · 6. *scaberrima*
    – Stems and leaf rhachis hairy with spreading bristly hairs 0.5–2 mm long; midribs of leaflets
    pubescent with short only slightly thickened hairs; inflorescences with long yellow bristly
    hairs · 21
21. Stems and leaf rhachis less conspicuously hairy; leaflets mostly c. 5 × 1–1.5 mm, mostly not
    curved and scarcely falcate · 3. *thymodora*
    – Stems and leaf rhachis very conspicuously yellow hairy; leaflets mostly c. 7 × 1–2 mm,
    curved and distinctly falcate · 5. *sp. A*
22. Leaflets larger, 5–17 × 2–7 mm; bracteoles lanceolate-falcate; standard 1.4–2 cm long · · ·
    · 8. *carsonii*
    – Leaflets smaller, 4–9.5 × 1.5–2.5 mm; bracteoles ovate; standard 1.6–1.7 cm long · 7. *longiloba*

1. **Kotschya africana** Endl., Nov. Stirp. Dec. **1**: 4 (1839); Iconogr. Gen. Pl.: t. 125 (1841). —
Dewit & Duvigneaud in Bull. Soc. Roy. Bot. Belgique **86**: 212 (1954); in F.C.B. **5**: 332
(1954). —White, F.F.N.R.: 158 (1962). —Verdcourt in Kew Bull. **24**: 21 (1970); in F.T.E.A.,
Leguminosae, Pap.: 412 (1971); in Kirkia **9**: 450 (1974). —Lock, Leg. Afr. Check-list: 115
(1989). Type from Sudan.
  Smithia kotschyi Benth. in Miquel, Pl. Jungh.: 211 (1852). —J.G. Baker in F.T.A. **2**: 153
(1871). —E.G. Baker, Legum. Trop. Africa: 309 (1929). Type as above.
  Damapana africana (Endl.) Kuntze, Revis. Gen. Pl. **1**: 179 (1891).
  Smithia africana (Endl.) Taub. in Engler, Pflanzenw. Ost-Afrikas **C**: 215 (1895).

Erect shrub 1–4.5(6) m tall. Stems covered with glandular bristly tubercular-based
hairs. Leaves 20–42-foliolate, leaflets 2–10(12.5) × 0.5–3.5(4) mm, oblong or oblong-
lanceolate, slightly falcate and acute at the apex, obliquely or minutely emarginate at
the base, glabrous or pubescent and with some tubercular-based hairs on the
submarginal main nerve beneath, margins denticulate-ciliate; petiole and rhachis
together 2–7.6 cm long; petiolules minute; stipules 3–7.5 × 1.2–3.5 mm, ovate-
lanceolate, acute, margined or covered with tubercular-based hairs, scarious, veined,
persistent. Inflorescences axillary, dense, 1–few-flowered, not strobilate; peduncle
3–5 mm long or very short or obsolete; rhachis 2–6 cm long, mostly with bristly yellow
hairs; pedicels 4–6 mm long; sterile bracts often several and conspicuous; bracts
3.5–8 × 1.3–3 mm, ovate, acute or acuminate, glabrescent or covered with bulbous-
based hairs outside; bracteoles 3–7 × 1.2–2 mm, ovate-lanceolate, acute or acuminate,
pubescent. Calyx pubescent and bristly-hairy, 10–22 mm long, the upper lip
emarginate or 2-fid for 0.5–3 mm, the lower lip deeply divided for just over half its
length into 3 diverging lanceolate-acuminate lobes, the central one usually longest
and narrowest. Standard orange-yellow, often veined with reddish-purple, 9.5–18 ×
4.5–10 mm, obovate or rounded, emarginate, glabrous; wings orange-yellow, hairy at
apex; keel greenish-yellow, often veined with pale red, hairy at apex. Ovary 6–9-
ovulate, pubescent. Fruit 4–9-jointed; articles 3–4 × 2–3.2 mm, pubescent. Seeds
ochraceous or reddish-brown, 2–2.8 × 1.8–2.2 × 1–1.3 mm, depressed half-globose-
reniform; hilum minute, eccentric.

1. Leaflets 3.5–4 mm wide · iv) var. *latifoliola*
– Leaflets mostly narrower, usually 2–3 mm wide · 2
2. Calyx 16–22 mm long · i) var. *africana*
– Calyx 10–15(18) mm long · 3
3. Calyx with hairs mostly shorter, 1.3–1.5(1.8) mm long; inflorescences 1–6 cm long · · · · ·
· ii) var. *bequaertii*
– Calyx very densely covered with long bristly yellow hairs 2.5 mm long and usually with a
silvery pubescence of fine hairs longer than those in other varieties; inflorescences short,
1–2 cm long · iii) var. *ringoetii*

i) Var. **africana** —Verdcourt in Kirkia **9**: 451 (1974).
  Smithia goetzei sensu Hutchinson in Botanist in Southern Africa: 508 (1946) non Harms.

Inflorescences 5–6 cm long. Calyx 16–20(22) mm long, sparsely to densely
covered with rather short bristly yellow hairs and very short pubescence. Leaflets
usually 2–3 mm wide.

Zambia. N: 68 km north of Kasama, fl. 18.vii.1930, *Hutchinson & Gillett* 3817 (BM; COI; K;
LISC; SRGH).
  Also in Ethiopia, Sudan, Dem. Rep. Congo and Tanzania. Riverine forest and swamp forest
edges; 1410 m.
  The distinctions between var. *africana* and var. *bequaertii* are slight but extremes are easily
nameable and in certain parts of Africa it is clearly desirable to keep them distinct. All three
sheets seen from Zambia are intermediate in nature, and judgement on the varieties should not
be made merely from Zambian material alone.

ii) Var. **bequaertii** (De Wild.) Verdc. in Kew Bull. **24**: 22 (1970); in F.T.E.A., Leguminosae, Pap.:
413 (1971); in Kirkia **9**: 451 (1974). Type from Dem. Rep. Congo.
  Smithia riparia R.E. Fr., Wiss. Ergebn. Schwed. Rhod.-Kongo-Exped.: 86, fig. 7 (1914). —
E.G. Baker, Legum. Trop. Africa: 308 (1929). Type: Zambia, Lake Bangweulu, near
Kasoma (Kasomo), *Fries* 658 (UPS, holotype).

Smithia bequaertii De Wild. in Rev. Zool. Bot. Africaines **13**: B23 (1925). —E.G. Baker, Legum. Trop. Africa: 309 (1929).

Smithia kotschyi sensu Brenan, Check-list For. Trees Shrubs Tang. Terr.: 442 (1949) pro parte non Benth. sensu stricto.

Kotschya africana sensu Dewit & P.A. Duvign. in Bull. Soc. Roy. Bot. Belgique **86**: 212 (1954) pro parte.

Inflorescences 1–6 cm long. Calyx mostly short, 13–15(18) mm long, covered with short hairs. Leaflets usually 2–3 mm wide.

Zambia. N: Mbala Distr., Kapata Village, Kituta Bay, fl. & fr. 20.v.1936, *B.D. Burtt* 6180 (BM; EA; K). **Malawi**. N: Viphya (Vipya), Luwawa, fl. & fr. 20.vii.1962, *Chapman* 1668 (K; SRGH). **Mozambique**. N: Maniamba, fl. & fr. 27.viii.1934, *Torre* 256 (BM; COI; K; LISC).

Also in Ethiopia, Dem. Rep. Congo, Burundi, Uganda, Kenya, Tanzania, and probably also in Madagascar (see note). Lake shores, where it often forms a distinct thicket zone with *Phragmites*, etc. inside sand beaches, pan and stream edges, etc.; 810–2280 m.

Dewit & Duvigneaud sank *Smithia bequaertii* into *Kotschya africana* completely, but I agree with the annotations of J.B. Gillett that there is undoubtedly a difference between extremes though not correlated with geography. Unfortunately the type of *Smithia bequaertii* is at the upper end of the range of size for the small-flowered variant. *S. chamaechrista* Benth. is very similar indeed to this variety and probably not to be distinguished; it was described from Madagascar. In East Africa *Kotschya uguenensis* is easily distinguishable from *Kotschya africana* and has a quite different distribution. In the north eastern part of Zambia on the Nyika Plateau intermediates occur between the two species and these also grade gradually into a larger-leafletted form treated here as a distinct variety, var. *latifoliola*. *Fanshawe* 9745 (K) (Zambia E: Nyika Plateau, fl. 26.vi.1966) and *Lawton* 446 (K) (Zambia E: Mafinga Mt., fl. 26.viii.1958) are examples of such intermediates.

iii) Var. **ringoetii** (De Wild.) Dewit & P.A. Duvign. in Bull. Soc. Roy. Bot. Belgique **86**: 212 (1954); in F.C.B. **5**: 333 (1954). —Verdcourt in Kew Bull. **24**: 23 (1970); in F.T.E.A., Leguminosae, Pap.: 414 (1971); in Kirkia **9**: 452 (1974). Type from Dem. Rep. Congo (Katanga Province).

Smithia ringoetii De Wild. in Repert. Spec. Nov. Regni Veg. **13**: 115 (1914). —E.G. Baker, Legum. Trop. Africa: 308 (1929).

Smithia kotschyi sensu Brenan, Check-list For. Trees Shrubs Tang. Terr.: 442 (1949) pro parte non Benth. sensu stricto.

Inflorescences mostly very short, 1–2 cm long. Calyx 10–15 mm long, very densely covered with long bristly yellow hairs up to 2.5 mm long and usually with a silvery pubescence of finer hairs longer than those in the other two varieties. Leaflets usually 2–3 mm wide.

Zambia. N: NE of Mweru Wantipa (Mweru-wa-Ntipa), Bulaya, Chishela (Chishyela) Dambo, fl. 16.viii.1962, *Tyrer* 508 (BM; SRGH). W: Mufulira, fl. 23.v.1934, *Eyles* 8318 (K; SRGH).

Also in Dem. Rep. Congo and Tanzania. Dambos, margins of swamps and rivers; 1200 m.

iv) Var. **latifoliola** Verdc. in Kew Bull. **24**: 24 (1970); in Kirkia **9**: 453 (1974). Type: Zambia, near top of Kangampande Mt., *White* 2557 (FHO; K, holotype).

Kotschya uguenensis sensu White, F.F.N.R.: 158 (1962) pro parte non (Taub.) F. White.

Inflorescences mostly c. 2 cm long. Calyx c. 1.2 cm long with subequal lobes, covered with rather short hairs. Leaflets mostly 3.5–4 mm wide.

Zambia. E: Nyika Plateau, fl. 7.vi.1962, *Verboom* 631 (K).

Almost certainly also in northern Malawi. Rocky slopes and seepage channels on fringe of evergreen forest; 2100–2150 m.

It has already been pointed out that the Nyika Plateau provides the meeting point for *Kotschya africana* and *Kotschya uguenensis*. An equally valid treatment would be to consider the above variety and the intermediates mentioned there as a variant of *Kotschya uguenensis*. A specimen from Mozambique, *Torre* 257 (N: Lichinga (Vila Cabral), fl. 5.xi.1934, (BM; COI; K; LISC)) on the other hand seems to be a broad-leaved form of var. *bequaertii* with the central lower calyx tooth longer than the rest as in that variety. *Mendonça* 683 (Mandimba Distr., R. Luculumesi, fl. & fr. 8.x.1942 (BM; LISC)) is similar.

2. **Kotschya uguenensis** (Taub.) F. White, F.F.N.R.: 455 (1962). —Verdcourt in Kew Bull. **24**: 24 (1970); in F.T.E.A., Leguminosae, Pap.: 414 (1971); in Kirkia **9**: 453 (1974). —Lock, Leg. Afr. Check-list: 118 (1989). Syntypes from Tanzania.

Smithia uguenensis Taub. in Engler, Pflanzenw. Ost-Afrikas **C**: 215 (1895). —Harms in Engler, Pflanzenw. Afrikas [Veg. Erde 9] **3** (1): 615, fig. 229 (1915). —E.G. Baker, Legum. Trop. Africa: 307 (1929). —Brenan, Check-list For. Trees Shrubs Tang. Terr.: 443 (1949).

Erect shrub (1)1.8–4.5(6) m tall, stated to smell unpleasant. Stems covered with sticky spreading bristly yellow tubercular-based hairs. Leaves 30–44-foliolate; leaflets 3–8.5 × 0.7–1.8 mm, linear-oblong, acute and slightly falcate at the apex, obliquely rounded and minutely emarginate at the base, with short tubercular-based hairs on the margin and on the almost marginal nerve beneath; petiole and rhachis together 2–4.5 cm long; petiolules minute; stipules 5–7 × 1.5–2 mm, ovate-oblong, acute, pubescent and ciliate, persistent. Inflorescences axillary, dense, distinctly strobilate, with the flowers at right-angles to the axis giving an oblong appearance; peduncles very short or up to 7 mm long; rhachis 1.5–4 cm long, densely hairy; pedicels 7–10 mm long; bracts 5–8 × 2.5–3 mm, ovate or broadly elliptic, margined with long yellow bristly hairs; bracteoles 5–6 × 1.5–2 mm, broadly lanceolate, bluntly acute, pubescent and margined with long yellow bristly hairs. Calyx covered with short and long golden-yellow hairs, 1.2–1.65 cm long, the upper lip divided for c. 3 mm into 2 obliquely acute or rounded oblong lobes, the lower lip divided to c. half-way into 3 obtuse or bluntly acute lobes, the outer 2 oblong, the inner oblong or lanceolate, often narrowest. Standard pale to bright yellow, 12–13(15) × 6.5–9 mm, rounded-oblong, rounded or very slightly mucronate, glabrous save for very few bristles at extreme tip; wings and keel pale or greenish-yellow. Ovary 3–4-ovuled, pubescent.   Fruit of 3–4 articles, each article 3 × 2 mm, rounded-oblong, compressed, pubescent. Seeds buff-brown, 2.2 × 1.5 × 1.2 mm, ellipsoid-reniform; hilum minute, eccentric.

Malawi. N: Nyasa–Tanganyika Plateau, south of Namitawa, fl. & fr. ix.1902, *McClounie* 131 (K). **Mozambique**. N: Lichinga (Vila Cabral), fl. vii.1934, *Torre* 260 (BM; COI; K; LISC).
Forming impenetrable thickets (fide McClounie); on the margins of water courses; 1200–1700 m.
The problem of the intermediates between this species and *Kotschya africana* has already been discussed under var. *bequaertii* and var. *latifoliola* of that species.

3. **Kotschya thymodora** (Baker f.) Wild in Kirkia **4**: 159 (1964). —Verdcourt in Kew Bull. **24**: 26 (1970); in F.T.E.A., Leguminosae, Pap.: 416 (1971); in Kirkia **9**: 454 (1974). —Lock, Leg. Afr. Check-list: 118 (1989).   Type: Zimbabwe, Chimanimani (Melsetter), 23.ix.1906, *Swynnerton* 655 (BM, holotype; K; SRGH).
Smithia thymodora Baker f. in J. Linn. Soc., Bot. **40**: 56 (1911); Legum. Trop. Africa: 308 (1929). —Rattray in Kirkia **2**: 88 (1961).

Erect much branched shrub 1.8–4.5 m tall. Stems very sticky and glandular, covered with shorter pubescence and also with longer yellow tubercular-based hairs. Leaves (16)22–32(36)-foliolate; leaflets 2–8 × 0.8–2.5 mm, narrowly oblong-lanceolate or linear-oblong, acute and oblique or very slightly falcate at the apex, obliquely rounded at the base, with short tuberculate hairs on the slightly serrulate margins and on the almost marginal main nerve beneath; petiole and rhachis together (10)15–45 mm long; petiolules minute; stipules 3–5 × 1.5–2 mm, ovate-lanceolate, pubescent and ciliate, persistent.   Inflorescences axillary, dense, distinctly strobilate, the flowers mostly held at right-angles to the axis; peduncle short, 0–4 mm long; rhachis 10–35 mm long, densely yellow bristly hairy; pedicels 4 mm long; bracts 3–7 × 1.5–4.5 mm, ovate, broadly elliptic or somewhat obovate, sometimes slightly falcate, conspicuously covered, particularly on or near the margins, with long yellow bristly hairs; bracteoles 3–4 × 1.2–1.5 mm, lanceolate or elliptic, acute or obtuse, pubescent and margined with bristly hairs. Calyx covered with short and rather longer yellow tubercular-based hairs, 5.5–10 mm long, the upper lip divided for 1.5–2 mm into 2 very obtuse rounded lobes, the lower lip divided for c. one-third to half its length into 3 oblong obtuse lobes, the inner usually the narrowest. Standard golden-yellow, flushed red, 10–12(14) × 6–8(11) mm, rounded oblong, glabrous; wings and keel yellow.   Ovary 2–5-ovuled, pubescent.   Fruit of 2–5 articles, each article compressed ellipsoid, 3.3 × 2.5 mm, pubescent with fine tubercular-based yellow hairs and slightly reticulate.   Seeds blackish-red, 2.8 × 2 × 1 mm, compressed-ellipsoid; hilum minute, eccentric.

Subsp. **thymodora** —Drummond in Kirkia **8**: 223 (1972). —Verdcourt in Kirkia **9**: 455 (1974). —Lock, Leg. Afr. Check-list: 118 (1989).

Ovary 2-ovulate.

Zimbabwe. E: Nyanga Distr., Pungwe R. Source, fl. 20.x.1946, *Wild* 1431 (K; SRGH).
Mozambique. MS: Catandica (Vila Gouveia), Serra de Chôa, fl. 3.vii.1941, *Torre* 2983 (BM; K; LISC).

Also in South Africa (Transvaal). Montane grassland and thicket, forest edges, etc., also in *Uapaca* formations in foothills; 1000–2200 m.

Subsp. **septentrionalis** Verdc. in Kew Bull. **24**: 26 (1970); in F.T.E.A., Leguminosae, Pap.: 416 (1971); in Kirkia **9**: 455 (1974). —Lock, Leg. Afr. Check-list: 118 (1989). Type from Tanzania.

Ovary (3)4–5-ovulate.

Malawi. N: Nyika Plateau, 10 km from Kasaramba Viewpoint, fl. 8.vii.1970, *Brummitt* 11873 (K). Also in Tanzania. Montane grassland, slopes with *Protea*, *Philippia*, etc.; 1830–2350 m.

It could well be argued that *Kotschya uguenensis*, *Kotschya scaberrima* and the two subspecies of *Kotschya thymodora* could all be treated as subspecies of *Kotschya uguenensis* and I have discussed this in Kew Bull. **24**: 28 (1970). It is only in northern Malawi and contiguous parts of Zambia that real difficulty will be met with in giving names.

4. **Kotschya recurvifolia** (Taub.) F. White, F.F.N.R.: 455, 157 (1962). —Verdcourt in Kew Bull. **24**: 28 (1970); in F.T.E.A., Leguminosae, Pap.: 417, fig. 59/1–11 (1971); in Kirkia **9**: 456 (1974). —Lock, Leg. Afr. Check-list: 117 (1989). Type from Tanzania.
    Smithia recurvifolia Taub. in Engler, Pflanzenw. Ost-Afrikas **C**: 215 (1895). —E.G. Baker, Legum. Trop. Africa: 307 (1929). —Brenan, Check-list For. Trees Shrubs Tang. Terr.: 443 (1949); in Mem. New York Bot. Gard. **8**: 254 (1953).
    Smithia congesta Baker in Bull. Misc. Inform., Kew **1897**: 259 (1897). —E.G. Baker, Legum. Trop. Africa: 308 (1929). Type: Malawi, between Khondowe (Kondowe) and Karonga, July 1896, *Whyte* (K, holotype).
    Smithia drepanophylla Baker in Bull. Misc. Inform., Kew **1897**: 260 (1897). —E.G. Baker, Legum. Trop. Africa: 308 (1929). Type: Malawi: Misuku (Masuku) Plateau, *Whyte* (K, holotype).

Erect much branched somewhat aromatic sticky shrub 0.2–3.6 (4.5) m tall. Stems pubescent and with longer whitish or golden bristly tubercular-based hairs, sometimes also covered with persistent stipules. Leaves densely congested or laxer, 8–18-foliolate; leaflets 3–9(11) × 0.7–2.5 mm, narrowly oblong-falcate, always characteristically and distinctly curved at the apex, acute at the apex, cuneate or oblique at the base, quite glabrous or ciliate, sometimes minutely ciliate-denticulate, otherwise glabrous save for bristles on the submarginal main nerve; basal nerves 4–5; petiole and rhachis together typically very short, 3–15(23) mm long; petiolules minute; stipules 3–7 × 1–2.5 mm, triangular-lanceolate, acute, scarious, ciliate, persistent. Inflorescences axillary, dense, few- to several-flowered, 1.5–3.5 cm long, subsessile; pedicels 2–7 mm long; rhachis of inflorescence with long golden bristly hairs 1–2 mm long; bracts 5–7 × (2.5)4–5 mm, ovate, with long golden bristly hairs on the margins; bracteoles 2–6 × 1–1.5 mm, ovate to lanceolate, with long hairs chiefly on the margins. Calyx pubescent and ciliate with golden hairs, 5–12 mm long, 2-lipped; lips oblong, upper shortly to deeply emarginate or divided into diverging lobes, the lower divided for two-thirds of its length into 3 oblong-rounded or lanceolate acute lobes. Standard pale to golden-yellow or whitish, 9–12(16) × 7–8(11.5) mm, elliptic to rounded or obovate-oblong, glabrous; wings golden-yellow; keel greenish. Fruit of 1–3 articles, each article straight above, strongly curved beneath, 3.2 × 2.8 mm, pilose. Seeds greenish to dark reddish-brown, 1.8–2.5 × 1.2–2 × 0.5–1 mm, oblong or reniform.

Tab. 3.6.**29**. KOTSCHYA RECURVIFOLIA subsp. RECURVIFOLIA. 1, flowering branch (× ²/₃); 2, detail of leafy shoot (× 4); 3, flower (× 2); 4, bracteoles (× 4); 5, upper calyx lip (× 4); 6, lower calyx lip (× 4); 7, standard (× 4); 8, wing (× 4); 9, keel (× 4); 10, androecium (× 4); 11, gynoecium (× 4); 1–11 from *Sanders* 103; 12, fruit (× 4); 13, seed (× 4), 12 & 13 from *Dale* in *Forest Dept*. 2684. Drawn by Derek Erasmus. From F.T.E.A.

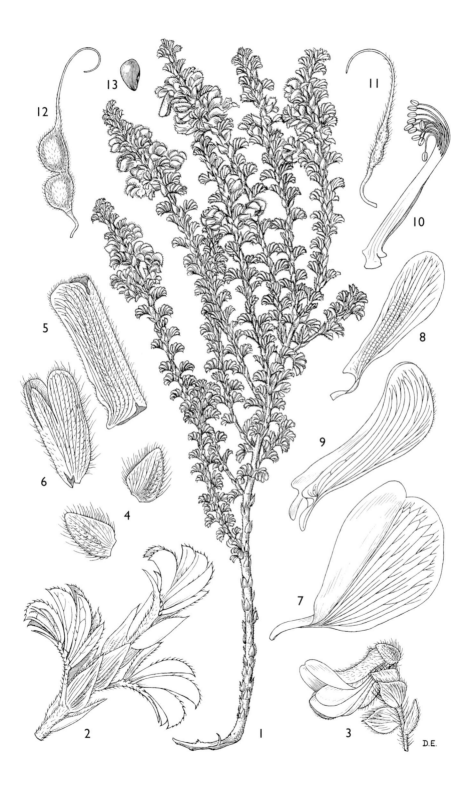

Subsp. **recurvifolia** —Verdcourt in Kirkia **9**: 457 (1974). —Lock, Leg. Afr. Check-list: 117 (1989). TAB. 3.6.**29**.

Foliage congested, the internodes mostly short and much of the stem hidden. Stems and leaves less or not glandular, sparsely to densely covered with often less tubercular hairs. Leaves shorter, typically 5–10 mm long. Leaflets few, typically 4–6 pairs, smaller and more falcate, typically 6 × c. 1 mm. Standard usually 9–12(14) mm long.

Zambia. E: Nyika Plateau, fl. 7.vi.1962, *Verboom* 638 (K; LISC). **Malawi.** N: Nyika Plateau, near top of Nganda, fl. 13.iii.1976, *E. Phillips* 1686 (K; MO).

Also in Tanzania. Mostly in montane grassland and evergreen forest margins, also in secondary forest and streamsides and seepage channels of granite outcrops; 1590–2600 m.

Other subspecies occur in East Africa and Ethiopia. The leaves in the Flora Zambesiaca material are mostly glabrous but quite hairy in one or two specimens which may be hybrids with *K. thymodora* subsp. *septentrionalis* e.g. *Pawek* 13056 (N: Misuku Hills, edge of Mugesse (Mughesse) Forest, 15.ix.1977 (K)).

5. **Kotschya sp. A** —Verdcourt in Kew Bull. **24**: 31 (1970); in Kirkia **9**: 457 (1974).

Bushy herb to 0.9 m tall, with erect stems having no main branches; young stems very densely covered with rather long yellow tubercular-based upwardly directed bristly hairs. Leaves 16–20-foliolate; leaflets 2–8 × 1–2 mm, narrowly oblong-falcate, acute and very curved at the apex, with a tapering bristle-like mucro, obliquely rounded at the base, glabrescent save for cilia on the lower margin and long bristle-like yellow hairs beneath on the almost marginal main nerve; petiole and rhachis together 1.5–3 cm long; petiolules obsolete; stipules 5 × 2 mm, triangular, striate, ciliate, persistent. Inflorescences axillary, almost strobilate, 10–15 mm long, densely golden hairy; peduncles 5–10 mm long, densely golden hairy; pedicels c. 6 mm long; bracts c. 6 × 2 mm, elliptic, nervose, persistent, ciliate with long yellow bristly hairs and some shorter ones on the main vein and base; bracteoles 6 × 1.5 mm, lanceolate, ciliate like the bracts. Calyx covered with rather long golden hairs, c. 9 mm long, the upper lip divided for about one fifth of its length into 2 obliquely rounded lobes, the lower lip deeply 3-lobed, the lobes 6 mm long, narrowly oblong or oblong-elliptic, obtuse, the inner the narrowest. Standard yellow, 11–12 mm long and wide, almost round, glabrous; wings and keel yellow. Ovary 2-ovuled, pubescent. Fruits not seen.

Zimbabwe. E: Chimanimani Distr., Orange Grove, 26.iv.1947, *Wild* 1947 (K; SRGH). Montane streamsides; 1500 m.

This plant comes from the area occupied by *Kotschya thymodora* but cannot be readily referred to that species; the whole aspect is different. Two further specimens which seem to be the same species have the indumentum of *Kotschya scaberrima* (Zimbabwe E: Mt. Pene, Mermaids' Grotto, *Corby* 1996 (SRGH) and Chimanimani (Melsetter), Kasipiti, fl. 1.vi.1969, *Loveridge* 1569 (K; LISC; SRGH)). A further specimen in this complex (Zimbabwe E: Chimanimani (Melsetter), Adams Ridge, fl. 23.vi.1968, *Chase* 8531) has a calyx just over 1 cm long and resembles *K. uguenensis* but has a 2-ovuled ovary. The leaflets tend to those of *sp. A* and have the indumentum of *K. scaberrima*.

6. **Kotschya scaberrima** (Taub.) Wild in Kirkia **4**: 159 (1964). —Verdcourt in Kew Bull. **24**: 32 (1970); in Kirkia **9**: 458 (1974). —Lock, Leg. Afr. Check-list: 117 (1989). Type: Malawi, Shire Highlands, *Buchanan* 934 (B†; K) neotype chosen by Wild & Gillett 1964.

    Smithia scaberrima Taub. in Engler, Pflanzenw. Ost-Afrikas **C**: 215 (1895). —E.G. Baker, Legum. Trop. Africa: 308 (1929). —Brenan in Mem. New York Bot. Gard. **8**: 254 (1953). Syntypes: Malawi, Shire Highlands, *Buchanan* 457* (B†); 934 (B†; K); and near Blantyre; *Last* s.n. (K).

Shrub 0.9–3 m tall. Stems covered with tubercles which render them scabrid; these tubercles correspond to the tubercular-based hairs of other species but the hair part is greatly reduced in this species; the older stems are purple-brown, ridged, lenticellate and glabrescent. Leaves 12–26-foliolate; leaflets 3–7.5 × 1–1.8 mm, obliquely oblong-elliptic or linear-oblong, obliquely acute at the apex but scarcely falcate, obliquely rounded at the base, with sparse tubercles along the margins and main nerve similar to those on the stem; main nerve almost marginal; petiole and

---

* *Buchanan* 487 at Kew is probably an isosyntype, the 5 being an error.

rhachis 1–3 cm long with hairs similar to the stem hairs; petiolules very short; stipules 2 × 1.5 mm, triangular, ciliate, striate, persistent. Inflorescences axillary, usually ± strobilate, 15–25 mm long, bristly yellow hairy; peduncles 5 mm long, similarly hairy; pedicels 4–5.5 mm long; bracts 4 × 2 mm, obliquely ovate, acuminate, with short tubercle-like hairs; bracteoles 3–3.5 × 1.2 mm, lanceolate or ovate-lanceolate, with similar hairs. Calyx covered with very short tubercular-based hairs and some fine pubescence as well, 7–9(11) mm long, the upper lip very shortly 2-fid or divided for about one-quarter of its length into 2 oblong rounded lobes, the lower lip deeply 3-lobed, the lobes 3–3.5 mm long, oblong or elliptic, blunt, the inner the narrowest. Standard yellow, 12–13 × 6–10 mm, round or elliptic, glabrous; wings and keel yellow. Ovary 2-ovuled, pubescent. Fruit of 1–2 articles, each article 3 × 2 mm, 1.2 mm thick, compressed oblong-ellipsoid, pubescent. Seed dark brown, 2.4 × 1.8 × 1.2 mm, ellipsoid-reniform; hilum minute, very eccentric.

Zimbabwe. E: Chimanimani Distr., Tarka Forest Reserve, lower slopes of Mt. Pene, fl. ix.1968, *Goldsmith* 124/68 (K; LISC; SRGH). **Malawi**. S: Mulanje Mt., slopes of Thuchila (Tuchila) River Valley, fl. 12.vii.1956, *G. Jackson* 1873 (K). **Mozambique**. N: Ribáuè Distr., Serra de Ribáuè, Mepáluè, fr. 28.i.1964, *Torre & Paiva* 10319 (LISC). Z: Serra do Gurué, fl. vi.1943, *Torre* 5698 (LISC).

Submontane rocky slopes, grassland with *Xerophyta* and *Helichrysum* above *Widdringtonia* forest; 1200–2000 m.

The specimen from Mt. Pene closely resembles *K. scaberrima* (Taub.) Wild and throws doubt on the distinctness of *K. thymodora* once more. *Simão* 1443 (Serra da Gorongosa, Pico Gogogo, fl. 10.vii.1947 (LISC; LMA)) is somewhat intermediate. See also note at the end of 5. *sp. A.*

7. **Kotschya longiloba** Verdc. in Kew Bull. **24**: 32 (1970); in Kirkia **9**: 459 (1974). —Lock, Leg. Afr. Check-list: 116 (1989). Type: Zambia, near Nsama, 31.v.1950, *Bullock* 2907 (EA; K, holotype).

Subshrubby herb 1.5–1.8 m tall; branches glandular, densely covered with very sticky yellow bristly tubercular-based hairs. Leaves 20–30-foliolate; leaflets 4–9.5 × 1.5–2.5 mm, narrowly oblong, obtuse or obliquely subacute at the apex, obliquely rounded or slightly emarginate at the base, puberulous; margins ciliate, glandular; main nerve submarginal, 3–4 basal nerves not very marked; petiole and rhachis 2–3.5 cm long; petiolules very short; stipules 3.5 × 1.5 mm, ovate-lanceolate, scarious, ciliate, veined, persistent. Inflorescences axillary, not strobiliform, 2–4 cm long, densely covered with sticky golden bristly hairs; peduncles 5–10 mm long; pedicels 7–9 mm long; bracts 4–5 × 2.5 mm, ovate acuminate, margins with bristly yellow hairs; bracteoles 3–4 × 2.5–3 mm, ovate, acute, margined with similar hairs. Calyx covered with tubercular-based hairs, 1.5–2 cm long, 2-lipped; the upper lip oblong, divided into 2 obtuse oblong lobes 6–8 mm long, lower lip 3-fid, the lobes 10–12 mm long, narrowly oblong, obtuse, diverging. Standard yellow or orange, 1.6–1.7 × 1.2–1.3 cm, broadly elliptic or rounded, glabrous save for a few apical setae; wing and keel yellow. Ovary 2–3-ovulate, pilose. Fruit not seen.

Zambia. N: Kaputa Distr., Mweru Wantipa, floodplain below Mpundu, 1050 m, fl. 11.iv.1957, *Richards* 9161 (K).

Known only from this part of Zambia. Floodplains and at foot of rocky slopes, in wet black soil; 1050 m.

8. **Kotschya carsonii** (Baker) Dewit & P.A. Duvign. in Bull. Soc. Roy. Bot. Belgique **86**: 213 (1954); in F.C.B. **5**: 340 (1954). —White, F.F.N.R.: 157 (1962). —Torre in C.F.A. **3**: 208 (1966). —Verdcourt in Kew Bull. **24**: 33 (1970); in F.T.E.A., Leguminosae, Pap.: 420 (1971); in Kirkia **9**: 460 (1974). —Lock, Leg. Afr. Check-list: 115 (1989). Type: Zambia, Fwambo, 1890, *Carson s.n.* (K, holotype).
      Smithia carsonii Baker in Bull. Misc. Inform., Kew **1893**: 156 (1893). —E.G. Baker, Legum. Trop. Africa: 309 (1929).
      Smithia harmsiana De Wild. in Ann. Mus. Congo, Sér. IV, Bot. [Études Fl. Katanga] **1**: 52, t. 22 (1902). —E.G. Baker, Legum. Trop. Africa: 309 (1929). Type from Dem. Rep. Congo.
      Smithia setosissima Harms in Bot. Jahrb. Syst. **45**: 314 (1910). —E.G. Baker, Legum. Trop. Africa: 309 in adnot. (1929). Type from Dem. Rep. Congo.
      Kotschya carsonii forma *multifoliolata* Dewit & P.A. Duvign. in Bull. Soc. Roy. Bot. Belgique **86**: 213 (1954); in F.C.B. **5**: 342 (1954). Type from Dem. Rep. Congo.

Erect shrub 0.3–1.8(?4.5) m tall. Stems glandular, sticky, densely covered with ±

golden bristly tubercular-based hairs. Leaves 14–26(30)-foliolate; leaflets 5–17 × 2–6.5 mm, elliptic or ± oblong, eccentrically acute and mucronulate at the apex, obliquely rounded at the base, sometimes denticulate and glandular-ciliate, glandular-pubescent to practically glabrous on both surfaces; basal nerves 4–6, the principal one submarginal and bristly beneath; venation rather prominent beneath; petiole and rhachis together 1–6.5 cm long; petiolules 0.5 mm long; stipules 7–15 × 2.5–4 mm, ovate-lanceolate, acute, scarious, pubescent, persistent. Inflorescences axillary, dense, 3–5 cm long, subsessile or with a peduncle up to 1 cm long; pedicels 10–12 mm long; bracts 6–12 × 3–4 mm, ovate or lanceolate, acuminate, pubescent and setose; bracteoles free, 4–9 × 2–2.5 mm, ovate or lanceolate-falcate, acuminate, pubescent and setose. Calyx pubescent and setose, 15–22 mm long, 2-lipped, the upper divided for two-thirds to three-quarters of its length (one-third in subsp. *reflexa*) into 2 rounded lobes, the lower for over three-quarters of its length into 3 acute lanceolate lobes. Standard orange-yellow, (14)15–20 × (8)10–15 mm, rounded, glabrous save for a few hairs on the main nerve; wings and keel yellow. Fruit of 2(3) articles, 5 × 3.5–5 mm, each article oblong, pubescent. Seeds chestnut-brown, 3.5 × 2.5 × 1.5 mm, elliptic-oblong, slightly beaked on the outside of the small eccentric hilum.

Subsp. **carsonii** —Verdcourt in Kirkia **9**: 461 (1974). —Lock, Leg. Afr. Check-list: 115 (1989).

Ovary 2-ovulate; upper lip of calyx divided for about two-fifths to three-quarters of its length; leaflets 2–6.5 mm wide; standard 15–20 mm long.

Zambia. N: Kawambwa, fl. 21.vi.1957, *E.A. Robinson* 2324 (K). W: Mwinilunga, Lisombo River, fl. 10.vi.1963, *Loveridge* 906 (K; LISC; PRE; SRGH). E: Nyika Plateau, fl. 7.vi.1962, *Verboom* 637 (K). **Malawi**. N: Viphya (Vipya) Plateau, Mtangatanga Forest Reserve, fl. 25.ix.1972, *Pawek* 5808 (K). **Mozambique**. N: Lichinga (Vila Cabral), fl. 12.vi.1934, *Torre* 258 (BM; COI; K; LISC).

Also in Dem. Rep. Congo, southern Tanzania and Angola. *Brachystegia* woodland, bushland, moorland, peaty swamps and forest edges; 1050–1800 m.

Subspecies *reflexa* (Portères) Verdc. occurs in Guinea and Ivory Coast. It has a 3-ovulate ovary, and the upper lip of the calyx is divided for c. one-third of its length.

9. **Kotschya bullockii** Verdc. in Kew Bull. **24**: 36, fig. 4 (1970); in F.T.E.A., Leguminosae, Pap.: 420, fig. 60 (1971); in Kirkia **9**: 461 (1974). —Lock, Leg. Afr. Check-list: 115 (1989). TAB. 3.6.**30**. Type from Tanzania.

Shrub or small tree 1.8–3.6 m tall. Stems densely covered with short upwardly directed yellow very bulbous-based stiff hairs, later roughened with stipule-bases and eventually peeling to reveal a reddish almost powdery surface beneath. Leaves 24–40-foliolate; leaflets 2.5–11 × 0.8–2 mm, linear-oblong, slightly but distinctly falcate, acute at the apex, oblique at the base, somewhat coriaceous, rather glossy, ciliate and with stronger bulbous-based cilia along the completely marginal main nerve; other basal nerves obscure; petiole and rhachis together 2–6 cm long, densely covered with bulbous-based hairs similar to those on the stem; petiolules very short; stipules 5–6 × 2.5 mm, triangular, scarious, veined, at least the bases persistent. Inflorescences axillary, short, lax, 1–4-flowered; peduncle c. 5 mm long; rhachis c. 3 cm long; pedicels 5–7 mm long, covered with hairs similar to those on the stem; bracts 6–7 × 2 mm, ovate-lanceolate, shortly ciliolate; bracteoles 8–9 × 2 mm, lanceolate, acuminate, covered with appressed bulbous-based hairs. Calyx with similar hairs, 2-lipped; lips c. 2 cm long, oblong, the upper shortly 2-fid for c. 3 mm into subacute lobes, the lower divided for 5–6 mm into 3 subacute lobes. Standard pinkish-yellow with purple lines or more usually blue, 24 × 12–13 mm, oblong, emarginate, glabrous; wings and keel presumably the same colour as the standard, with a few bristly hairs at the apex. Ovary 2-ovulate, pubescent. Fruit with stipe 4 mm long, of 2 articles, each article 6 × 4 mm, elliptic, pilose with short and long hairs. Seeds chestnut-brown, 4.7 × 3 × 1.7 mm, ellipsoid-reniform, compressed.

Tab. 3.6.**30**. KOTSCHYA BULLOCKII. 1, flowering branch (× ²⁄₃); 2, portion of stem showing indumentum (× 4), 1 & 2 from *Eggeling* 6176; 3, leaf (× ²⁄₃); 4, leaflet (× 2), 3 & 4 from *Bullock* 3290; 5, flower (× 2); 6, bract (× 2); 7, upper calyx lip (× 2); 8, lower calyx lip (× 2); 9, standard (× 2); 10, wings (× 2); 11, keel (× 2); 12, androecium (× 2); 13, gynoecium (× 2); 14, young fruit (× 2), 5–14 from *Eggeling* 6176. Drawn by Derek Erasmus. From Kew Bull.

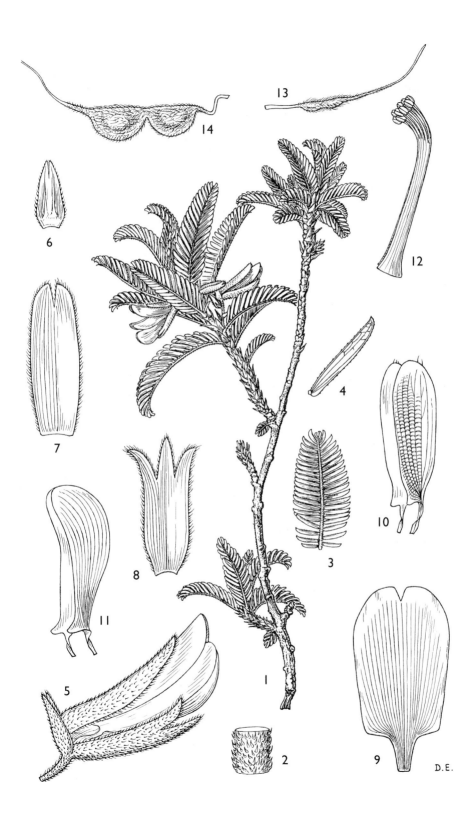

Zambia. N: Kawambwa, fl. 22.viii.1957, *Fanshawe* 3522 (EA; K).
Also in western Tanzania. Granite outcrops on dambo margins, with *Protea–Terminalia*, etc.,
also in watershed dambo systems; 1414 m.

10. **Kotschya suberifera** Verdc. in Kew Bull. **24**: 38, fig. 5 (1970); in Kirkia **9**: 462 (1974). —
    Lock, Leg. Afr. Check-list: 118 (1989).   TAB. 3.6.**31**.   Type: Zambia, Mwinilunga Distr.,
    Kalene Hill, *Angus* 554 (FHO, holotype).
      Kotschya sp. 1 of F. White, F.F.N.R.: 158 (1962).

Small tree c. 2 m tall; branches thickened, up to 2.5 cm in diameter covered with
a thick, corky, brown, corrugated bark; young branchlets similar but densely covered
with persistent stipules and tubercular-based hairs. Leaves congested at the apices of
the branchlets, 7–12-foliolate; leaflets 4–20 × 1.2–6.5 mm, obliquely oblong, falcate,
apiculate at the apex, very asymmetric, obliquely rounded at the base, ciliate and
with glandular hairs; main nerve submarginal; petiole and rhachis together c. 2–7.5
cm; petiolules 0.5 mm long; stipules 6–8 × 4–6 mm, triangular, striate, densely
covered outside with tubercular-based hairs, subpersistent. Inflorescences axillary,
2–4 cm long, 1–6-flowered, rather lax, densely sticky-pilose; peduncle 5–7 mm long;
pedicels 9–14 mm long; bracts 5–7 × 2.5 mm, lanceolate; bracteoles 5–6 × 1.5–2 mm,
lanceolate. Calyx glandular-pubescent, 2-lipped; upper lip 2-fid, 12 mm long, lobes
7–8 × 2–2.5 mm, lower lip 3-fid, 9–11 mm long, lateral lobes 6–7 × 2–2.5 mm, median
lobe 5–5.5 × 1.5–1.8 mm; all lobes narrowly elliptic-oblong. Standard dull wine-
coloured (when dry), 18 × 12 mm, obovate-elliptic, rounded at the apex, auriculate,
glabrous. Ovary 1-ovulate, 3 × 1 mm, densely white pilose. Fruit not seen.

Zambia. W: Mwinilunga Distr., Kalene Hill, fl. 25.ix.1952, *Angus* 554 (FHO).
Known only from this locality. Dominant over small area in hollow on top of hill in shallow
stony soil together with *Dissotis, Syzygium, Hymenocardia* and *Chrysophyllum*.

11. **Kotschya aeschynomenoides** (Welw. ex Baker) Dewit & P.A. Duvign. in Bull. Soc. Roy. Bot.
    Belgique **86**: 213 (1954); in F.C.B. **5**: 338, fig. 2. —Hepper in F.W.T.A., ed. 2, **1**: 581
    (1958). —White, F.F.N.R.: 157 (1962). —Torre in C.F.A. **3**: 206 (1966). —Verdcourt in Kew
    Bull. **24**: 39 (1970); in F.T.E.A., Leguminosae, Pap.: 421 (1971); in Kirkia **9**: 463 (1974). —
    Lock, Leg. Afr. Check-list: 115 (1989). Syntypes from Angola.
      Smithia aeschynomenoides Welw. ex Baker in F.T.A. **2**: 153 (1871). —Harms in Engler,
    Pflanzenw. Afrikas [Veg. Erde 9] **3** (1): 616 (1915). —E.G. Baker, Legum. Trop. Africa: 306
    (1929). —Hutchinson, Botanist South. Africa: 532 (1946). —Brenan, Check-list For. Trees
    Shrubs Tang. Terr.: 441 (1949).
      Damapana aeschynomenoides (Welw. ex Baker) Kuntze, Revis. Revis. Gen. Pl. **1**: 179 (1891).
    —Hiern, Cat. Afr. Pl. Welw. **1**: 237 (1896).
      Smithia volkensii Taub. in Engler, Pflanzenw. Ost-Afrikas **C**: 215 (1895). —E.G. Baker,
    Legum. Trop. Africa: 305 (1929). —Brenan, Check-list For. Trees Shrubs Tang. Terr.: 442
    (1949). Type from Tanzania.
      Smithia sphaerocephala Baker in Bull. Misc. Inform., Kew **1897**: 260 (1897). —E.G. Baker,
    Legum. Trop. Africa: 305 (1929). Type: Malawi, Khondowe (Kondowe) to Karonga, *Whyte*
    (K, holotype).
      Smithia ruwenzoriensis Baker f. in J. Linn. Soc., Bot. **38**: 246 (1908); Legum. Trop. Africa:
    306 (1929). Type from Uganda.
      Smithia mildbraedii Harms in Mildbraed, Wiss. Ergebn. Deutsch. Zentr.-Afrika Exped.,
    Bot. **2**: 260 (1911). —E.G. Baker, Legum. Trop. Africa: 308 (1929). Type from Rwanda.

Erect, rarely decumbent, aromatic shrub (0.15)0.5–3 m tall. Stems sticky and
glandular, densely covered with bristly tubercular-based hairs. Leaves (12)16–34-
foliolate; leaflets 2–14(17) × 1–3.5 mm, oblong to oblong-lanceolate, obtuse and
mucronulate at the apex, oblique at the base, glabrous save for sparse glandular hairs
on the main nerve beneath; basal nerves 2–3, the main one submarginal; petiole and
rhachis together 1.3–7.5 cm long; petiolules 0.5 mm long; stipules 4–7 × 1–3 mm,
lanceolate, acute, scarious, glandular-ciliate, ± deciduous. Inflorescences axillary,
often curved, dense, strobilate; peduncle 5–15 mm long; rhachis 1–7 cm long;
pedicels 5–7 mm long; bracts free, 2–6 × 1–3 mm, falcate-lanceolate or ovate, acute,
asymmetric at the base, pubescent; bracteoles free, 2–6(7) × 0.7–2 mm, elliptic to
elliptic-lanceolate, pubescent. Calyx pubescent, 4–8 mm long, 2-lipped; lips elliptic
or oblong, the upper deeply emarginate, the lower deeply divided into 3 rounded or
acute ovate lobes. Standard white, pale lilac or bluish-violet, often veined mauve,

Tab. 3.6.**31**. KOTSCHYA SUBERIFERA. 1, habit (× 1); 2, apical portion of leaf (× 2); 3, stipules (× 3); 4, flower (× 3); 5, calyx (× 2); 6, standard (× 2); 7, wing (× 2); 8, keel petal (× 2); 9, stamen (× 8); 10, young fruit (× 2), 1–10 from *Angus* 554. Drawn by Mary Grierson. From Kew Bull.

8–12 × 4–9 mm, rounded, emarginate, glabrous save for a few hairs on the central nerve; claw conspicuous, wings white; keel white or bluish-violet. Fruit of 1–2 articles; articles 2–2.5 mm long and wide, semicircular, glabrous. Seeds dark brown, 2 × 1.7 × 1.2 mm, oblong-rounded.

Zambia. N: Mbala Distr., Lumi River Marsh close to Kawimbi Mission, fl. 20.v.1955, *Richards* 5775 (K; SRGH). W: Ndola, fl. 29.v.1953, *Fanshawe* 49 (EA; K). C: 51.2 km NE of Serenje Corner, fl. 25.vii.1930, *Hutchinson & Gillett* 4076 (BM; K). E: Nyika Plateau, fl. 7.vi.1962, *Verboom* 628 (K). **Malawi**. N: Nyika Plateau, fl. vii.1953, *Chapman* 119 (FHO; K).
Also in Sudan, Dem. Rep. Congo, Rwanda, Uganda, Kenya, Tanzania and Angola. *Brachystegia* woodland, evergreen forest margins, wet places such as seasonal streams, river banks, marshes, swamps, dambos, boggy drainage channels on granite outcrops, etc.; 1200–2280 m.
It must be admitted that as treated above this is a very variable species. The main variation is to be found in the length of the bracts, bracteoles, calyx and standard. Typical *K. aeschynomenoides* has short elliptic bracteoles 3 mm long, the standard 8–9 mm long and the calyx c. 5 mm long. *Smithia sphaerocephala* and *S. ruwenzoriensis* represent large-flowered variants, the former having bracteoles 3.5 mm long and the latter 4.5 mm long; in both the standard is 11–12 mm long. *S. volkensii* applies to the variant having longer lanceolate bracteoles, 5–6(7) mm long, and small standard, 7.5–8 mm long. Between these variants it is possible to find specimens with other combinations of characters and a division into subspecies with a good geographical or ecological basis is not obvious. *S. mildbraedii* is included in the above synonymy purely from the description – the leaflets are said to be up to 20-jugate; the petals were not described. It was not dealt with in the F.C.B. account.

12. **Kotschya strobilantha** (Welw. ex Baker) Dewit & P.A. Duvign. in Bull. Soc. Roy. Bot. Belgique **86**: 213 (1954); in F.C.B. **5**: 337, pl. 26 (1954). —White, F.F.N.R.: 157 (1962). — Torre in C.F.A. **3**: 207 (1966). —Verdcourt in Kew Bull. **24**: 42 (1970); in Kirkia **9**: 464 (1974). —Lock, Leg. Afr. Check-list: 118 (1989). TAB. **3.6.32**. Type from Angola.
  Smithia strobilantha Welw. ex Baker in F.T.A. **2**: 154 (1871). —Harms in Warburg, Kunene-Samb.-Exped. Baum: 262 (1903). —Fries, Wiss. Ergebn. Schwed. Rhod.-Kongo-Exped.: 88 (1914). —E.G. Baker, Legum. Trop. Africa: 307 (1929). —Brenan, Check-list For. Trees Shrubs Tang. Terr.: 442 (1949).
  Damapana strobilantha (Welw. ex Baker) Kuntze, Revis. Gen. Pl. **1**: 179 (1891). —Hiern, Cat. Afr. Pl. Welw. **1**: 237 (1896).

Erect, rarely prostrate shrub or subshrub 0.15–2 m tall. Young stems with bristly hairs which are somewhat thickened at the base, older soon glabrous usually with a rather peeling epidermis. Leaves sensitive, 6–8-foliolate; leaflets (5)9–18 × (1.5)2–4 mm, oblong or oblong-oblanceolate, slightly falcate, acute at the apex, obliquely rounded at the base, glabrous all over or with margins ciliate or ciliate-denticulate, 3–4 nerved from the base, petiole 5–16 mm long; rhachis 5–10 mm long; petiolules 0.5 mm long and wide; stipules joined at base for 3–4 mm, 8–16 × 1–2.5 mm, lanceolate, acute or acuminate, scarious, persistent, strongly veined, pubescent or glabrous, often dense and sheathing on the long shoots. Inflorescences axillary, dense, strobilate, subsessile, 1–2.5 cm long; pedicels 2–4 mm long with long setae; bracts 3–6 × (1.5)2–4 mm, ± semicircular, sometimes falcate, venose, pubescent to glabrescent and ciliate with long bristly hairs; bracteoles 3–5 × 1–2.5 mm, ovate, elliptic or oblong, ± falcate, acute, venose. Calyx glabrescent to densely pubescent and ciliate, 3–6 mm long, 2-lipped; the upper lip shortly 2-lobed, the lower lip deeply 3-lobed, the lobes ovate, rounded, subequal, venose. Standard blue, white or pink, sometimes veined with darker colour, 6–7 × 3–3.5 mm, obovate-cuneate or ± violin-shaped, emarginate and with 1–2 bristly hairs at the apex; wings blue or pink, sparsely ciliate at the apex. Fruit of 1–2 articles; articles 3 × 2–2.5 mm, ellipsoid or oblong, compressed, pubescent. Seeds pale yellowish to brown, 2.2 × 1.8 × 1 mm, reniform; hilum small, very eccentric.

Tab. 3.6.32. KOTSCHYA STROBILANTHA. 1, flowering branch (× ½), from *Mullenders* 920; 2, stipules (× 3); 3, open flower, bract and bracteoles (× 5); 4, open flower, longitudinal section (× 5); 5, calyx, spread out, internal face (× 3); 6, standard, spread out, internal face (× 5); 7, wing, external face (× 5); 8, keel, spread out, internal face (× 5); 9, androecium and gynoecium (× 5), 2–9 from *Quarré* 1814; 10, fruit (× 5), from *Mullenders* 920. Drawn by J.M. Lerinckx. From Fl. Congo Belge. Reproduced with permission of Jardin Botanique National de Belgique.

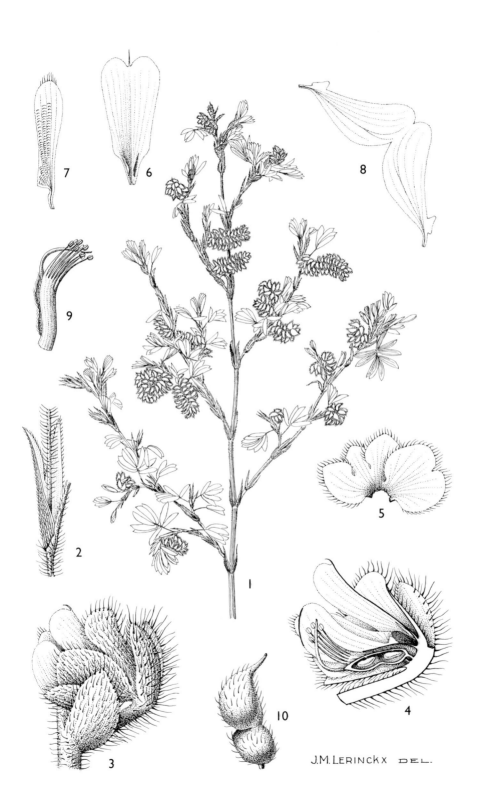

J.M. LERINCKX DEL.

Var. **strobilantha** —Verdcourt in Kew Bull. **24**: 42 (1970). —Drummond in Kirkia **8**: 223 (1972). —Verdcourt in Kirkia **9**: 465 (1974). —Gonçalves in Garcia de Orta, Sér. Bot. **5**: 93 (1982).

Leaflets 9–12 mm long; shrub to c. 2 m tall.

Zambia. B: Kaoma Distr., Luampa River, fl. & fr. *Verboom* 1042 (K; SRGH). N: Luwingu, fl. 29.v.1964, *Fanshawe* 8705 (K). W: Solwezi Boma, fl. 4.vi.1930, *Milne-Redhead* 416 (PRE; K). C: 32.8 km from Lusaka on Great East Road, fl. 13.vi.1957, *Angus* 1624 (FHO; K). S: Choma, fl. 11.vii.1930, *Hutchinson & Gillett* 3534 (BM; K). **Zimbabwe**. N: Gokwe South Distr., Charama, near the headwaters of the Manyoni River, fl. 28.iii.1962, *Bingham* 191 (K; PRE; SRGH). C: Harare (Salisbury), fl. 8.v.1927, *Eyles* 4938 (K; SRGH). E: Odzani River Valley, 1914, *Teague* 188 (BOL; K). **Malawi**. N: Mzimba Distr., Marymount, Mzuzu towards Lunyangwa, fl. 11.ix.1966, *Pawek* 77 (SRGH). C: Lilongwe Distr., Dzalanyama Forest Reserve, 13 km from Sinyala Gate on road to Choulongwe (Chaulongwe) Falls, fl. 26.vi.1970, *Brummitt* 11700 (K). **Mozambique**. N: Maniamba Distr., Mepoche, fl. 15.ix.1934, *Torre* 250 (BM; COI; K; LISC). T: Furancungo, fl. 25.iii.1941, *Torre* 3333 (BM; K; LISC).

Also in Burundi, Dem. Rep. Congo and Angola. *Brachystegia, Julbernardia, Uapaca* and other mixed woodlands, bushland, grassland with scattered trees, swampy grassland, dambo margins, sometimes in dry rocky places or on sandy soil; 1050–1800 m.

The abundant material of this common species shows a good deal of variation particularly in the indumentum of the various parts of the inflorescence. Two diffuse varieties, one with an almost glabrous inflorescence and the other with a densely hairy inflorescence might be separated but intermediates are not infrequent. Var. *kundelunguensis* Dewit & P.A. Duvign., a smaller shrublet with uniformly small leaves, occurs in the Dem. Rep. Congo and at least one Zambian specimen is somewhat similar though not identical (N: Shiwa Ngandu, fl. 4.vi.1956, *E.A. Robinson* 1573).

13. **Kotschya prittwitzii** (Harms) Verdc. in Kew Bull. **24**: 45 (1970); in F.T.E.A., Leguminosae, Pap.: 426 (1971; in Kirkia **9**: 466 (1974). —Lock, Leg. Afr. Check-list: 117 (1989). Type from Tanzania.

Smithia prittwitzii Harms in Bot. Jahrb. Syst. **45**: 314 (1910); in Engler, Pflanzenw. Afrikas [Veg. Erde 9] **3** (1): 616 (1915). —E.G. Baker, Legum. Trop. Africa: 307 (1929). —Brenan, Check-list For. Trees Shrubs Tang. Terr.: 443 (1949).

Smithia strobilantha sensu Hutchinson, Botanist South. Africa: 497 (1946) non Welw. ex Baker.

Erect subshrub or shrub c. 0.9–1.8 m tall. Stems pubescent when young with tubercular-based hairs; bark becoming fissured and somewhat flaky on the older stems. Leaves 9–10(12)-foliolate; leaflets 6–18.5 × 1.2–4 mm, oblanceolate or oblong, falcate, acute at the apex, cuneate, entire or ciliate-denticulate, glabrous or minutely pubescent; basal nerves 3–6; petiole and rhachis together 10–25 mm long; petiolules very short; stipules joined for 3–4 mm, sometimes dense and sheathing on the young shoots, 8–18 × 3.5–6 mm, triangular or lanceolate, acute or acuminate, membranous, pubescent, persistent. Inflorescences axillary, dense, few–several-flowered, subsessile, c. 10–15(20) mm long; pedicels 3–3.5(5) mm long; bracts 5 × 2 mm, ± ovate, striate, pubescent; bracteoles 5 × 2.5 mm, ovate, pubescent. Calyx glandular-pubescent, 8–10 mm long, 2-lipped; lips oblong, the upper divided for half its length into 2 acute lobes, the lower divided for two-thirds of its length into 3 lanceolate lobes. Standard yellow, somewhat hooded, 10.5–14.5 × 4–8 mm, obovate, hairy on midrib outside, otherwise glabrous; wings and keel yellow. Fruit of 1–2 articles; articles 3 × 2.5 mm, rounded, pilose. Seeds chestnut-brown, longest dimension c. 2–2.5 mm, trigonous or oblong-reniform, slightly beaked beyond the eccentric hilum.

Var. **prittwitzii** —Verdcourt in Kirkia **9**: 466 (1974).

Corolla 10–14.5 mm long. Staminal tube including filaments and anthers 12–15 mm long.

Zambia. N: Mbala Distr., Chilongowelo Escarpment, fl. 7.vi.1961, *Richards* 15216 (K). C: 8 km east of Chiwefwe, fl. 15.vii.1930, *Hutchinson & Gillett* 3680 (BM; COI; K; LISC).

Also in Tanzania. *Brachystegia* woodland, grassy and rocky places, bushland, also in damp places by rivers and lakes, riverine forest; 1500–1740 m.

Var. **parviflora** Verdc. in Kew Bull. **24**: 46 (1970); in Kirkia **9**: 467 (1974). Type: Zambia, Kawambwa, 23.viii.1957, *Fanshawe* 3545 (K, holotype).

Corolla 10 mm long. Staminal tube including filaments and anthers 9 mm long.

Zambia. N: Mansa (Fort Rosebery), fl. 3.v.1964, *Fanshawe* 8509 (K). W: Kitwe, fl. 22.v.1957, *Fanshawe* 3293 (K).
Not known elsewhere. *Brachystegia* woodland, riverine forest on sandy alluvium but sometimes in rocky places; 1200–1414 m.

14. **Kotschya imbricata** Verdc. in Kew Bull. **26**: 73, fig. 1 (1971); in Kirkia **9**: 467 (1974). —Lock, Leg. Afr. Check-list: 116 (1989). TAB. 3.6.**33**. Type: Zambia, Solwezi Distr., Kabompo Gorge road, *Mutimushi* 3138 (K, holotype; NDO).

Small shrub with many slender woody unbranched stems to 0.6 m tall. Stems covered with rather dense erect white hairs above, chestnut-brown and somewhat fissured at the base. Leaves 20–50-foliolate, fasciculate on very short side shoots; leaflets very small, 1–3 × 0.5–1 mm, oblong-ovate to narrowly oblong, acute and very shortly mucronulate at the apex, mostly very oblique at the base, glabrescent save for long marginal cilia, mostly several-nerved from the base, the midnerve ± central, venation yellowish and prominent beneath; petiole and rhachis together 6–25 mm long; petiolules obsolete; stipules 8 × 1.5 mm, lanceolate, scarious, striate, very persistent. Inflorescences axillary and falsely terminal, dense and cone-like, 10–15 × 10 mm, ovoid, sessile, scented; pedicels 1–2.5 mm long; bracts tightly imbricate, 5–6 × 5 mm, obliquely ovate, pilose; bracteoles 4 × 2 mm, ovate, pilose, slightly joined at the base, persistent, veined. Calyx densely pilose, 7–8 mm long, the upper lip of two oblong-lanceolate lobes 7 × 2 mm, joined for half their length, lower lip deeply 3-lobed, the outer lobes 2 × 2 mm, flat, lanceolate, the midlobe 7 × 2.5 mm when flattened out, boat-shaped. Standard crimson, 9.5 × 9 mm, broadly obovate-quadrate, truncate or concave at the apex and with a densely pilose mucro 2 mm long which continues from the midrib which is densely pilose outside in the upper half; wings and keel crimson. Ovary 2-ovuled, densely pilose. Fruit not seen.

Zambia. W: Solwezi Distr., Kabompo Gorge road, fl. 14.v.1969, *Mutimushi* 3138 (K; NDO).
Not known elsewhere. "Isenga" woodland with granite rocks; 1200 m.

15. **Kotschya strigosa** (Benth.) Dewit & P.A. Duvign. in Bull. Soc. Roy. Bot. Belgique **86**: 212 (1954); in F.C.B. **5**: 335 (1954). —Hepper in F.W.T.A., ed. 2, **1**: 581 (1958). —White, F.F.N.R.: 157 (1962). —Torre in C.F.A. **3**: 207 (1966). —Verdcourt in Kew Bull. **24**: 46 (1970); in F.T.E.A., Leguminosae, Pap.: 423 (1971); in Kirkia **9**: 468 (1974). —Lock, Leg. Afr. Check-list: 118 (1989). Type from Madagascar.
    Smithia strigosa Benth. in Miquel, Pl. Jungh.: 211 (1852). —J.G. Baker in F.T.A. **2**: 154 (1871). —Harms in Engler, Pflanzenw. Afrikas [Veg. Erde 9] **3** (1): 614 (1915). —E.G. Baker, Legum. Trop. Africa: 306 (1929). —Brenan, Check-list For. Trees Shrubs Tang. Terr.: 442 (1949).
    Damapana strigosa (Benth.) Kuntze, Revis. Gen. Pl. **1**: 179 (1891). —Hiern, Cat. Afr. Pl. Welw. **1**: 237 (1896).
    Sarcobotrya strigosa (Benth.) Viguier in Notul. Syst. (Paris) **14**: 169 (1952).

Erect or sprawling subshrub 0.3–2.4 m tall. Stems bristly with tubercular-based hairs. Leaves fasciculate on reduced shoots, 10–24-foliolate; leaflets 3–10 × 1–2 mm, linear-spathulate, often falcate, acute at the apex, cuneate at the base, ciliate and sparsely bristly with tubercular-based hairs on the main nerve beneath; basal nerves 2–4, the main one central or submarginal; petiole and rhachis together 12–22 mm long; petiolules minute; stipules 6–10 × 3–4 mm, lanceolate, acuminate, scarious, ciliate, glabrescent, persistent. Inflorescences axillary in upper leaves, very dense, subsessile, 15–20 mm long; pedicels 2–3 mm long; bracts 5–13 × 4–5 mm, ovate-triangular, long-cuspidate or acuminate, ciliate with long hairs, with conspicuous main nerve; bracteoles free, 5–7 × 2–3 mm, ovate or lanceolate, falcate, acuminate, ciliate with long hairs. Calyx ciliate, 6–7 mm long, 2-lipped; lips oblong, the upper deeply emarginate, the lower divided for one-half to two-thirds of its length into 3 ovate acuminate lobes. Standard bluish outside, deep blue inside with yellow at the base, 6–13 × 3–4 mm, squarish-obovate, glabrous; wings deep blue; keel hyaline with blue veins. Fruit of 1–2 articles; articles 3–4 × 2.5–4 mm, pilose. Seeds chestnut-brown, bluntly trigonous, 2.5 × 2 × 1 mm.

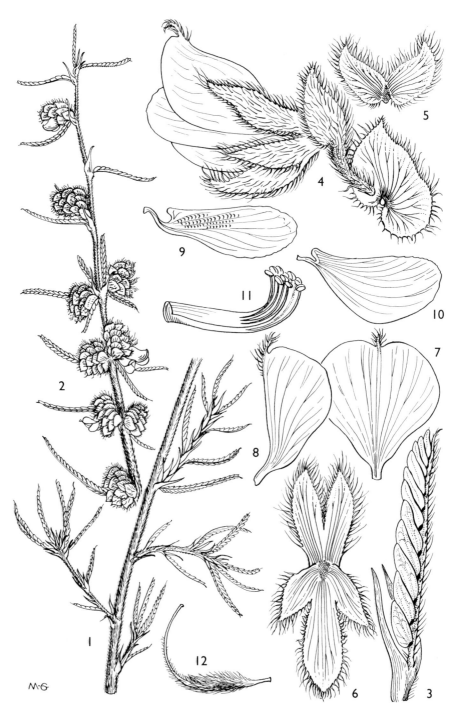

Tab. 3.6.**33**. KOTSCHYA IMBRICATA. 1, portion of stem with leafy branches (× 1); 2, portion
of flowering branch (× 1); 3, leaf (× 6); 4, flower with bract and bracteoles (× 5); 5, bracteoles
(× 5); 6, calyx (× 5); 7, standard, spread out, internal face (× 5); 8, standard, side view (× 5);
9, wing petal (× 5); 10, keel (× 5); 11, androecium (× 5); 12, gynoecium (× 5), 1–12 from
Mutimushi 3138. Drawn by Mary Grierson. From Kew Bull.

Var. **strigosa** —Drummond in Kirkia **8**: 223 (1972). —Verdcourt in Kirkia **9**: 468 (1974). —Gonçalves in Garcia de Orta, Sér. Bot. **5**: 93 (1982).

Bracts 7–8(10) mm long; inflorescences hairy; standard 6–9 mm long.

Zambia. N: Mbala (Abercorn), Lake Chila, fl. 3.v.1955, *Richards* 5503 (K). W: 4.8 km east of Mufulira, fl. 13.vi.1948, *Cruse* 366 (K). C: 14.4 km north of Lusaka, *Cole* 30 (K). S: Choma, fl. 15.iv.1963, *van Rensburg* 1932 (K). **Zimbabwe**. C: Shurugwi (Selukwe), Ferny Creek, 6.viii.1966, *Biegel* 1291 (SRGH). E: Chimanimani Distr., Chipinge (Chipinga) Commonage, fl. 3.iv.1957, *Chase* 6380 (K; PRE; SRGH). S: Bikita Distr., upper region of Dafana River, fl. 7.v.1969, *Biegel* 3045 (K; LISC; SRGH). **Malawi**. N: Mzimba to Lundazi, km 22.4, fl. 29.iv.1952, *White* 2511 (FHO; K; PRE). C: Chongoni Forest, Kangoli Seed Orchard, fl. 14.iv.1971, *Salubeni* 1537 (K; SRGH). S: Shire Highlands, *Buchanan* 1511 (K). **Mozambique**. N: Massangulo Mt., 65.6 km north of Mandimba, fl. 26.v.1961, *Leach & Rutherford-Smith* 11046 (K; LISC; PRE; SRGH). Z: Alto Mólocuè, road to Gurué, fl. 28.v.1937, *Torre* 1515 (COI; LISC). T: between Kazula (Casula) and Furancungo, 70 km from Kazula (Casula), fl. 9.vii.1949, *Barbosa & Carvalho* in Barbosa 3531 (K; LMA). MS: Encosta da Serra de Chôa, Catandica (Vila Gouveia), fl. 17.ix.1942, *Mendonça* 270 (BM; K; LISC).
    Also in Nigeria, W Cameroon, Dem. Rep. Congo, Burundi, Uganda, Tanzania, Angola and Madagascar. *Brachystegia, Julbernardia, Uapaca* and other woodlands, evergreen fringing forest, bushland, streamsides, lake shores, marsh edges, on peaty or sandy soil; 1050–1800 m.

Var. **grandiflora** Dewit & P.A. Duvign. in Bull. Soc. Roy. Bot. Belgique **86**: 212 (1954); in F.C.B. **5**: 336 (1954). —Verdcourt in Kew Bull. **24**: 49 (1970); in F.T.E.A., Leguminosae, Pap.: 424 (1971); in Kirkia **9**: 469 (1974). Type from Dem. Rep. Congo.

Bracts 10–13 mm long; inflorescences, particularly the margins of bracts and bracteoles very densely pilose with long hairs; standard 10–12 mm long.

Zambia. N: Mbala (Abercorn), pans on Old Kasama road, fl. 2.v.1955, *Richards* 5460 (K).
    Also in Burundi, Dem. Rep. Congo and Tanzania. *Brachystegia–Julbernardia* and *Parinari* woodland, also grassland, sometimes on termite mounds or in rough rocky ground; 1500–1800 m.
    A number of Tanzania and Dem. Rep. Congo specimens are intermediate between the two varieties.

16. **Kotschya speciosa** (Hutch.) Hepper in Kew Bull. **11**: 124 (1956); in F.W.T.A., ed. 2, **1**: 581 (1958). —White, F.F.N.R.: 157 in adnot. (1962). —Verdcourt in Kew Bull. **24**: 49 (1970); in F.T.E.A., Leguminosae, Pap.: 424 (1971). —Drummond in Kirkia **8**: 223 (1972). —Verdcourt in Kirkia **9**: 469 (1974). —Lock, Leg. Afr. Check-list: 117 (1989). Type from Nigeria.
    Smithia speciosa Hutch. in Bull. Misc. Inform., Kew **1921**: 365 (1921). —Hutchinson & Dalziel, F.W.T.A. **1**: 416 (1928). —E.G. Baker, Legum. Trop. Africa: 306 (1929). —Brenan, Check-list For. Trees Shrubs Tang. Terr.: 443 (1949).

Erect shrub or subshrub 0.9–1.5 m tall. Stems densely covered with bristly tubercular-based hairs. Leaves subfasciculate on very short reduced shoots, 12–18-foliolate; leaflets 1.3–8 × 0.5–2 mm, oblong-elliptic, lanceolate or oblanceolate, sometimes slightly falcate, acute at the apex, cuneate or oblique at the base, glabrous or sparsely ciliate, 3-nerved from the base, the main nerve subcentral or submarginal, never truly marginal; petiole and rhachis together 0.5–2.7 cm long; petiolules very short; stipules 7–10 × 1.5–3 mm, lanceolate, acuminate, scarious, pubescent, persistent. Inflorescences axillary, dense, subsessile, 10–25 mm long; pedicels 2–2.5 mm long; bracts 6–8 × 4 mm, ovate, falcate, acuminate, striate, with a thickened often eccentric midrib, ciliate; bracteoles similar, 4–6 × 2–3 mm, narrowly ovate, hairy. Calyx densely hairy, 9–13 mm long, 2-lipped; the upper lip oblong, divided to c. half-way into 2 acute lobes, the lower elliptic, divided half-way into 3 ovate-lanceolate acute lobes. Standard blue with yellow mark at the base, 14 × 8–9(10.5) mm, obovate, glabrous; wings blue, white at the base; keel blue. Ovary 2-locular, pilose. Ripe fruit and seeds not seen.

Zambia. N: Mbala Distr., Old Kasama road, Pans, fl. 2.v.1955, *Richards* 5474 (K). **Zimbabwe**. N: Mvurwi (Umvukwe) Mts., 8 km north of Banket, fl. 23.iv.1948, *Rodin* 4410 (K; PRE; UC). W: Matopos, fl. xi.1902, *Eyles* 1117 (PRE). C: Harare (Salisbury), fl. 10.iv.1940, *Verdoorn* 2047 (K; PRE). **Mozambique**. T: Angónia, between Zóbuè and Vila Coutinho, 1962, *Gomes e Sousa* 4766 (COI; K; LMA; PRE) (somewhat intermediate with *Kotschya strigosa*).

Also in Nigeria, Cameroon and Tanzania. *Brachystegia* woodland, bushland, grassland, sandy lake shores, often on granite kopjes in the south; 1300–1680 m.

This is very similar to *Kotschya strigosa* but has considerably larger petals; the supposed differences in the bracts and bracteoles are mostly illusory. In habit it is closest to *Kotschya coalescens* but clearly distinguished by the bracteoles not being joined. It seems unwise at this stage to consider these as subspecies of *Kotschya strigosa* as they are easily identifiable. The name *Smithia eylesii* found on some Zimbabwean specimens has never been published.

17. **Kotschya coalescens** Dewit & P.A. Duvign. in Bull. Soc. Roy. Bot. Belgique **86**: 212 (1954); in F.C.B. **5**: 336 (1954). —White, F.F.N.R.: 157 in adnot. (1962). —Verdcourt in Kew Bull. **24**: 50 (1970); in F.T.E.A., Leguminosae, Pap.: 425 (1971); in Kirkia **9**: 470 (1974). —Lock, Leg. Afr. Check-list: 115 (1989). Type from Dem. Rep. Congo.

Smithia strobilantha sensu Hutchinson, Botanist South. Africa: 497 (1946) non Welw. ex Baker.

Erect or rambling perennial herb or subshrub, 0.75–1.8 m tall. Stems covered with bristly tubercular-based hairs. Leaves subfasciculate on very short reduced shoots, 10–18-foliolate; leaflets 2–6 × 0.7–2 mm, narrowly oblong or lanceolate, somewhat falcate, rounded or acute at the apex, cuneate at the base, setulose-ciliate, punctulate; basal nerves 3–4; petiole and rhachis together 5–21 mm long; petiolules very short; stipules 5–7 × 1–2.5 mm, ovate-lanceolate, acuminate, scarious, pubescent, persistent. Inflorescences axillary, few-flowered in axils of normal leaves or leaves reduced to stipules, dense, subsessile, 2 cm long; pedicels 2–3 mm long; bracts 4–5 × 2–2.5 mm, ovate, acute, pubescent; bracteoles joined for three-quarters of their length, 4–5 × 2 mm, ovate, acute, pubescent. Calyx pubescent or densely hairy, 10–13 mm long, 2-lipped; lips divided for one-half to two-thirds of their length, the upper oblong, divided into 2 obtuse or subacute lobes, the lower almost round, divided into 3 obtuse to acute ovate or lanceolate lobes. Standard blue, 10–12 × 6–8 mm, obovate-oblong or rounded, glabrous or with some hairs on the margins and main nerve; wing blue, hairy at the apex; keel blue. Fruit of 1–2(3) articles; articles 3–3.5 mm long and wide, rounded, pubescent. Seeds grey-brown, 2.5 × 2 × 1.2 mm, rounded-reniform.

Zambia. W: Solwezi Distr., stream just north of Mutanda Bridge, fl. 23.vi.1930, *Milne-Redhead* 587 (K). C: Mkushi Distr., river east of Chiwefwe, fl. 15.vii.1930, *Hutchinson & Gillett* 3695 (BM, COI; K; LISC; SRGH).

Also in Dem. Rep. Congo and Tanzania. Moist sandy dambos, stream and river banks in moist grassland and bushland; c. 1500 m.

18. **Kotschya eurycalyx** (Harms) Dewit & P.A. Duvign. in Bull. Soc. Roy. Bot. Belgique **86**: 213 (1954); in F.C.B. **5**: 343 (1954). —Torre in C.F.A. **3**: 209 (1966). —Verdcourt in Kew Bull. **24**: 50 (1970); in F.T.E.A., Leguminosae, Pap.: 425 (1971); in Kirkia **9**: 471 (1974). —Lock, Leg. Afr. Check-list: 115 (1989). Type from Angola.

Smithia eurycalyx Harms in Bot. Jahrb. Syst. **45**: 312 (1910); in Engler, Pflanzenw. Afrikas [Veg. Erde 9] **3** (1): 616 (1915). —E.G. Baker, Legum. Trop. Africa: 309 (1929).

Erect or decumbent herb or subshrub 12–50 cm tall from a woody rootstock. Stems with tubercular-based bristly hairs but later glabrous and sometimes with peeling epidermis. Leaves 5–10-foliolate; leaflets 4–14 × 1.5–6 mm, obliquely elliptic-oblong, slightly to distinctly falcate, acute or subacute at the apex, obliquely subcordate at the base, ciliate with spaced marginal hairs or practically glabrous, 5–6-nerved from the base, the venation prominent on both sides; main nerve eccentric; petiole and rhachis together 3–6 cm long; petiolules minute; stipules 10–20 × 1–3 mm, narrowly lanceolate, acuminate, scarious, ribbed, ciliate, persistent. Inflorescences axillary in the upper leaves, together forming a dense head, individual parts several–many-flowered, very dense, strobilate, subsessile, 1.5–2.5 cm long; pedicels 4–6 mm long; bracts 5–15 × 3–4.5 mm, obliquely ovate or ovate-lanceolate, acuminate, striate, ciliate; bracteoles 4–8 × 2–3 mm, lanceolate, striate. Calyx leaf-like and venose, pubescent, 10–14 mm long, 2-lipped; upper lip divided into 2 almost free obovate-oblong lobes 13–14 mm long, each 7.5–9.5 mm wide; lower lip deeply divided into 3 lobes, the outer pair 10–12 × 5.5–6.5(9) mm, ovate or obovate, but the central one small, 6–8 × 2–2.5 mm, elliptic; all lobes ciliate. Standard bright blue, 8–12 × 8.5–10 mm, round or squarish-obovate, pubescent along the midrib and ciliate along the upper margin; wings blue, ciliate at apex; keel greenish. Fruit of 1–2 articles, 4–5 × 4–4.5 mm, each article kettle-drum-shaped, densely pilose. Seeds chestnut-brown, 3.5 × 3 × 1.3 mm, oblong-reniform.

Subsp. **venulosa** Verdc. in Kew Bull. **24**: 51 (1970); in F.T.E.A., Leguminosae, Pap.: 426 (1971); in Kirkia **9**: 472 (1974). —Lock, Leg. Afr. Check-list: 116 (1989). Type: Zambia, Mbala (Abercorn), Uningi (Ningi) Pans, *Richards* 18089 (K, holotype).

Calyx lobes more pointed than in typical variety and with transverse venation between the marginal nerve and leaf margin evident.

Zambia. N: Mbala Distr., Kawimbe, fl. 2.vi.1957, *Richards* 9972 (K). **Malawi**. N: Nyika National Park, fl. 26.iv.1973, *Pawek* 6580 (K).
Also in southern Tanzania. Grassland, rocky outcrops and roadsides; 1470–1900 m.
The typical subsp. is known only from the Dem. Rep. Congo and Angola. Further work may indicate that *Kotschya eurycalyx* and *Kotschya schweinfurthii* (Taub.) Dewit & P.A. Duvign. are best combined into one species subdivided into several subspecies. In general the two are easily separated and the eastern subsp. *venulosa* is clearly not identical with the types of either. An occasional sheet, e.g. *Letouzey* 6803, from Cameroon is somewhat intermediate. I have not seen enough typical *Kotschya eurycalyx* to decide against Harms' original judgement.

19. **Kotschya capitulifera** (Welw. ex Baker) Dewit & P.A. Duvign. in Bull. Soc. Roy. Bot. Belgique **86**: 212 (1954); in F.C.B. **5**: 334 (1954). —Torre in C.F.A. **3**: 205 (1966). —Verdcourt in Kew Bull. **24**: 53 (1970); in F.T.E.A., Leguminosae, Pap.: 428 (1971); in Kirkia **9**: 472 (1974). —Lock, Leg. Afr. Check-list: 115 (1989). Type from Angola.
    Smithia capitulifera Welw. ex Baker in F.T.A. **2**: 152 (1871). —Oliver in Trans. Linn. Soc., London **29**: 58, t. 33 (1872). —Harms in Engler, Pflanzenw. Afrikas [Veg. Erde 9] **3** (1): 616 (1915). —E.G. Baker, Legum. Trop. Africa: 305 (1929).
    Damapana capitulifera (Welw. ex Baker) Kuntze, Revis. Gen. Pl. **1**: 179 (1891). —Hiern, Cat. Afr. Pl. Welw. **1**: 236 (1896).
    Smithia burttii Baker f. in J. Bot. **73**: 78 (1935). Type from Tanzania.
    Kotschya capitulifera var. *robusta* Dewit & P.A. Duvign. in Bull. Soc. Roy. Bot. Belgique **86**: 212 (1954); in F.C.B. **5**: 334 (1954). Type from the Dem. Rep. Congo.

Annual ascending or erect unbranched to much branched aromatic herb, 5–45 cm tall, ± decumbent at the base. Stems glabrous or covered with both bristly and short curved hairs. Leaves 6–20-foliolate; leaflets 2–6 × 0.5–1.5 mm, oblong, elliptic, obliquely obovate-oblong or oblong-lanceolate, acute, terminated by a bristle, practically glabrous or with a few long bristly hairs on the nerves and margins; petiole 2–6 mm long; rhachis 4–19 mm long; stipules 3–4 × 1 mm, lanceolate, acute or acuminate, scarious, persistent. Inflorescences small terminal dense heads, 5–18 mm in diameter; peduncles obsolete; pedicels 2 mm long; bracts 2–4 × 2–3 mm, ovate, glabrous or pubescent; bracteoles 2–5 × 0.5–1 mm, falcate, elliptic or lanceolate, glabrous or serrulate with long cilia. Calyx 2-lipped; lips scarious and ciliate, 2.5–4 mm long, the upper lip shortly divided into 2 acute or acuminate lobes, the lower shortly divided into 3 obtuse or acute, ovate, elliptic or lanceolate lobes. Standard blue, pink, white or violet, 4–5.5(11) × 3.5–4(6) mm, ± obovate, glabrous; wings and keel similarly coloured. Fruit 2.5 mm long, of 2 articles; articles 1.2 × 1.2 mm, kettle-drum-shaped, glabrous. Seeds dark reddish-brown, longest dimension c. 0.8 mm, oblong-reniform, slightly produced at one end of hilum.

Var. **capitulifera** —Drummond in Kirkia **8**: 223 (1972). —Verdcourt in Kirkia **9**: 473 (1974).

Rootstock slender; standard 4–5 mm long.

Zambia. N: Mbala (Abercorn), road to Uningi Pans, fl. 17.iv.1963, *Richards* 18104 (K). W: 11 km north of Chingola, banks of River Kafue, fl. 4.v.1960, *E.A. Robinson* 3709 (K). S: Choma Distr., 120 km north of Choma, Kabulamwanda, fl. 21.vi.1955, *E.A. Robinson* 1233 (K). **Zimbabwe**. W: Matobo Distr., Farm Besna Kobila, fl. ii.1957, *Miller* 4130 (K; LISC; SRGH). S: Masvingo Distr., Mutirikwi (Mtilikwe) C.L., fl. 19.iii.1972, *Wild* 7911 (K; LISC; SRGH).
    Also in Burundi, Kenya, Tanzania and Angola. Grassland, particularly short seasonally waterlogged grassland on sandy soil, swamps, lateritic dambos, sometimes in crevices of rocky outcrops; 900–1690 m.
    As in the case of many annuals, there is much variation in habit from very short and unbranched to robust and branched.

Var. *grandiflora* Verdc. with flowers twice as large and stouter rootstock occurs in one place in Tanzania.

77. SMITHIA Aiton

Smithia Aiton, Hort. Kew. ed. 1, **3**: 496, t. 13 (1789) *nom. conserv.* —Verdcourt in Kirkia **9**: 445 (1974).

Erect or decumbent herbs or subshrubs. Leaves pinnately 6–12-foliolate; leaflets opposite, ± asymmetric at the base, with a single centrally placed main nerve; stipules membranous, persistent, prolonged below the point of attachment into a biauriculate appendage, the one auricle short and rounded, the other longer, linear; stipels absent. Inflorescences axillary, usually dense, ± umbel-like scorpioid cymes; bracts entire, scarious, brownish, shorter than the flowers and soon falling; bracteoles present, scarious, inserted below the calyx, persistent. Calyx scarious, 2-lipped; lips entire or slightly toothed. Corolla small, usually red or blue. Standard rounded or obovate, often emarginate; wings oblong, obliquely appendaged at the base, with a series of small pockets, free; keel petals with a lateral appendage almost equalling the claw. Stamens usually in 2 groups of 5, alternately long and short, joined for c. two-thirds of their length; anthers uniform. Intrastaminal disk present. Ovary linear, shortly stipitate, 2–9-ovuled; style glabrous, inflexed; stigma terminal. Fruit stipitate, folded like a concertina, included in the persistent accrescent calyx, articulate; segments rounded, indehiscent, smooth or tuberculate. Seeds mostly compressed-reniform; hilum small, eccentric; rim-aril not developed.

A genus of c. 30 species in the Old World tropics, mainly in Asia and Madagascar, only one in the Flora Zambesiaca area.

Smithia elliotii Baker f., Legum. Trop. Africa: 304 (1929). —Brenan in Mem. New York Bot. Gard. **8**: 254 (1953). —Dewit & Duvigneaud in F.C.B. **5**: 346 (1954). —Hepper in F.W.T.A., ed. 2, **1**: 582 (1958). —White, F.F.N.R.: 105 (1962). —Verdcourt in Kew Bull. **24**: 16 (1970); in F.T.E.A., Leguminosae, Pap.: 408, fig. 58 (1971); in Kirkia **9**: 446 (1974). —Lock, Leg. Afr. Check-list: 120 (1989). Type from Uganda.

Decumbent herb, 0.4–1.8 m long. Stems bristly to glabrescent. Leaves sensitive, 10–28-foliolate; leaflets 3–15 × 1.2–5 mm, linear-oblong, rounded or subacute but apiculate at the apex, obliquely rounded at the base, ciliate, bristly on the nerves and punctate beneath; petiole 2–5 mm long; rhachis 1.5–5.5 cm long; petiolules 0.5 mm long; stipules 15–33 × 1.5–4 mm, glabrous. Inflorescences dense, subumbellate, 8–15 mm long, up to 25 mm wide, above a peduncle 1–4 cm long; pedicels 2–5 mm long; bracts 4–5 × 1–2 mm, ovate-lanceolate, glabrous save for sparse cilia on the margins; bracteoles 2–4 × 2 mm, oblong or elliptic, glabrous or with a few hairs on the main nerve. Calyx densely to very sparsely covered with yellow bristles, striate, c. 7–9 mm long, somewhat accrescent in fruit and then c. 10–12 mm long; tube 2.5 mm long; upper lip bidentate, c. 5–6.5 mm wide; lower lip tridentate, 2–4 mm wide. Corolla rose, mauve or blue; standard c. 10 × 4–8 mm, broadly obovate, glabrous. Fruit 4–7-jointed, each article c. 2–3 mm long and wide, tuberculate. Seeds dark brown, c. 1.5 mm across, 0.5 mm thick, rounded-reniform, compressed.

Var. **elliotii** —Verdcourt in Kirkia **9**: 446 (1974). TAB. 3.6.**34**.

Calyx densely covered with long yellow bristly hairs.

Tab. 3.6.**34**. SMITHIA ELLIOTII var. ELLIOTII. 1, flowering branch (× 1); 2, stipules (× 3); 3, leaflet, lower and upper surface (× 5); 4, flower (× 5); 5, accrescent calyx (× 5); 6, standard and androecium (× 5); 7, wing (× 5); 8, keel (× 5); 9, gynoecium (× 5); 10, young fruit (× 5); 11, segment of fruit (× 10); 12, seed (× 10), 1–12 from *Thomas* 2215. Drawn by Margaret Stones. From F.T.E.A.

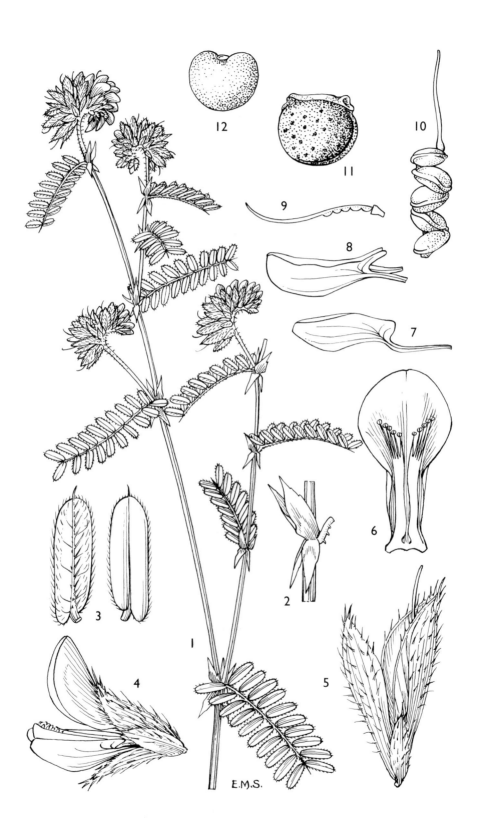

E.M.S.

Zambia. N: near Mbala (Abercorn), Kawimbe Mission, fl. 20.v.1955, *Richards* 5805 (K). W: Solwezi, Lualaba, fl. 13.vi.1962, *Holmes* 1483 (K). C: Lukanga Swamp, fl. 21.v. 1972, *Verboom* 3210 (K; SRGH). **Malawi**. N: Nkhata Bay Distr., South Viphya (Vipya), Luwawa fl. 23.vi.1952, *G. Jackson* 852 (K). C: Ntchisi (Nchisi) Mt., fl. 6.v.1963, *Verboom* 864 (K; LISC). S: Zomba Plateau, fl. 31.v.1946, *Brass* 16118 (K; NY). **Mozambique**. Z: Gurué, montes de Gurué, near waterfalls, fl. 7.iv.1943, *Torre* 5088 (BM; K; LISC).

Also in Nigeria, Cameroon and Dem. Rep. Congo, widespread in east Africa; also in Madagascar. Ditches, swamps, moist dambos, stream banks, edge of rain forest and swamp forest mostly in coarse grass and other dense herbage; 1500–1830 m.

Var. *sparse-strigosa* Verdc. occurs in Dem. Rep. Congo, Burundi and Uganda and may be distinguished by the calyx which is sparsely hairy with very short hairs and is not bristly.

78. HUMULARIA P.A. Duvign.

Humularia P.A. Duvign. in Bull. Soc. Roy. Bot. Belgique **86**: 145–205, figs. 1–9 (1954). — Gledhill in Bol. Soc. Brot. **42**: 305 (1968) (theoretical considerations). —Verdcourt in Kirkia **9**: 474 (1974).
Geissaspis auctt. afr. non Wight & Arn.

Erect or prostrate subshrubs, or less often small shrubs. Leaves pinnately 2–12-foliolate, the leaflets asymmetrical at the base, often with one part cuneiform and the other rounded or half-cordate, the main nerve eccentric or even marginal; stipules rounded, cordate or unequally auriculate at the base, often similar in size to the leaves, leaf-like or sometimes submembranous or subscarious, persistent or deciduous; stipels absent. Inflorescences axillary or terminal, in subcylindrical bracteate racemes, generally dense and scorpioid, less often lax and zigzag; bracts big, distichous, bilobed, densely imbricate, sometimes ± hiding the flowers and completely hiding the fruits, mostly membranous and coloured; bracteoles membranous, persistent, nervose. Calyx 2-lipped; lips almost free, the upper ± 2-fid, the lower 3-fid. Corolla small or medium-sized, yellow, orange or reddish, often lined with red. Standard ± pandurate, narrowed into a claw; wings spathulate, united by their basal appendages, and with a series of small pockets; keel petals obovate, very shortly joined near the apex, narrowed into a claw. Stamens arranged in 2 bundles of 5; anthers uniform. Ovary stipitate, 2-ovuled; style inflexed, glabrous; stigma terminal. Fruit somewhat woody, flattened or eventually biconvex, 1–2-jointed, straight or one joint folded back on the other, often beaked by the persistent style-base. Seeds rounded-reniform, smooth.

A genus of about 40 species confined to Africa, formerly included in the Asiatic genus *Geissaspis* Wight & Arn.; it passes gradually into *Aeschynomene* but the two are best kept separate.

The variation in this genus follows the pattern of several other singularly difficult genera which literally explode into a plethora of minor taxa in Central Africa (e.g. *Cryptosepalum*). Although *Humularia* is perhaps not so difficult as *Cryptosepalum*, still the division into species is very unsatisfactory. Duvigneaud in his monograph has had to make use of minor characters such as the number of leaflets in order to classify them at all. The problem is essentially a field one and must await further evidence. Whether (as has proved to be the solution in *Cryptosepalum*) a very drastic reduction of the number of taxa recognised is advisable will then perhaps be evident. I have followed the broad outlines of Duvigneaud's classification which was based on a careful consideration of the geographical distribution of the usable characters, but have felt it imperative to reduce the rank of some of his taxa and one of my own.

1. Bracts all completely divided to the base · · · · · · · · · · · · · · · · · · 12. *pseudaeschynomene*
 – Bracts divided for one-quarter to three-quarters of their length into two lobes, or sometimes the small basal ones only completely divided · 2
2. Stems prostrate; leaves 2–4-foliolate · 3
 – Stems ± erect · 4
3. Stipules slightly emarginate at the base · · · · · · · · · · · · · · · · · · · 1. *kapiriensis* var. *repens*
 – Stipules distinctly unequally biauriculate at the base · · · · · · · · · · · · · · · · · · · 9. *rosea*
4. Leaflets with main nerve internal, though sometimes close to the margin in *H. submarginalis*
 · 5
 – Leaflets with main nerve completely marginal · 12
5. Leaves 2(rarely 4)-foliolate; stipules conspicuously biauriculate at the base · · · · · · · · · 6

- Leaves usually with more than 2 leaflets, or if with only 2 then stipules not conspicuously biauriculate at the base · 7
6. Leaflets divided by the main nerve into two ± equal parts · · · · · · · · · · 5. *elisabethvilleana*
- Leaflets divided by the main nerve into two very unequal parts · · · · · · · 6. *drepanocephala*
7. Stipules not conspicuously biauriculate but often emarginate at the base · · · · · · · · · · · 8
- Stipules mostly conspicuously biauriculate at the base (but some small stipules need careful observation since the auricles are very small) · 11
8. Leaves mostly (2)4(rarely 6)-foliolate · 9
- Leaves 6–22-foliolate · 10
9. Stipules larger, 7–28 × 6–25 mm; stems typically glabrous · · · · 1. *kapiriensis* var. *kapiriensis*
- Stipules smaller, 5–22 × 5–15 mm; stems mostly pubescent · · · · · · · · · · · · · · 2. *bequaertii*
10. Bracts larger, 8–25 × 15–24 mm, glabrescent to hairy · · · · · · · · · · · · · · · · · 3. *welwitschii*
- Bracts smaller, 5–7.5 × 8–14 mm, glandular-hairy · 4. *kassneri*
11. Lamina of leaflet divided into two somewhat unequal parts of equal length · · 7. *apiculata*
- Lamina of leaflet divided into two very unequal parts, one much narrower than the other and also shorter, i.e. scarcely if at all reaching the apex · · · · · · · · · · · · · 8. *submarginalis*
12. Leaves 4–8-foliolate; leaflets obovate-oblong or oblong-falcate · · · · · · · · · · 10. *descampsii*
- Leaves 2-foliolate; leaflets obovate · 11. *minima*

Tab. 3.6.**35**. HUMULARIA. Leaflets (× 1). 1, H. ELISABETHVILLEANA, from *Bequaert* 275; 2, H. DREPANOCEPHALA, from *Ringoet* 488; 3, H. DREPANOCEPHALA var. HOMBLEI, from *Homblé* 355; 4, H. DESCAMPSII var. DECAMPSII, from *Quarré* 7667; 5. H. DESCAMPSII var. ABERCORNENSIS, from *Bullock* 2614. Drawn by M. Boutique (fig. 4 by P. Halliday). From Fl. Congo Belge. Reproduced with permission of Jardin Botanique National de Belgique.

1. **Humularia kapiriensis** (De Wild.) P.A. Duvign. in Bull. Soc. Roy. Bot. Belgique **86**: 180 (1954); in F.C.B. **5**: 308, pl. 24 (1954). —Verdcourt in Kirkia **9**: 476 (1974). —Lock, Leg. Afr. Check-list: 113 (1989). TAB. 3.6.**36**. Type from Dem. Rep. Congo.
   Geissaspis welwitschii var. *kapiriensis* De Wild. in Bull. Jard. Bot. État **4**: 124 (1914). —E.G. Baker, Legum. Trop. Africa: 314 (1929).

Erect or prostrate subshrub 0.6–2.1 m tall or long. Stems mostly much-branched, glabrous or densely covered with stiff tubercular-based hairs, often flattened. Leaves (2)4(6)-foliolate; leaflets 1–35 × 6–23 mm, elliptic, elliptic-oblong or obovate, obtuse or emarginate at the apex, often mucronulate, obliquely rounded at the base, glabrous or nearly so, entire, the main nerve oblique, dividing the leaf into unequal parts, the one about double the width of the other; venation prominent; up to 7 other basal nerves; petiole 6–30 mm long with similar indumentum to the stem; rhachis 0–2 cm long often prolonged as a bristle c. 2 mm long; stipules 7–28 × 6–25 mm, broadly elliptic, rounded at both ends or only slightly emarginate at the base, densely veined. Inflorescences mostly borne on shoots with leaves reduced to stipules, 4 cm long, sometimes branched; peduncle 5–12 mm long, glabrous or hairy; pedicels 0.5 mm long; bracts yellow-green, 14–18 × 2–24 mm, rounded, divided shortly or up to one-third their length into two rounded entire, glabrous or very sparsely ciliate venose lobes; bracteoles 3–6 × 1–3 mm, elliptic or oblong, glabrous or with a few cilia. Calyx lobes 10–11 × 3–4 mm, oblong-lanceolate, the upper entire, the lower curved, very shortly 3-fid. Standard yellow, 1–16 × 5–10 mm, panduriform or rectangular below with an enlarged rounded apex. Fruit of 1–2 articles; articles

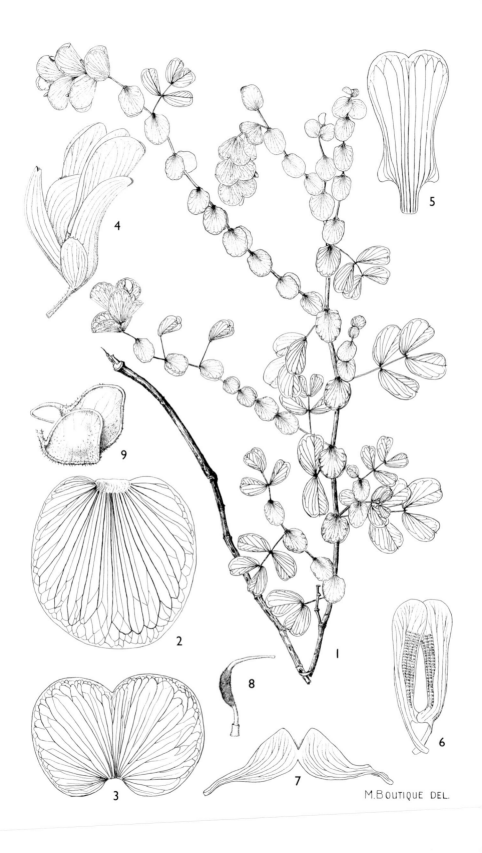

M. Boutique del.

three-quarters-elliptic, the upper margin straight, the lower strongly curved, 5–6 × 4–5.5 mm, pubescent with tubercular-based hairs, mostly bent back on each other. Seeds very dark red-brown, 3.2 × 2.8 × 1.5 mm, compressed ellipsoid, the minute round hilum very eccentric with seed somewhat beaked beyond it.

Var. **kapiriensis** —Verdcourt in Kirkia **9**: 477 (1974).

Stems erect, glabrous to glandular-pubescent or hairy. Inflorescences on leafy shoots or shoots with leaves reduced to stipules, not branched.

Zambia. W: River Matonchi, below dam, fr. 21.x.1937, *Milne-Redhead* 2881 (K).
Evergreen vegetation, swampy grassland; 1500 m.
Milne-Redhead 2881 has been annotated by Duvigneaud as *H. kapiriensis* but he cited it as a probable hybrid with *H. bequaertii*. It is, I think, sufficiently close to the type to accept as the species itself. *Loveridge* 887 (Mwinilunga, Lisombo River, fl. 10.vi.1963 (K; LISC; SRGH)) differs in having the leaflets, stipules and bracts minutely closely glandular-denticulate.

Var. **repens** Verdc. in Kew Bull. **27**: 440 (1972); in Kirkia **9**: 477 (475 *reptans* in key sphalm.) (1974). Type: Zambia, 30 km west of Mwinilunga, *E.A. Robinson* 3652 (K, holotype).

Stems prostrate, densely covered with stiff tubercular-based short hairs. Inflorescences on shoots with leaves reduced to stipules, often rather extensive and branched.

Zambia. W: Mwinilunga, fr. 16.v.1969, *Mutimushi* 3447 (K).
Not known elsewhere. Dry sandy plateau grassland; c. 1500 m.

2. **Humularia bequaertii** (De Wild.) P.A. Duvign. in Bull. Soc. Roy. Bot. Belgique **86**: 183 (1954); in F.C.B. **5**: 310 (1954). —Verdcourt in Kirkia **9**: 477 (1974). —Lock, Leg. Afr. Check-list: 112 (1989). Type from Dem. Rep. Congo.
Geissaspis bequaertii De Wild. in Repert. Spec. Nov. Regni Veg. **11**: 523 (1913). —E.G. Baker, Legum. Trop. Africa: 313 (1929).

Subshrub 0.6–1.5 m tall. Stems erect, mostly much-branched, often flattened, purplish-brown, glabrous to very glandular with tubercular-based hairs or glandular-setulose particularly on younger stems. Leaves 4–6-foliolate; leaflets 7–35 × 3.5–21 mm, obovate or elliptic, obtuse but mucronulate at the apex, obliquely cuneate to rounded at the base, often pruinose, glabrous to glandular-pubescent, entire to distinctly denticulate-ciliate; the main nerve oblique with venation strongly reticulate especially beneath and with 3–6 other basal nerves; petiole 5–30 mm long and rhachis 5–20 mm long, both glabrous to densely pubescent or with a few tubercular-based hairs, the rhachis prolonged as a bristle 1–10 mm long; stipules often pruinose, 5–22 × 5–15 mm, oblong, elliptic, ovate or almost round, rounded at both ends or slightly subcordate at the base, densely veined with numerous approximate basal nerves; indumentum similar to the leaflets. Inflorescences often borne on branches where leaves are reduced to stipules, 1.5–4 cm long, densely strobilate; peduncle 5–15 mm long, glandular-pubescent; pedicels 2.5–3.5 mm long, similarly pubescent; bracts green or purplish then turning reddish-yellow, 0.7–2 × 1.4–3 cm, rounded, divided for one-fifth to half of their length into 2 rounded or elliptic lobes which are venose, sparsely to densely ciliate round the margins, glabrous or pubescent with tubercular-based hairs; the outermost bracts may be only 5 mm long and wide and divided almost to the base; bracteoles 3–5 × 1–2.5 mm, narrowly elliptic, ciliate. Calyx lobes 10–11 × 4.5–6 mm, oblong-elliptic, the

Tab. 3.6.**36**. HUMULARIA KAPIRIENSIS. 1, flowering branch (× ¹/₂), from *Homblé* 1231; 2, stipule (× 2), from *Quarré* 6144; 3, bract (× 2); 4, open flower (× 2); 5, standard, external face (× 2); 6, wings, external face (× 2); 7, keel, spread out, internal face (× 2); 8, gynoecium (× 2), 3–8 from *Homblé* 1231; 9, fruit (× 3), from *Gilbert* 12333. Drawn by M. Boutique. From Fl. Congo Belge. Reproduced with permission of Jardin Botanique National de Belgique.

upper entire or slightly 2-fid, the lower shortly but distinctly 3-fid. Standard yellow or orange, 9–14 × 4–7.5 mm, oblong with a slight median waist. Fruit of 1–2 articles; articles kettle-drum-shaped, the upper margin straight, the lower semi-elliptic, 5–6 mm long and wide, finely glandular-pubescent. Seeds chestnut-brown, 4 mm long and wide, 1.8 mm thick, obliquely rounded-cordate or round in outline; the small hilum distinctly eccentric.

Var. **bequaertii** —Verdcourt in Kirkia **9**: 478 (1974).

Bracts covered with short tubercular-based hairs.

Zambia. W: Solwezi, fl. 8.vi.1930, *Milne-Redhead* 453 (K).
Also in Dem. Rep. Congo. *Brachystegia* woodland; c. 1500 m.

Var. **purpureocoerulea** (P.A. Duvign.) Verdc. in Kew Bull. **27**: 440 (1972); in Kirkia **9**: 478 (1974). Type from Dem. Rep. Congo.
    Humularia purpureocoerulea P.A. Duvign. in Bull. Soc. Roy. Bot. Belgique **86**: 180 (1954); in F.C.B. **5**: 308 (1954).

Bracts glabrous or margins with scattered cilia.

Zambia. N: Mbala Distr., on road to Kambole, fl. 30.v.1936, *B.D. Burtt* 6122 (EA; K). W: Kasempa, fl. 19.viii.1961, *Fanshawe* 6686 (K; LISC).
Also in Dem. Rep. Congo and Tanzania. *Brachystegia, Protea, Julbernardia, Uapaca* woodland, also in grassy bushland and swampland; 1050–1660 m.
    The *B.D. Burtt* specimen cited above is not typical but according to Duvigneaud, who kept the taxa as separate species, the pubescent young stems indicate introgressive hybridization with *H. bequaertii*.
    There is certainly much variation in the indumentum. *Scott-Elliot* 8305 from "Stevenson Road" cited by Duvigneaud as *H. rubrofarinacea* is clearly the same as *B.D. Burtt* 6122.

3. **Humularia welwitschii** (Taub.) P.A. Duvign. in Bull. Soc. Roy. Bot. Belgique **86**: 182 (1954). —Torre in C.F.A. **3**: 210 (1966). —Verdcourt in Kirkia **9**: 479 (1974). —Lock, Leg. Afr. Check-list: 114 (1989). Type from Angola.
    Smithia welwitschii Taub. in Bot. Jahrb. Syst. **23**: 190 (1896).
    Damapana welwitschii (Taub.) Hiern, Cat. Afr. Pl. Welw. **1**: 238 (1896).
    Geissaspis welwitschii (Taub.) Baker f. in J. Bot. **46**: 113 (1908). —De Wildeman in Bull. Jard. Bot. État **4**: 124 (1914). —Harms in Engler, Pflanzenw. Afrikas [Veg. Erde 9] **3** (1): 118, fig. 300 (1915). —E.G. Baker, Legum. Trop. Africa: 314 (1929).

Erect shrub or subshrub 0.15–1 m tall. Stems mostly robust and woody, branched, subcylindrical, glabrous, pubescent or densely scabrous with tubercular-based hairs, viscid-glandular. Leaves 6–16(22)-foliolate; leaflets 9–45 × 5–30 mm, elliptic or elliptic-oblong, often oblique, obtuse to acute and mucronulate at the apex, obliquely unequally rounded at the base, glabrous or glandular-serrate; the main nerve oblique dividing the lamina into unequal parts, the one 1.5–2 times as broad as the other; venation strongly reticulate, there being 5 other basal nerves; petiole 10–15 mm long; rhachis 2.5–7 cm long, with similar indumentum to the stem, the rhachis prolonged into a bristle 3 mm long; stipules 6–20 × 4–15 mm, elliptic or elliptic-oblong, rounded at base and apex, glabrous to densely scabrid with tubercular-based hairs, densely veined with numerous approximate basal nerves. Inflorescences often borne on branches with leaves reduced to stipules, 4–6 cm long, mostly densely strobilate; peduncle 1–2 cm long, scabrid-pubescent; pedicels 5 mm long, pubescent; bracts yellow, 8–25 × 15–24 mm, rounded, divided for one-sixth to half of their length into two rounded lobes which are venose, glabrous to densely covered with tubercular-based hairs; bracteoles 4–6 × 1.5–2.5 mm, oblong to ovate-lanceolate, ciliate and with tubercular-based hairs. Calyx lobes 10–11 × 5 mm, elliptic-oblong, the upper shortly 2-fid, the lower shortly distinctly 3-fid. Standard yellow, 12–17 × 7–12 mm, rectangular with upper half broadly widened and rounded. Fruit of 1–2 articles, each article half-elliptic, the upper margin straight, the lower very strongly curved, 5–6 mm long and wide, glandular-pubescent. Seeds dark reddish-brown, 4 mm long and wide, 2 mm thick, irregularly oblong in outline; the small hilum eccentric.

Var. **welwitschii** —Verdcourt in Kirkia **9**: 480 (1974).

Stems, stipules, bracts, etc. mostly glabrous or sometimes pubescent or ciliate (stems often densely glandular-pubescent in the Flora Zambesiaca material).

Zambia. B: Mongu, fl. 29.ix.1962, *Fanshawe* 7055 (K). W: Kasempa, fl. 18.viii.1961, *Fanshawe* 6682 (K) (suffrutescent form).
Also in Angola. Floodplains and semi-dambo grassland, also woodland on Kalahari Sand; 1050 m.
Gilges 171 (Zambia B: Zambezi (Balovale), fl. 11.viii.1952 (K; PRE)) is cited by Duvigneaud as *H. megalophylla* but appears to be a suffrutescent form near var. *welwitschii*. All Flora Zambesiaca material has the stems more hairy than in typical var. *welwitschii*.

Var. **lundaensis** (P.A. Duvign.) Verdc. in Kew Bull. **27**: 440 (1972); in Kirkia **9**: 480 (1974). Type from Dem. Rep. Congo.
  Smithia megalophylla Harms in Bot. Jahrb. Syst. **26**: 292 (1899); in Warburg, Kunene-Samb.-Exped. Baum: 262 (1903); in Engler, Pflanzenw. Afrikas [Veg. Erde 9] **3** (1): 618 (1915). Type from Angola.
  Geissaspis megalophylla (Harms) Baker f. in J. Bot. **46**: 114 (1908); Legum. Trop. Africa: 314 (1929).
  Geissaspis castroi Baker f. in Bol. Soc. Brot., sér. 2, **8**: 106 (1933). Type from Angola.
  Humularia lundaensis P.A. Duvign. in Bull. Soc. Roy. Bot. Belgique **86**: 187 (1954); in F.C.B. **5**: 312 (1954). Type from Dem. Rep. Congo.
  Humularia megalophylla (Harms) P.A. Duvign. in Bull. Soc. Roy. Bot. Belgique **86**: 186 (1954). —Torre in C.F.A. **3**: 211 (1966).

Stems, stipules, bracts, etc. glandular-pubescent with tubercular-based hairs.

Zambia. W: 32 km north of Mwinilunga, fl. 29.v.1960, *Angus* 2313 (FHO; K; SRGH).
Also in Angola and Dem. Rep. Congo. *Cryptosepalum* woodland on sand, *Monotes* scrubland, grassy plains, edaphic grassland; 1500 m.
I have chosen the name *lundaensis* in preference to *megalophylla* since the indumentum varies considerably in Angolan material referred to that name. It does tend to have larger more numerous leaflets and to be more robust in which case some may prefer to keep it separate. Taking a broad view, species 1–3 could be considered variants of one variable species.

4. **Humularia kassneri** (De Wild.) P.A. Duvign. in Bull. Soc. Roy. Bot. Belgique **86**: 185 (1954); in F.C.B. **5**: 313 (1954). —Verdcourt in Kirkia **9**: 481 (1974). —Lock, Leg. Afr. Check-list: 113 (1989). Type from Dem. Rep. Congo.
  Geissaspis kassneri De Wild. in Bull. Jard. Bot. État **4**: 112 (1914). —E.G. Baker, Legum. Trop. Africa: 314 (1929).

Subshrub 0.8–0.9 m tall. Stems erect, densely covered with glandular hairs. Leaves 6–14(18)-foliolate; leaflets 7–18 × 5–9 mm, obliquely elliptic, obtuse to truncate at the apex, obliquely cuneate at the base, glabrous to hairy on upper surface, hairy beneath, the hairs appressed and tubercular-based, the margins ciliate with longer but similar hairs, the main nerve oblique, with 4–5 other main basal nerves; venation reticulate; petiole 5–12 mm long; rhachis 1–6 cm long, both with similar indumentum to the stems, the rhachis prolonged beyond the terminal leaflets as a bristle 1.5 mm long; stipules 5–10 × 4–7 mm, ovate, obtuse at the apex, rounded or subcordate at the base, hairy outside and glabrous inside, long ciliate, eventually deciduous. Inflorescences mostly on branches with reduced leaves, 10–15 mm long, densely to loosely strobilate, hairy; peduncle 5 mm long, hairy; pedicels 1.5 mm long; bracts 5–7.5 × 8–14 mm, rounded, divided for about one-third to half of their length into ovate lobes, pubescent on one or both sides with tubercular-based glandular hairs, ciliate-denticulate with similar hairs; bracteoles (2)4–6 × 0.5–1.5 mm, lanceolate, ciliate. Calyx lobes 8–9 mm long, the upper broadly lanceolate, shortly but distinctly 3-fid, the lower oblong-elliptic, ± entire. Standard orange, 10–12 × 4.5–8 mm, oblong with a very slight waist. Fruit of 1 article, 5 mm long and wide, glandular and densely hairy.

Var. **kassneri** —Verdcourt in Kirkia **9**: 481 (1974).

Leaflets in 3–6(7) pairs, glabrous above, hairy on main nerve beneath, denticulate-ciliate with tubercular-based hairs. Bracteoles 4–6 × 0.5–1 mm.

Zambia. N: Mporokoso Distr., Kawambwa, fl. 7.i.1960, *Richards* 12083 (K).
Also in Dem. Rep. Congo. Swamp edges in *Brachystegia* woodland; 1200 m.
Var. *kibaraensis* P.A. Duvign. has bracts glandular hairy on both faces and 5–9 pairs of leaflets;
var. *vanderystii* (DeWild.) P.A. Duvign. has finely ciliate leaflets and bracteoles under 2 mm long;
and var. *perpilosa* P.A. Duvign. has leaflets hairy on both faces, otherwise as in var. *vanderystii*.
These three varieties occur in Dem. Rep. Congo.

5. **Humularia elisabethvilleana** (De Wild.) P.A. Duvign. in Bull. Soc. Roy. Bot. Belgique **86**: 190,
   fig. 4/J (1954); in F.C.B. **5**: 316, fig. 21A (1954). —Verdcourt in F.T.E.A., Leguminosae,
   Pap.: 430 (1971); in Kirkia **9**: 481 (1974). —Lock, Leg. Afr. Check-list: 113 (1989). TAB.
   3.6.35, fig. 1. Type from Dem. Rep. Congo.
      Geissaspis elisabethvilleana De Wild. in Repert. Spec. Nov. Regni Veg. **11**: 523 (1913); in
   Ann. Mus. Congo, Sér. IV, Bot. [Études Fl. Katanga] **2** (1): 68 (1913).

Erect subshrub 1–2 m tall. Stems glabrous or with a very few scattered tubercular-
based hairs. Leaves 2-foliolate; leaflets subsessile, 10–31 × 8–25 mm, regularly or
somewhat obliquely obcordate, the main nerve dividing the blade into unequal or
subequal parts, the larger part at the most twice as broad as the narrower part,
cuneate, glabrous, entire; venation prominent and reticulate on both surfaces;
petiole 4–8 mm long; stipules 7–23 × 5–16 mm, elliptic or obovate-oblong, rounded
or obtuse at the apex, unequally auriculate at the base, leaf-like, entire, glabrous,
persistent. Inflorescences 10–20-flowered; peduncle 3–4 mm long, sparsely
pubescent with tubercular-based hairs; rhachis 2–5 cm long; pedicels 2 mm long;
bracts green or purplish, 10–14 × 11–19 mm, split for c. one-third of their length into
2 obtuse or rounded and apiculate entire or sparsely ciliolate-denticulate lobes,
glabrous; bracteoles 2–3 × 0.5–1 mm, lanceolate, entire, glabrous. Calyx 7 mm long,
glabrous. Standard orange, 8–9 × 3–4 mm, panduriform. Fruit of 1 article, 5 mm
long and wide, glabrous. Seeds not seen.

Zambia. W: 56 km north of Kasempa on road to Solwezi, fl. 22.iii.1961, *Drummond &*
Rutherford-Smith 7192 (K; LISC; PRE; SRGH).
Also in Burundi, Dem. Rep. Congo and Tanzania. *Brachystegia* woodland; 1350–1500 m.

6. **Humularia drepanocephala** (Baker) P.A. Duvign. in Bull. Soc. Roy. Bot. Belgique **86**: 190, fig.
   4/G, H (1954); in F.C.B. **5**: 317, fig. 21B (1954). —Verdcourt in F.T.E.A., Leguminosae,
   Pap.: 431 (1971); in Kirkia **9**: 482 (1974). —Lock, Leg. Afr. Check-list: 112 (1989). TAB.
   3.6.35, fig. 2. Syntypes: Malawi, Nyika Plateau, 6000–7000 ft., *Whyte* (K); between Mpata
   and the commencement of the Tanganyika Plateau, 2000–3000 ft., July 1896, *Whyte* (K).
      Geissaspis drepanocephala Baker in Bull. Misc. Inform., Kew **1897**: 260 (1897). —E.G.
   Baker, Legum. Trop. Africa: 313 (1929). —Brenan, Check-list For. Trees Shrubs Tang.
   Terr.: 425 (1949); in Mem. New York Bot. Gard. **8**: 255 (1953).

Erect herb or subshrub 20–40 cm tall. Stems tufted, woody, ± densely pubescent
with hairs which are not or scarcely bulbous-based, but glabrescent towards the base;
rootstock massive, woody. Leaves 2-foliolate (rarely 4-foliolate); leaflets subsessile,
7–27 × 7–18 mm, obliquely obcordate, the main nerve dividing the blade into very
unequal parts, the smaller part two-thirds to the same height as and one-twelfth to
two-thirds the width of the larger part, cuneate, glabrous, entire or sparsely and very
finely denticulate, sometimes only at the apex; petiole 4–20 mm long; stipules 9–28 ×
6–20 mm, elliptic, obovate or deltoid, obtuse at the apex, unequally auriculate at the
base, leaf-like, entire or finely ciliate-denticulate, glabrous, persistent. Inflorescences
8–15-flowered; peduncle 2–7 mm long, densely hairy; rhachis 3–6 cm long; pedicels
3–4 mm long; bracts yellow-green, 11–17 × 10–18 mm, split for one-seventh to one-half
of their length into 2 obtuse to acuminate, apiculate, entire or denticulate lobes,
glabrous; bracteoles 2–3 × 0.5–1 mm, lanceolate, entire, glabrous. Calyx 7–9 mm
long, glabrous. Standard cream to orange-yellow, 9–11 × 4–4.5 mm, panduriform;
wings pale yellow. Fruit of 1 article, 5–6.5 × 4.5–5 mm, semicircular, glabrous.

1. Leaflets distinctly divided into 2 unequal parts by the midrib and separated by an
   emargination; narrower lobe one-quarter to one-third the width of the larger lobe · · · · ·
   · i) var. *drepanocephala*
   – Leaflets with narrow part very reduced, the midrib almost submarginal; emargination
   scarcely evident; narrower part one-twelfth to one-fifth the width of the larger part · · · · 2

2. Bracts denticulate · iii) var. *forcipiformis*
– Bracts not denticulate · 3
3. Leaflets and stipules not denticulate · · · · · · · · · · · · · · · · · ii) var. *homblei* forma *homblei*
– Leaflets and stipules denticulate · · · · · · · · · · · · · · · · · ii) var. *homblei* forma *denticulata*

i) Var. **drepanocephala** —Verdcourt in F.T.E.A., Leguminosae, Pap.: 433, fig. 61/1–12 (1971); in Kirkia **9**: 483 (1974). TAB. 3.6.**37**.

Leaflets mostly 15–25 mm long, entire or finely denticulate at the apex, the narrowest lobe c. one-third the width of the larger lobe and much shorter, separated by a mostly shallow emargination; petiole c. 5–20 mm long; stipules entire. Bracts divided for one-sixth to one-quarter of their length into 2 obtuse to acute lobes, entire.

　　Zambia. N: Mbala Distr., Ndundu, fl. 23.i.1960, *Richards* 12446 (K; SRGH). W: Ndola, fl. 29.ii.1954, *Fanshawe* 725 (K; SRGH). **Malawi**. N: 33 km WSW of Karonga, near Kayelekera, fl. 8.vi.1989, *Brummitt* 18434 (K).
　　Also in Dem. Rep. Congo and Tanzania. Grassland with scattered *Uapaca*, *Brachystegia* woodland; 1300–1870 m.

ii) Var. **homblei** (De Wild.) P.A. Duvign. in Bull. Soc. Roy. Bot. Belgique **86**: 191 (1954); in F.C.B. **5**: 318, fig. 21C (1954). —Verdcourt in Kirkia **9**: 483 (1974). TAB. 3.6.**35**, fig. 3. Type from Dem. Rep. Congo.
　　Geissaspis homblei De Wild. in Repert. Spec. Nov. Regni Veg. **11**: 522 (1913); in Ann. Mus. Congo, Sér. IV, Bot. [Études Fl. Katanga] **2** (1): 69 (1913). —E.G. Baker, Legum. Trop. Africa: 313 (1929).

Leaflets mostly 14–22 mm long, entire or distinctly denticulate, the narrowest lobe very reduced, scarsely one-tenth to one-fifth the width of the larger lobe and much shorter, passing gradually into it, not emarginate; petiole 5–16 mm long; stipules entire. Bracts divided for c. one-third of their length into 2 obtuse or subacute lobes, entire.

Forma **homblei** P.A. Duvign. in Bull. Soc. Roy. Bot. Belgique **86**: 192 (1954). —Verdcourt in Kirkia **9**: 484 (1974).

Stipules and leaflets entire.

　　Zambia. W: Mufulira, fl. 17.iv.1941, *Cruse* 256 (K). C: Serenje Distr., near Kanona, south of Mulembo R. (Vale of Mlembo), fl. 31.v.1961, *Symoens* 9498 (K). **Malawi**. N: Nyika Road, km 32, fl. iii.1953, *Chapman* 122 (BM).
　　Also in Dem. Rep. Congo and Tanzania. Plateau woodland, grassland; 1260–1800 m.
　　There are numerous intermediates with var. *drepanocephala*.

Forma **denticulata** P.A. Duvign. in Bull. Soc. Roy. Bot. Belgique **86**: 192 (1954); in F.C.B. **5**: 318 (1954). —Verdcourt in Kirkia **9**: 484 (1974). Type from Dem. Rep. Congo.

Stipules and leaflets denticulate.

　　Zambia. W: Kitwe, fl. 10.iv.1967, *Fanshawe* 10023 (K).
　　Also in Dem. Rep. Congo and Tanzania. *Brachystegia* woodland; 1200 m.

iii) Var. **forcipiformis** P.A. Duvign. in Bull. Soc. Roy. Bot. Belgique **86**: 192 (1954); in F.C.B. **5**: 318 (1954). —Verdcourt in Kirkia **9**: 484 (1974). Type from Dem. Rep. Congo.

Leaflets mostly 13–22 mm long, entire or finely denticulate, the narrower lobe one-twelfth to one-fifth of the width of the larger lobe, sometimes practically obsolete, mostly distinctly shorter and running into it without or with only a faint emargination; petiole 6–8 mm long; stipules denticulate. Bracts divided for one-seventh to half their length into 2 obtuse or acuminate lobes, denticulate with tubercular-based hairs.

　　Zambia. N: Great North Road, 46.4 km south of Shiwa Ngandu turnoff, fl. & fr. 29.iii.1961, *Angus* 2577 (FHO; K). W: Kitwe, fl. 28.ii.1959, *Mutimushi* 31 (K). E: recorded from Lundazi but

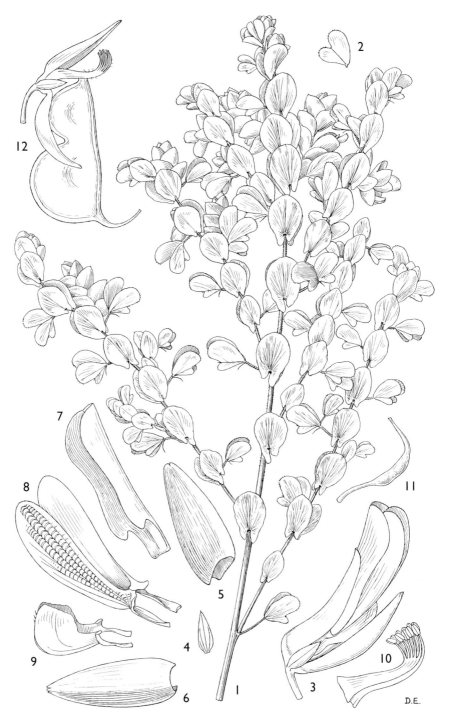

Tab. 3.6.**37**. HUMULARIA DREPANOCEPHALA var. DREPANOCEPHALA.   1, flowering branch ($\times\,^2/_3$); 2, leaflet ($\times$ 1); 3, flower ($\times$ 4); 4, bracteole ($\times$ 4); 5, upper calyx lip ($\times$ 4); 6, lower calyx lip ($\times$ 4); 7, standard ($\times$ 4); 8, wings ($\times$ 4); 9, keel ($\times$ 4); 10, androecium ($\times$ 4); 11, gynoecium ($\times$ 4), 1–11 from *Richards* 11875; 12, fruit ($\times$ 4), from *Milne-Redhead & Taylor* 9125A. Drawn by Derek Erasmus. From F.T.E.A.

no specimen retained. **Malawi**. N: Chitipa Distr., Misuku Hills, 19 km from Mugesse (Mughesse), fl. 7.iv.1969, *Pawek* 1964 (K). C: Ntchisi (Nchisi) Mt., fl. 20.ii.1959, *Robson & Steele* 1686 (BM; K; LISC; SRGH). Also in Dem. Rep. Congo and southern Tanzania. *Brachystegia* woodland and also near streams; 1200–1450 m.

7. **Humularia apiculata** (De Wild.) P.A. Duvign. in Bull. Soc. Roy. Bot. Belgique **86**: 195 (1954); in F.C.B. **5**: 324 (1954). —Brummitt in Wye Coll. Malawi Proj. Rep.: 65 (1973). — Verdcourt in Kirkia **9**: 484 (1974). —Lock, Leg. Afr. Check-list: 112 (1989). Type from the Dem. Rep. Congo.
  Geissaspis apiculata De Wild. in Bull. Jard. Bot. État **4**: 104 (1914). —E.G. Baker, Legum. Trop. Africa: 316 (1929). —Hutchinson, Botanist South. Africa: 503 (1946).
  Geissaspis bakeriana De Wild. in Bull. Jard. Bot. État **4**: 105 (1914). —E.G. Baker, Legum. Trop. Africa: 315 (1929). Type: Malawi, between Mpata and the commencement of the Tanganyika Plateau, *Whyte* (K, holotype).
  Geissaspis luentensis De Wild. in Bull. Jard. Bot. État **4**: 116 (1914). —E.G. Baker, Legum. Trop. Africa: 315 (1929). Type from the Dem. Rep. Congo.
  Humularia luentensis (De Wild.) P.A. Duvign. in Bull. Soc. Roy. Bot. Belgique **86**: 195 (1954); in F.C.B. **5**: 224 (1954).
  Humularia bakeriana (De Wild.) P.A. Duvign. in Bull. Soc. Roy. Bot. Belgique **86**: 193 (1954).
  Humularia katangensis var. *glabrescens* P.A. Duvign. in Bull. Soc. Roy. Bot. Belgique **86**: 199 (1954). Type: Zambia, Mbala Distr., hills near Mwambeshi R, *B.D. Burtt* 6324 (EA; K, holotype).

Erect subshrub 0.3–2 m tall. Stems usually many from a woody base, branched, glabrous or scabrid with tubercular-based hairs. Leaves 4–6(8)-foliolate; leaflets 6–23 × 35–12 mm, obovate, obovate-elliptic or narrowly obovate-oblong, truncate, emarginate or rounded at the apex, apiculate, obliquely rounded at the base, margins sparsely ciliate-denticulate to distinctly denticulate or in some variants quite smooth and glabrous, sometimes glaucescent; main nerve distinctly lateral, dividing the lamina into two unequal parts, one 1.5–3 times wider than the other and often of unequal length; other basal nerves 2–4; venation reticulate; petiole 5–9 mm long, glabrous or with tubercular-based hairs; rhachis 5–10 mm long, glabrous or similarly hairy, ending in a bristle 1.5–5 mm long; stipules mostly yellow with pinkish-white blotch, leaf-like or submembranous, 3–14 × 2–6 mm, lanceolate, broadly elliptic or elliptic-oblong, acuminate or apiculate at the apex, equally to unequally distinctly or obscurely biauriculate at the base, entire to closely ciliate-denticulate on the margins, mostly glabrous, usually deciduous. Inflorescences borne on leafy shoots, densely strobilate, 8–16-flowered, 2–6 cm long; peduncle 5–7.5 mm long, pedicels 3 mm long; bracts pale green and pink, often glaucous, 7–15 × 9–20 mm, divided into 2 lobes for one-third to half their length (or in outer ones almost their entire length); lobes ovate, obtuse or acuminate, entirely to closely denticulate, glabrous or ciliate; bracteoles 2–3 × 0.5–1 mm, lanceolate, ciliate-denticulate. Calyx lobes 7 × 2.5–3 mm, ovate-oblong, the upper one entire, the lower shortly 3-fid. Standard yellow or orange, 8–9 × 3–3.5 mm, panduriform. Fruit of 1 kettle-drum-shaped article, 4–5 mm long and wide, the upper margin straight, the lower margin very rounded. Seeds very dark blackish-purple, 3.8 × 3.5 × 1.3 mm, obliquely reniform, pitted; hilum small eccentric.

Zambia. N: road to Isoka Village from Mbala road, fl. 31.i.1955, *Richards* 4304 (K). W: Chingola, fl. 13.xi.1964, *van Rensburg* 2985 (K; SRGH). C: 35.2 km NE of Serenje Corner, fr. 16.vii.1930, *Hutchinson & Gillett* 3719 (BM; K). **Malawi**. N: Chitipa Distr., 8 km northwest of Mzengapakweru (Muzengapakweru), fr. 12.ix.1972, *Synge* WC452 (K). Also in Dem. Rep. Congo. *Brachystegia* woodland and montane grassland; 900–2100 m (Whyte's generalised label gives 2000–3000 ft. but this seems very low). I have taken a broad view of this species. The material now available scarcely allows Duvigneaud's scheme to be maintained.

8. **Humularia submarginalis** Verdc. in Kew Bull. **27**: 440 (1972); in Kirkia **9**: 486 (1974). —Lock, Leg. Afr. Check-list: 114 (1989). TAB. 3.6.**38**. Type: Zambia, Mansa (Fort Rosebery), *Fanshawe* 8545 (K, holotype).

Shrub 45–90 cm tall. Stems at first scabrid with short tubercular-based hairs which

Tab. 3.6.**38**. HUMULARIA SUBMARGINALIS. 1, flowering branch (× 1); 2, inflorescence with
subtending leaf (× 1⅓); 3, flower (× 5); 4, standard (× 6); 5, wing (× 6); 6, keel (× 6);
7, androecium (× 6); 8, gynoecium (× 6); 9, fruit (× 4), 1–9 from *Fanshawe* 8545. Drawn by
Victoria Friis. From Kew Bull.

wear off on the older stems. Leaves 4–6-foliolate; leaflets 9–23 × 3–10 mm, narrowly elliptic or oblanceolate, unequal and apiculate at the apex, very obliquely rounded at the base, margins ciliate-denticulate and also often with some short hairs on the main and secondary nerves; main nerve very distinctly lateral to practically marginal, dividing the blade into two very unequal parts, the narrower being considerably shorter to almost the same length as the wider, usually rather thick in texture; other basal nerves 3–5; venation reticulate beneath; petiole 4–8 mm long; rhachis 3–15 mm long, ending in a distinct bristle, all with similar indumentum to the stem; stipules leaf-like, 7–14 × 5–8 mm, ovate, acute to shortly acuminate at the apex, unequally biauriculate at the base, ciliate-denticulate, mostly with some scattered hairs, ± persistent, densely venose.   Inflorescences on leafy shoots, densely strobilate, sometimes branched, 1.5–4 cm long; peduncle 5–10 mm long; pedicels 3–4 mm long; bracts turning reddish, 7–13 × 8–22 mm, ovate, divided into two lobes for about one-third of their length; lobes ovate, acuminate, ciliate-denticulate and mostly pubescent; bracteoles 3 × 1 mm, lanceolate, sparsely ciliate. Calyx lobes glabrous, 8–9 × 2–4 mm, elliptic-oblong, the upper almost entire, the lower shortly 3-fid. Standard yellow, 8 × 4 mm, oblong with a slight waist. Fruit of 1 kettle-drum-shaped article, 7.5 × 6 mm, the upper margin straight, the lower margin very rounded. Ripe seeds not seen.

Zambia. N: near Luwingu, south of the road to Mansa (Fort Rosebery), fl. 3.vi.1962, *Symoens* 9543 (K).
Wooded grassland and *Marquesia–Brachystegia* woodland; 1300 m.
This has been misidentified as *H. ciliato-denticulata* (De Wild.) P.A. Duvign.

9. **Humularia rosea** (De Wild.) P.A. Duvign. in Bull. Soc. Roy. Bot. Belgique **86**: 196 (1954); in F.C.B. **5**: 322 (1954). —Verdcourt in Kirkia **9**: 487 (1974). —Lock, Leg. Afr. Check-list: 114 (1989). Type from Dem. Rep. Congo.
  Geissaspis rosea De Wild. in Repert. Spec. Nov. Regni Veg. **11**: 524 (1913). —E.G. Baker, Legum. Trop. Africa: 315 (1929).
  Geissaspis rosea var. *divergentiloba* De Wild., Pl. Bequaert. **4**: 56 (1926). —E.G. Baker, Legum. Trop. Africa: 315 (1929). Type from Dem. Rep. Congo.
  Geissaspis incognita De Wild. in Repert. Spec. Nov. Regni Veg. **11**: 524 (1913). —E.G. Baker, Legum. Trop. Africa: 316 (1929). Type from Dem. Rep. Congo.
  Geissaspis incognita var. *latifoliolata* De Wild., Pl. Bequaert. **4**: 50 (1926). —E.G. Baker, Legum. Trop. Africa: 316 (1929). Type from Dem. Rep. Congo.
  Geissaspis subscabra De Wild. in Bull. Jard. Bot. État **4**: 123 (1914). —E.G. Baker, Legum. Trop. Africa: 317 (1929). Type from Dem. Rep. Congo.
  Geissaspis robynsii De Wild., Pl. Bequaert. **4**: 54 (1926). —E.G. Baker, Legum. Trop. Africa: 316 (1929). Type from Dem. Rep. Congo.

Subshrub with slender rampant stems 0.5–1 m long, mostly purplish-brown, pubescent bifariously or all round with tubercular-based hairs but scarcely scabrid. Leaves 2–4-foliolate; leaflets 15–35 × 8–18 mm, obovate-elliptic, obliquely emarginate at the apex, obliquely truncate at the base, glabrous, entire or obscurely ciliate; main nerve distinctly lateral with 3–4 other basal nerves; venation reticulate beneath; petiole 6–13 mm long, rhachis 5–10 mm long, both similarly hairy to the stem; stipules 6–18 × 2.5–9 mm, ovate or elliptic, rounded to acuminate at the apex, unequally biauriculate at the base, entire, glabrous or ciliate, denticulate, persistent, densely veined. Inflorescence often on leafless shoots, densely strobilate, 1.5–3 cm long; peduncle 1–15 mm long, pubescent; pedicels 3 mm long; bracts green or yellow, often turning reddish, 7–15 × 8–15 mm, rounded, divided into two lobes for one-quarter to half of their length, or outer ones for their entire length; lobes ovate, obtuse, glabrous, ciliate-denticulate; bracteoles 1.6–2.5 × 0.8–1 mm, lanceolate, sparsely ciliate. Calyx lobes glabrous, 7–8 × 3 mm, elliptic-oblong, the upper 2-fid, the lower 3-fid. Standard yellow or orange, 6–10 × 5–8.5 mm, panduriform. Fruit of 1 kettle-drum-shaped article, 5.5 mm long and wide, the upper margin straight, the lower margin very rounded. Seeds chestnut-brown, compressed, 3.2–3.5 mm long and wide, 1.5 mm thick, almost round in outline; hilum small, eccentric.

Var. **rosea** —Verdcourt in Kirkia **9**: 488 (1974).
  Humularia rosea var. *denticulata* P.A. Duvign. in Bull. Soc. Roy. Bot. Belgique **86**: 199 (1954). Type: Zambia, Mbala Distr., between Malombe and Mululwe rivers, dambo on way to Mungomba, fl. 4.vi.1936, *B.D. Burtt* 6436.

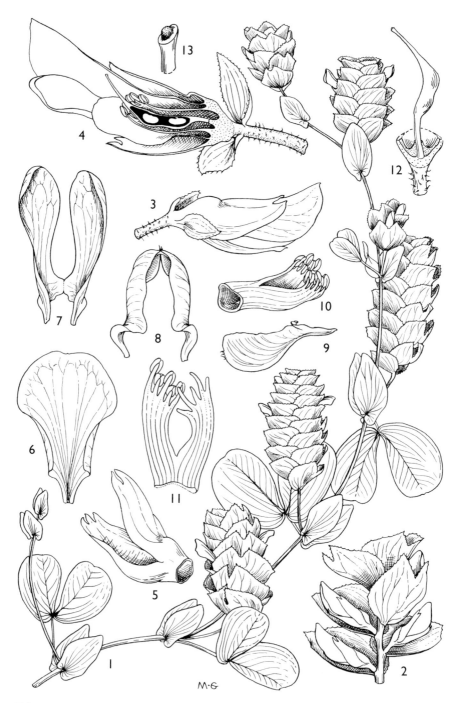

Tab. 3.6.**39**. HUMULARIA ROSEA var. REPTANS. 1, portion of stem with flowering shoots (× 1); 2, part of inflorescence with bracts partially removed (× 2); 3, flower (× 4); 4, flower, longitudinal section (× 5); 5, calyx (× 4); 6, standard (× 4); 7, wings (× 4); 8, keel (× 4); 9, keel petal, side view, flattened (× 4); 10, androecium (× 5); 11, androecium, cut open and flattened (× 5); 12, gynoecium (× 5); 13, style apex and stigma (× 26), 1–13 from *Richards* 21432. Drawn by Mary Grierson. From Kew Bull.

Leaves 4-foliolate.

Zambia. N: Mbala Distr., Kambole waterfall, fl. 30.i.1964, *Richards* 18907 (K). W: Kitwe, fl. 12.v.1955, *Fanshawe* 2276 (K).
Also in Dem. Rep. Congo. *Brachystegia* woodland, woodland/dambo edges; 1350–1800 m.
Var. *denticulata* P.A. Duvign. has the stipules and some leaflets rather obscurely ciliate-denticulate. It and *Richards* 18907, cited above, have rather shorter broader leaflets than the material from the Zambian Copperbelt.

Var. **reptans** (Verdc.) Verdc. in Kew Bull. **27**: 441 (1972); in Kirkia **9**: 488 (1974). TAB. 3.6.**39**.
    Type from Tanzania.
        Humularia reptans Verdc. in Kew Bull. **24**: 54, fig. 6 (1970); in F.T.E.A., Leguminosae, Pap.: 430 (1971).

Leaves 2-foliolate.

Zambia. N: Kalambo Falls, fl. 16.iv.1966, *Richards* 21432 (K).
Also in southern Tanzania. *Brachystegia* woodland; 1350 m.

10. **Humularia descampsii** (De Wild. & T. Durand) P.A. Duvign. in Bull. Soc. Roy. Bot. Belgique
    86: 200, fig. 4/K, 5/D (1954); in F.C.B. **5**: 328, pl. 25, fig. 21E (1954). —Verdcourt in Kirkia
    9: 488 (1974). —Lock, Leg. Afr. Check-list: 112 (1989). TAB. 3.6.**40**. Type from Dem. Rep.
    Congo.
        Geissaspis descampsii De Wild. & T. Durand in Bull. Soc. Roy. Bot. Belgique **39**: 65 (1900).
    —E.G. Baker, Legum. Trop. Africa: 317 (1929). —Brenan in Mem. New York Bot. Gard. **8**:
    255 (1953).

Subshrub 0.4–1.5 m tall. Stems erect, robust, woody, sparsely to densely covered with stiff tubercular-based hairs, scabrid, mostly dark brown. Leaves 4(8)-foliolate; leaflets 3–30 × 2–8 mm, obovate-oblong or oblong-falcate, obliquely rounded at both ends, glabrous or pilose, finely denticulate on the margins; main nerve completely marginal, ending in a small point not far below the apex of the leaflet; 4–6 additional nerves radiating from the base; venation strongly reticulate; petiole 5–10 mm long and rhachis 2–9 mm long, both scabrid-pubescent like the stems, the rhachis prolonged beyond the ultimate pair of leaflets for 2.5–7 mm; petiolules very short; stipules 5–27 × 4–15 mm, rounded to ovate, leaf-like, obtuse to acute at the apex, very unequally bluntly bilobed at the base, densely veined with numerous approximate basal nerves, glabrous or pilose, denticulate, persistent. Inflorescences borne on leafy shoots, densely strobilate, 2–5 cm long; bracts greenish-yellow, becoming brown, 10–18 × 10–15(23) mm, rounded, divided into two lobes for half to two-thirds of their length; lobes ovate, acute or acuminate, denticulate, glabrous or pilose; bracteoles 2–5 × 1–3 mm, lanceolate, ciliate; peduncle 1–3(5) mm long, scabrid-pubescent; pedicels 1 mm long. Calyx lobes sparsely pubescent or glabrescent, 7–10 × 3 mm, oblong-lanceolate, the upper entire, the lower 3-fid. Standard yellow or orange, 8–9 × 3–4 mm, oblong with a slight waist. Fruit of 1 rounded kettle-drum-shaped article, 4–5.5 × 4–4.5 mm, the upper margin ± straight, the lower margin strongly curved, glabrous, nervose. Seeds dark crimson-brown, 3 × 2.5 × 1.5 mm, reniform; the small hilum eccentric.

Var. **descampsii** —Duvigneaud in Bull. Soc. Roy. Bot. Belgique **86**: 201 (1954); in F.C.B. **5**: 328
    (1954). —Verdcourt in Kirkia **9**: 489 (1974). TAB. 3.6.**35**, fig. 4.

Stems very densely scabrid. Leaflets 4(6), 10–15 × 4–6 mm, glabrous; stipules mostly ± obtuse, glabrous. Bracts rather small, 10–12 × 12–15 mm, with some hairs near the base.

Forma **pilosa** P.A. Duvign. in Bull. Soc. Roy. Bot. Belgique **86**: 201 (1954); in F.C.B. **5**: 329
    (1954). —Verdcourt in Kirkia **9**: 489 (1974). Type from Dem. Rep. Congo.

Differs from typical form in having leaflets, stipules, bracts, etc. all densely covered with rather short tubercular-based hairs.

Zambia. N: Mansa (Fort Rosebery), fl. 14.xi.1964, *Mutimushi* 1102 (K).

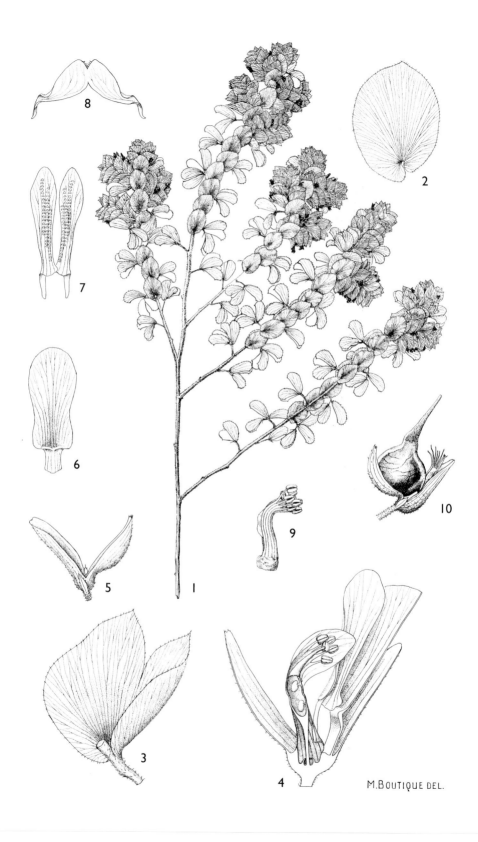

M.BOUTIQUE DEL.

Also in Dem. Rep. Congo. *Brachystegia* woodland.

Var. **abercornensis** P.A. Duvign. in Bull. Soc. Roy. Bot. Belgique **86**: 202 (1954). —Verdcourt in Kirkia **9**: 489 (1974). TAB. 3.6.**35**, fig. 5. Type: Zambia, Mbala, *Bullock* 2614 (EA; K, holotype).
Humularia descampsii var. *nyassica* P.A. Duvign. in Bull. Soc. Roy. Bot. Belgique **86**: 202 (1954). Type: Malawi, Mzimba to Kasungu, viii.1946, *Brass* 17387 (K, holotype; NY).
Geissaspis ?descampsii sensu Brenan in Mem. New York Bot. Gard. **8**: 255 (1953).

Stems densely scabrid. Leaflets 4–6, 12–26 × 4–8 mm, pubescent with tubercular-based hairs beneath, particulary when young; stipules similarly hairy. Bracts with lobes acuminate, mostly also similarly hairy.

Zambia. N: Mbala Distr., Uningi (Ningi) Pans, fl. 11.ii.1960, *Richards* 12452 (K; SRGH). W: Mwinilunga, fr. 10.vi.1974, *Chisumpa* 162 (K; NDO) (glabrous atypical form). **Malawi.** N: Mzimba Distr., Viphya (Vipya) Mts., fl. 28.iii.1954, *G. Jackson* 1284 (BM; K). C: 22.4 km north of Dwangwa (Dwanga) River, fl. 8.vi.1938, *Pole Evans & Erens* 638 (K; PRE).
Not known elsewhere. Grassland with scattered trees, *Brachystegia* woodland, sometimes in rocky places or on termite mounds; 480–1800 m.
The leaflets, stipules and bracts are often almost glabrous but the variety is well-defined. I am unable to separate var. *nyassica* P.A. Duvign. and probably var. *acuta* P.A. Duvign. described from the Dem. Rep. Congo should not be separated either.

11. **Humularia minima** (Hutch.) P.A. Duvign. in Bull. Soc. Roy. Bot. Belgique **86**: 204 (1954). —Verdcourt in Kirkia **9**: 490 (1974). —Lock, Leg. Afr. Check-list: 113 (1989). Type: Zambia, c. 12.8 km NW of Mbala (Abercorn), vii.1930, *Hutchinson & Gillett* 4006 (BM; K, holotype; LISC; SRGH).
Geissaspis minima Hutch., Botanist Southern Africa: 526 (1946).

Small woody subshrub 0.3–0.5 m tall with several diverging branched stems from a woody rootstock; young stems with a very few short stout bristles but otherwise glabrous or rather densely covered with very short tubercular-based hairs, purplish and usually with a whitish bloom. Leaves 2-foliolate; leaflets 4.5–13 × 3–10 mm, obovate or obovate-rhombic, broadly rounded at the apex, oblique at the base, glabrous, main nerve marginal and ending in a small point about half-way up the inner side of the leaflet; 3–4 additional nerves radiating from the base; venation strongly reticulate; petiole 3–10 mm long, prolonged between the leaflet pair as a bristle c. 1 mm long and with 4–5 short bristles similar to those on the stems; petiolules very short; stipules 2.5–10.5 × 1.5–9 mm, oblong-elliptic, very rounded at the apex, unequally bluntly bilobed at the base, densely veined, with numerous approximate basal nerves. Inflorescences densely strobilate, 1–4.5 cm long; peduncle 3–6 mm long; pedicels 2.5–3 mm long; bracts pale green tinged yellow, brown or red, 6–12 × 6–18 mm, obovate, round or rhombic, divided into two rounded lobes for one-third to three-quarters or sometimes their entire length; bracteoles then, 1.5–4 × 0.8–1.3 mm, narrowly ovate to lanceolate, often with a few tubercular-based short marginal cilia giving a faintly toothed appearance. Calyx lobes 5–7.5 × 3–3.5 mm, oblong or elliptic-ovate to ovate, the upper entire, the lower very shortly 3-fid. Standard bright yellow, apricot- or orange-coloured, 7–9 × 4–4.5 mm, violin-shaped. Fruit of 1 round or somewhat squarish article, 4–5 mm long and wide, the upper margin ± straight, lower strongly curved, slightly granular in texture. Seeds chestnut-brown, 3 × 2.5 × 1.8 mm, irregularly quadrangular or rounded-reniform, obscurely obtusely beaked at one end beyond the small round eccentric hilum.

Tab. 3.6.**40**. HUMULARIA DESCAMPSII. 1, flowering branch (× ¹/₂); 2, stipule (× 2); 3, bracts (× 3), 1–3 from *Dubois* 1369; 4, flower, longitudinal section (× 5), from *van den Brande* 31; 5, calyx and bracteoles (× 3); 6, standard, internal face (× 3); 7, wings, external face (× 3); 8, keel, spread out, internal face (× 3); 9, androecium and gynoecium (× 3), 5–9 from *Dubois* 1369; 10, fruit (× 3), from *van den Brande* 31. Drawn by M. Boutique. From Fl. Congo Belge. Reproduced with permission of Jardin Botanique National de Belgique.

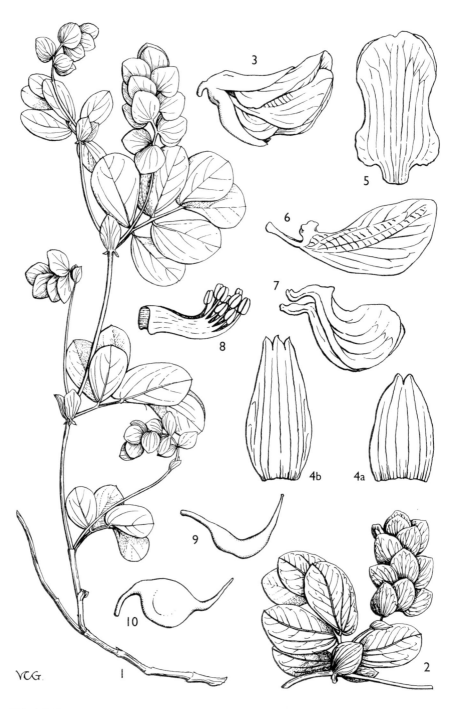

Tab. 3.6.**41**. HUMULARIA PSEUDAESCHYNOMENE. 1, stem (× 1); 2, inflorescence (× 1⅓); 3, flower (× 4); 4a, upper calyx lip, 4b, lower calyx lip (× 6); 5, standard (× 6); 6, wing (× 6); 7, keel (× 6); 8, androecium (× 6); 9, gynoecium (× 6); 10, fruit (× 4), 1–10 from *White* 3349. Drawn by Victoria Friis. From Kew Bull.

Subsp. **minima** —Verdcourt in Kirkia **9**: 491 (1974). —Lock, Leg. Afr. Check-list: 114 (1989).

Leaflets 4.5–9.5 × 3–7 mm; stipules 2.5–6 × 1.5–4 mm; bracts 6–12 × 6–13 mm; standard 7 mm long.

Zambia. N: Mbala Distr., between Isanya (Issanya) Estate and Wambeshi River, fl. 19.iv.1936, *B.D. Burtt* 6375 (EA; K).
Known only from the Mbala area being one of the remarkable series of endemics from there. *Brachystegia* bushland and open woodland, grassland and also open ground on sandy soil; 1680–1800 m.

Subsp. **flabelliformis** (P.A. Duvign.) Verdc. in Kew Bull. **27**: 441 (1972); in Kirkia **9**: 491 (1974).
—Lock, Leg. Afr. Check-list: 114 (1989). Type: Zambia, Mwinilunga Distr., just south of Kalalima R. (River Kamwezhi), *Milne-Redhead* 3798 (K, holotype).
Humularia flabelliformis P.A. Duvign. in Bull. Soc. Roy. Bot. Belgique **86**: 204, fig. 4/A–B (1954).

Leaflets 7–13 × 5–10 mm; stipules 3.5–10.5 × 3–9 mm; bracts 8–11 × 11–18 mm; standard 9 mm long.

Zambia. W: Mwinilunga Distr., c. 27.2 km south of Mwinilunga on the Kabompo road, fr. 6.vi.1963, *Loveridge* 838 (K; SRGH).
Not known elsewhere. *Brachystegia* woodland on Kalahari Sand; 1442 m.

12. **Humularia pseudaeschynomene** Verdc. in Kew Bull. **27**: 443 (1972); in Kirkia **9**: 491 (1974).
—Lock, Leg. Afr. Check-list: 114 (1989). TAB. 3.6.**41**. Type: Zambia, Mwinilunga Distr., north of Kalene Hill Mission, 24.ix.1952, *White* 3349 (FHO; K, holotype).

Rhizomatous suffrutex with creeping stems 15–40 cm long which are burnt back every year, glabrous or densely covered with rather short tubercular-based glandular hairs. Leaves (2)4–6-foliolate, glaucous; leaflets 12–27 × 7–18 mm, elliptic or elliptic-obovate, rounded to truncate and slightly mucronulate at the apex, cuneate or rounded at the base, the apical ones obliquely so, glabrous but sometimes with small yellow dots of resinous secretion which are readily detachable and perhaps not normal, entire; main nerve central or slightly eccentric but with 3–4 nerves to one side and only one on the other; venation distinctly raised and reticulate; petiole 5–10 mm long, glabrous or hairy with similar indumentum to that on the stems; petiolules c. 1 mm long; rhachis 7–14 mm long; stipules 5–9 × 3–6 mm, oblong-elliptic to obovate, obtuse at the apex, variable at the base on the same shoot, either completely unappendaged or with 2 distinct basal symmetrical appendages or only 1, mostly c. 2 mm long, glabrous, veined, entire. Inflorescences rather more loosely strobilate than in most species, 1.5–3 cm long, axillary or sometimes borne on axillary branches on which leaves are very reduced or represented only by stipules; peduncles 8–35 mm long, glabrous or hairy; pedicels 1.5 mm long; bracts divided completely to the base, lobes 5–8 × 4.5–7 mm, round or broadly elliptic, rounded at both ends, glabrous, entire with venation prominently reticulate; bracteoles not seen, ?early deciduous. Calyx lobes glabrous, 5.5 × 2.5 mm, elliptic, the upper entire, the lower distinctly 3-fid. Standard yellow, 7 × 4.5 mm, panduriform, emarginate. Ovary glabrous, 2-ovuled. Fruit not seen.

Zambia. W: Mwinilunga Distr., north of Kalene Hill Mission, fl. 24.ix.1952, *White* 3349 (FHO; K).
Not known elsewhere. Watershed grassland on Kalahari Sand, Congo–Zambezi divide; 1200–1500 m.

79. CYCLOCARPA Afzel. ex Baker

Cyclocarpa Afzel. ex Baker in F.T.A. **2**: 151 (1871). —Verdcourt in Kirkia **9**: 444 (1974).

Erect or spreading annual herbs, mostly branched and tufted. Leaves alternate, sensitive to touch, paripinnate; stipules membranous, spurred; stipels absent.

Inflorescences umbel-like racemes, 1–4-flowered; bracts and bracteoles small, thin. Flowers small, yellow. Calyx deciduous, 2-lipped; lips almost entire. Standard obovate, emarginate, shortly clawed; wings oblong, shortly clawed, finely denticulate above, marked with a few small pockets; keel obtuse, spurred, finely denticulate above. Stamens all joined, the tube split completely unilaterally or for half its length to form 2 bundles of 5, or with the vexillary filament free; anthers uniform. Ovary linear, falcate, subsessile, many-ovuled; style curved, inflexed; stigma small, terminal. Fruit many-jointed, linear, compressed, coiled into a ring or a spiral, minutely roughened on the margins; articles numerous, separating, the nerve persistent after they have fallen. Seeds broadly irregularly reniform; hilum minute, eccentric.

A monotypic genus widely distributed but very local in the tropics of the Old World.

Cyclocarpa stellaris Afzel. ex Baker in F.T.A. **2**: 151 (1871). —Léonard in F.C.B. **5**: 241 (1954). —Hepper in F.W.T.A., ed. 2, 1: 580 (1958); in Webbia **19**: 613, fig. 13 (1965). —Verdcourt in F.T.E.A., Leguminosae, Pap.: 406, fig. 57 (1971). —Drummond in Kirkia 8: 219 (1972). —Verdcourt in Kirkia **9**: 444 (1974). —Lock, Leg. Afr. Check-list: 111 (1989). TAB. 3.6.**42**. Type from Sierra Leone.

Glabrous herb 3.5–50 cm tall or long. Leaves 4–10-foliolate; leaflets 4–12 × 2–5 mm, oblong, elliptic or obovate-oblong, acute to rounded and mucronulate at the apex, obliquely rounded at the base, very finely serrulate; petiole and rhachis together 5–9 mm long; petiolules 0.5 mm long; stipules 3–10 × 0.5–1.5 mm, lanceolate, persistent. Inflorescences with a rhachis under 5 mm long; peduncle 1 mm long; pedicels 1.5–3 mm long; bracts 1–2.5 × 0.3–0.5 mm, lanceolate, acuminate, persistent; bracteoles 1.2 × 0.7 mm, ovate-lanceolate. Upper calyx lip 3–3.5 × 1.5–2

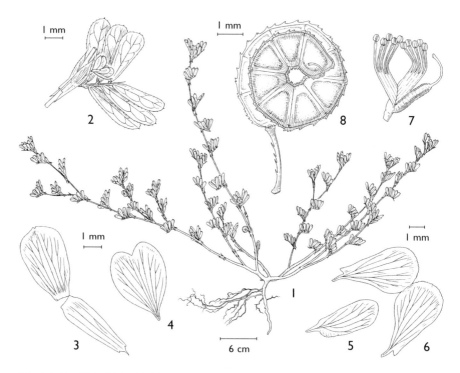

Tab. 3.6.**42**. CYCLOCARPA STELLARIS. 1, habit; 2, leaf with stipules and axillary shoot; 3, calyx; 4, standard; 5, wing; 6, keel petals; 7, androecium and gynoecium, 1–7 from *Jordan* 368; 8, fruit, from *Faulkner* 1266. From F.T.E.A.

mm, obovate-oblong, blunt; lower lip 3–4 × 1 mm, lanceolate, acuminate. Standard pale yellow, 3–4 × 1.5–2 mm. Fruit of 8–11 articles arranged in a spiral 4–5 mm in diameter, each article 1.5–2 × 1.5 mm, trapezoidal, with a rounded keeled outer margin, venose, dehiscent. Seeds olive- to dark brown, 1–1.2 × 1 × 0.25 mm.

Zambia. W: Kitwe, fl. & fr. 25.iii.1955, *Fanshawe* 2221 (K). S: Choma Distr., c. 4.8 km NE of Mapanza, fl. & fr. 10.iv.1955, *E.A. Robinson* 1227 (K; SRGH). **Zimbabwe**. S: Chivi Distr., Rundi (Lundi) River, near turn-off to Hippo Pools, fl. & fr. 4.v.1961, *Drummond & Rutherford-Smith* 7661 (K; LISC; PRE; SRGH). **Malawi**. S: Chiradzulu, fl. & fr. 21.x.1905, *Cameron* 99 (K). **Mozambique**. N: between Corrane and Nampula, fl. & fr. 11.iv.1937, *Torre* 1330 (COI; LISC).

Also from Guinea-Bissau to Gabon and Dem. Rep. Congo, also in Tanzania, and extending to Laos, Borneo, north Australia and Queensland. In damp places, cracks in rocks, seepage zones, pond edges, dambos; 210–1280 m.

80. ZORNIA J.F. Gmel.

Zornia J.F. Gmel., Syst. Nat. **2**: 1076, 1096 (1791). —Verdcourt in Kirkia **9**: 497 (1974).

Erect or prostrate, perennial or annual herbs. Leaves digitately 2- or 4-foliolate; leaflets mostly glandular-punctate; stipules well developed, produced below the point of insertion, mostly similarly punctate; stipels absent. Inflorescence spicate, terminal or axillary; bracts paired, stipule-like, mostly similarly punctate, ciliate, enclosing the flowers when in bud. Calyx hyaline, ciliate, persistent, 5-lobed, the upper pair joined and the laterals smaller than the other 3. Corolla medium-sized, mostly yellow or orange-yellow, often striated with purple; standard almost round, clawed; wings and keel clawed. Stamens monadelphous, united into a column below, which at maturity is circumscissile and the upper part is shed with the petals, the lower tubular part remaining to protect the young ovary; 5 small versatile anthers alternate with 5 larger sub-basifixed anthers. Ovary subsessile, with (2)5–8 ovules; style curved with minute terminal stigma. Fruit sessile, 2–15-jointed; the articles glabrous to pilose with or without stiff bristles, mostly prominently nerved, sometimes glandular. Seeds ovoid, compressed, mostly black or dark brown, without appendages.

1. Leaflets usually 2; plants annual or perennial ·2
 – Leaflets usually 4; plants perennial ·3
2. Plants normally annual; shoots erect or decumbent; upper leaflets lanceolate, acute, sparsely and obscurely glandular-punctate on lower surface, especially near the margin; bracts acute, without or with a few inconspicuous pellucid glands; flowers much shorter than the bracts; articles with bristles glochidiate and retrorsely hispid · · · · · 1. *glochidiata*
 – Plants normally perennial; shoots usually decumbent; bracts sparsely pellucid-punctate; articles with bristles bearing spreading hairs · 2. *pratensis*
3. Articles of fruits devoid of pubescent bristles, mostly glandular, glabrous or pubescent particularly on the margins; stems usually glabrous or glabrescent; leaflets mostly shorter and more elliptic · 4. *capensis*
 – Articles of fruits covered with glabrous or retrorsely pubescent bristles; stems usually hairy ·4
4. Leaflets up to 2.6 cm long, mostly obovate to broadly elliptic; bristles of articles 0.5–4 mm long · 3. *setosa*
 – Leaflets up to 4.6 cm long, predominantly narrowly elliptic to linear; bristles of articles c. 1 mm long · 5. *milneana*

1. **Zornia glochidiata** C. Rchb. ex DC., Prodr. **2**: 316 (1825). —Léonard & Milne-Redhead in F.C.B. **5**: 356 (1954). —Milne-Redhead in Bol. Soc. Brot., sér. 2, **28**: 87 (1954); in F.W.T.A., ed. 2, **1**: 575 (1958). —Mohlenbrock in Webbia **16**: 108 (1961). —Torre in C.F.A. **3**: 215 (1966). —Schreiber in Merxmüller, Prodr. Fl. SW. Afrika, fam. 60: 125 (1970). —Milne-Redhead in F.T.E.A., Leguminosae, Pap.: 444 (1971). —Drummond in Kirkia **8**: 229 (1972). —Verdcourt in Kirkia **9**: 498 (1974). —Gonçalves in Garcia de Orta, Sér. Bot. **5**: 123 (1982). —Lock, Leg. Afr. Check-list: 122 (1989). Type from Senegal.
 Zornia diphylla sensu auct., e.g. Baker in F.T.A. **2**: 158 (1871) pro major parte, Robyns, Fl. Sperm. Parc Nat. Alb. **1**: 325 (1948) pro parte non (L.) Pers.

Erect or decumbent annual herb (rarely perennial) (4)10–70 cm tall. Stems sometimes slightly woody at the base, glabrous, glabrescent or puberulous. Leaves 2-foliolate; upper leaflets 6–45 × 2–14 mm, lanceolate to ovate-lanceolate, acute and mucronulate at the apex, rounded at the base, glabrescent or pubescent beneath and sparsely glandular-punctate; lower leaflets mostly relatively broader; petiole 4–22 mm long; petiolules c. 1 mm long; stipules 5–18 mm long including the 1–7 mm long spur, 1–3 mm wide, narrowly lanceolate or lanceolate, glabrous and very sparsely punctate. Inflorescences 4–20 cm long; peduncle 1–5 cm long; bracts 5–13 × 2–7 mm, ovate or elliptic, acute, ciliate but otherwise glabrous, with no or very few glands. Calyx tube 1.5 mm long; lobes 0.5–1.5 mm long, glabrous. Standard white, yellow or orange, veined with red or pink, or mauve or crimson often paler within and yellow at base, 4.5–6 × 3–4.5 mm; wings yellow, pink or crimson; keel mostly greenish, sometimes orange to vinaceous at the apex. Fruits 11–17 mm long of 3–5 articles; each article 2–3 mm long and broad, not glandular but covered with numerous barbellate spreading bristles 0.8–2.5 mm long. Seeds brown, 1.5 × 1.1 × 0.5 mm, compressed reniform; hilum small, eccentric, the rim raised at each end.

Botswana. N: Ngamiland Distr., Okavango, near Tsao (Tsau), fl. & fr. 18.iii.1961, *Richards* 14791 (K). SW: 16 km NE of Mamuno, Olifants Kloof, 13.ii.1970, *R.C. Brown* 35 (K; SRGH). SE: Mochudi, fl. & fr. 15.iv.1967, *Mitchison* A38 (K). **Zambia**. B: Sesheke, fl. & fr. iv.1910, *Macaulay* (*Gairdner*) 527 (K). N: Mbala Distr., old road to Kasakalawe (Cascalawa) from Chemba Village, fl. & fr. 16.ii.1960, *Richards* 12487 (K). W: Ndola, fl. & fr. 29.i.1954, *Fanshawe* 732 (K; SRGH). C: Lusaka, Gloucester Road, fl. & fr. 19.iii.1961, *Best* 285 (K; SRGH). E: Great East Road, near Petauke, Mvuvye River, fl. & fr. 5.xii.1958, *Robson* 840 (BM; K; LISC; PRE; SRGH). S: 8 km south of Mapanza Mission, fl. & fr. 20.iii.1955, *E.A. Robinson* 1133 (K; SRGH). **Zimbabwe**. N: Mazowe Distr., Wengi River, Mtoroshanga–Concession Road, fl. & fr. 2.iii.1963, *Corby* 1253 (K; SRGH). W: Bulawayo, fl. & fr. 15.v.1923, *Walters* 3269 (K; SRGH). C: Chegutu Distr., Poole Farm, fl. & fr. 30.i.1955, *Hornby* 3370 (COI; K; PRE; SRGH). E: Mutare (Umtali), Golf Club, fr. 4.iv.1966, *Corby* 1588 (K; PRE; SRGH). S: Beitbridge Distr., between Tshiturapadsi (Chiturupazi) and Chikwarakwara, fl. & fr. 24.ii.1961, *Wild* 5374 (K; PRE; SRGH). **Malawi**. N: Mzimba Distr., Champhila (Champira), 19.iv.1974, *Pawek* 8394 (K; MAL; MO). C: Dedza Distr., near foot of escarpment, 9 km west of Golomoti on Dedza road, fl. & fr. 27.iii.1977, *Brummitt, Seyani & Patel* 15012 (K). S: near Kasupe, fl. 15.ii.1964, *Salubeni* 257 (K; SRGH). **Mozambique**. N: c. 5 km from Montepuez towards Nairoto (Nantulo), fl. & fr. 8.iv.1964, *Torre & Paiva* 11733 (LISC). T: between Lupata and Tete (Tette), fl. ii.1859, *Kirk* (K). GI: Nhachengue (Inhachengo), fl. & fr. 26.ii.1955, *Exell, Mendonça & Wild* 639 (BM; LISC; SRGH).

Widespread from Senegal and Eritrea to South Africa; also in Madagascar. Grassland, clearings, roadsides, etc. in open *Brachystegia, Julbernardia, Acacia* or *Colophospermum* woodland, *Brachystegia–Hyparrhenia* mosaics, waste places, riverbanks, old cultivations, mostly on sandy soil, shallow bare soil in rocky places, inland sand dunes, also recorded from semi-evergreen forest of *Milicia, Adansonia* and *Acacia;* 100–1500 m.

This and other species were long confused with *Zornia diphylla* (L.) Pers. but the identity of this is explained by Dandy & Milne-Redhead in Kew Bull. **17**: 73–74 (1963). They conclude that the name *Zornia diphylla* must be restricted to the plant hitherto known as *Zornia conjugata* (Willd.) Sm., which far from being a pantropical weed, seems to be unknown outside its natural area of distribution, namely Sri Lanka and southern India.

2. **Zornia pratensis** Milne-Redh. in Bull. Jard. Bot. État **24**: 127 (1954); in Bol. Soc. Brot., sér. 2, **28**: 93 (1954). —Léonard & Milne-Redhead in F.C.B. **5**: 358 (1954). —Mohlenbrock in Webbia **16**: 74 (1961). —Torre in C.F.A. **3**: 216 (1966). —Milne-Redhead in F.T.E.A., Leguminosae, Pap.: 445 (1971). —Verdcourt in Kirkia **9**: 499 (1974). —Lock, Leg. Afr. Check-list: 122 (1989). Type from Uganda.

Decumbent or, less often, erect perennial herb 5–40 cm long or tall. Stems often many from the woody stock, branched, glabrous or somewhat pubescent. Leaves 2-foliolate; leaflets 10–27(40) × 1–5(10) mm, obliquely linear-elliptic, linear or lanceolate, acute at the apex, cuneate at the base, glabrescent or sparsely hairy, glandular-punctate, particularly near the margins; petiole 5–20 mm long, glabrous; petiolules c. 1 mm long; stipules 9–18 mm long including the 3–6 mm long spur, 1.5–3 mm wide, lanceolate, glabrous and obscurely pellucid-punctate. Inflorescences 5–25 cm long; peduncle 2–6 cm long; bracts 5–12 × 2–4(6) mm, elliptic, ovate-lanceolate or lanceolate, acute, ciliate but otherwise glabrous, fairly densely to sparsely pellucid-punctate. Calyx tube 2 mm long; lobes 0.5–1.5 mm long, glabrous save for midnerve of the lowest lobe. Standard yellow or orange, marked with red or brown, or sometimes pink, 7–11 × 6–10 mm; wings yellow marked with red; keel yellow or green. Fruits 10–17 mm long of 4–6 articles, each article 1.5–2.5 × 1.5–2 mm, sometimes slightly glandular, puberulous or rarely glabrous, covered with plumose or glabrous bristles 0.2–0.5(1) mm or 2–3 mm long. Seeds dark brown, compressed, 1.8 × 1.1 × 0.5 mm, irregularly reniform, somewhat beaked beyond the small eccentric hilum.

Subsp. **pratensis** —Milne-Redhead in Bol. Soc. Brot., sér. 2, **28**: 94 (1954). —Léonard & Milne-Redhead in F.C.B. **5**: 358, fig. 25B (1954). —Torre in C.F.A. **3**: 216 (1966). —Milne-Redhead in F.T.E.A., Leguminosae, Pap.: 445 (1971). —Drummond in Kirkia **8**: 229 (1972). —Verdcourt in Kirkia **9**: 500 (1974). —Lock, Leg. Afr. Check-list: 122 (1989). *Zornia diphylla* sensu Robyns, Fl. Sperm. Parc Nat. Alb. **1**: 325 (1948) pro parte non (L.) Pers.

Bristles of articles up to 1 mm long, usually shorter or scarcely developed.

Var. **pratensis** —Verdcourt in Kirkia **9**: 500 (1974).

Bristles of the articles with short spreading hairs.

Botswana. SW: 25 km northwest of the entry to Khutse Game Reserve via Salajwe along track to Kikao, fr. 7.iii.1977, *Skarpe* S157 (K). **Zambia**. E: 24 km from Chipata (Fort Jameson) on Malawi road, Kachebere R.C. Mission, fl. & fr. 25.ix.1960, *Wright* 278 (K). **Zimbabwe**. N: Mwami (Miami), fl. & fr. 4.x.1946, *Wild* 1325 (K; SRGH). **Malawi**. C: near Kasungu Hill, fl. & fr. 14.i.1959, *Robson & Jackson* 1146 (BM; K; LISC; PRE; SRGH).
Also in the Sudan, Ethiopia, Dem. Rep. Congo, Rwanda, Burundi, East Africa and Angola. Grassland, open bushland, dambo margins, also in *Acacia* woodland; 1100–1350 m.

Var. **glabrior** Milne-Redh. in Kew Bull. **25**: 178 (1971); in F.T.E.A., Leguminosae, Pap.: 445 (1971). —Verdcourt in Kirkia **9**: 500 (1974). —Gonçalves in Garcia de Orta, Sér. Bot. **5**: 123 (1982). Type from Tanzania.

Bristles of the articles c. 1 mm long, glabrous; faces of articles glabrous save for sparse gland dots and marginal cilia.

Malawi. N: Nkhata Bay Distr., Viphya Plateau, 61 km SW of Mzuzu, fl. 25.ix.1975, *Pawek* 10181 (K; MAL; MO; SRGH; UC) (articles not present). C: Dedza Distr., Ngondonda (Ngononda) to Bembeke road, fr. 15.xi.1967, *Salubeni* 894 (K; LISC; SRGH). **Mozambique**. T: Moatize, toward Vila Coutinho, 17 km after the crossing of the roads Zóbuè–Moatize and Moatize–Vila Coutinho, fl. & fr. 31.i.1966, *Correia* 437 (LISC).
Also in Tanzania. Burnt grassland, also open *Brachystegia–Piliostigma* woodland; 350–1600 m.

Subsp. **barbata** J. Léonard & Milne-Redh. in Bull. Jard. Bot. État **24**: 127 (1954); in Bol. Soc. Brot., sér. 2, **28**: 96 (1954). —Léonard & Milne-Redhead in F.C.B. **5**: 359, fig. 25C (1954). —Milne-Redhead in F.T.E.A., Leguminosae, Pap.: 446 (1971). —Verdcourt in Kirkia **9**: 500 (1974). —Lock, Leg. Afr. Check-list: 122 (1989). Type from Dem. Rep. Congo.
Zornia diphylla sensu Hutchinson, Botanist South. Africa: 526 (1946) non (L.) Pers.
Zornia setifera Mohlenbr. in Webbia **16**: 69 (1961). Type as for subsp. *barbata*.

Bristles of articles 2–3 mm long.

Zambia. N: Mbala Distr., Nkali Dambo, fl. & fr. 21.viii.1956, *Richards* 5879 (K; SRGH). W: Chingola, fl. & fr. 24.ix.1955, *Fanshawe* 2459 (K).
Also in Dem. Rep. Congo and south western Tanzania. Open woodland, grassland with scattered trees, dry grassland and dambos, especially those subjected to severe burning; 1300–1740 m.
The specimen *Hutchinson & Gillett* 4020 collected near Mbala and cited by Milne-Redhead (loc. cit. 96) as subsp. *pratensis* is, judging by its ovary and on geographical grounds, better referred to this subspecies. It is difficult to decide who is correct concerning the status of this plant — it is certainly very distinctive in appearance but the differences seem to be restricted to an extra 1–2 mm in the length of the fruit bristles.

3. **Zornia setosa** Baker f., Legum. Trop. Africa: 324 (1929). —Milne-Redhead in Bol. Soc. Brot., sér. 2, **28**: 99 (1954). —Léonard & Milne-Redhead in F.C.B. **5**: 353 (1954). —Mohlenbrock in Webbia **16**: 41 (1961). —Milne-Redhead in F.T.E.A., Leguminosae, Pap.: 448 (1971). —Verdcourt in Kirkia **9**: 501 (1974). —Lock, Leg. Afr. Check-list: 123 (1989). Type: Malawi, Angoniland, 1901, *Purves* 83 (K, holotype).

Prostrate or erect perennial herb with pubescent stems 3–50 cm long or tall. Leaves (2–3)4-foliolate; leaflets elliptic to obovate or oblanceolate, the terminal pair usually the largest, and leaflets of lower leaves often larger than those of the upper leaves, 5–26 × 2–18 mm, obtuse, rounded (rarely acute) and ± mucronulate at the apex, cuneate at the base, densely glandular-punctate, glabrous or pubescent on both surfaces, mostly ciliate when young; petiole 2–16 mm long; petiolules 0.5–1 mm long; stipules 4–9 mm long including the 0.5–2.5 mm long spur, 2–3 mm wide, ovate, ovate-lanceolate or oblong-elliptic, glabrous or pubescent, densely glandular-punctate. Inflorescences 2–9 cm long; peduncle 7–35 mm long; bracts 5–10 × 2.5–4.5 mm, elliptic-obovate or elliptic, acute, punctate like the stipules, pubescent or glabrescent. Calyx glabrous save for base of lowest lobe; tube 2–2.5 mm long, lobes 1–2 mm long. Standard yellow, orange or pink to purple, mostly flushed or veined with darker colour, 7–9 × 7–8 mm; keel green. Fruits 10–18 mm long, of 3–7 articles; each article 2–3 × 2–2.5 mm, puberulous or glabrous, glandular or not, covered with plumose or glabrous bristles 0.5–4 mm long. Seeds dark purple-brown, 1.8 × 1.4 × 0.8 mm, rounded reniform, slightly beaked beyond the small eccentric hilum.

Subsp. **setosa** —Milne-Redhead in Bol. Soc. Brot., sér. 2, **28**: 99 (1954); in F.T.E.A., Leguminosae, Pap.: 448 (1971). —Verdcourt in Kirkia **9**: 502 (1974). —Gonçalves in Garcia de Orta, Sér. Bot. **5**: 123 (1982). —Lock, Leg. Afr. Check-list: 123 (1989).

Articles puberulous; bristles plumose, 3–4 mm long; bracts pubescent or glabrous.

Zambia. N: Mweru Wantipa (Mweru-wa-Ntipa), Chocha airfield, fl. & fr. 15.vii.1957, *Whellan* 1376 (K; PRE; SRGH). W: Solwezi, fl. & fr. 12.x.1953, *Fanshawe* 399 (K). E: Chipata Distr., Msekera Agriculture Station, fl. 3.ix.1962, *Verboom* 674 (K; PRE; SRGH). **Malawi**. C: Dedza, fl. & fr. 20.x.1956, *G. Jackson* 2071 (K). **Mozambique**. T: Tsangano Distr., near Vila Mouzinho, fl. & fr. 15.x.1943, *Torre* 6040 (BM; K; LISC).
Also in Dem. Rep. Congo and south western Tanzania (one gathering only known). Grassland subject to severe burning, dambos, open *Brachystegia* woodland on sandy soil, also disturbed ground; 900–1650 m.

Subsp. **obovata** (Baker f.) J. Léonard & Milne-Redh. in F.C.B. **5**: 353, pl. 27 (1954); in Bol. Soc. Brot., sér. 2, **28**: 100 (1954). —Milne-Redhead in F.T.E.A., Leguminosae, Pap.: 448, fig. 64 (1971). —Verdcourt in Kirkia **9**: 502 (1974). —Lock, Leg. Afr. Check-list: 123 (1989). TAB. 3.6.43. Type from Kenya.
Zornia tetraphylla var. *obovata* Baker f., Legum. Trop. Africa: 324 (1929).
Zornia tetraphylla sensu auct. mult. non Michx. nec (L.) Fawc. & Rendle.
Zornia obovata (Baker f.) Mohlenbr. in Webbia **16**: 41, fig. 25, 27 (1961).

Articles glandular, puberulous or glabrous; bristles glabrous or plumose (rarely retrorsely hispid) 0.5–1 mm long.

Zambia. N: Mbala, road turn off up to Red Locust Organisation, H.Q., fl. & fr. 4.iii.1955, *Richards* 4767 (K).
Also in Ethiopia, Dem. Rep. Congo, Rwanda and East Africa. Short grassland, etc. in very sandy often damp places; 1500–1680 m.

4. **Zornia capensis** Pers., Synops. Pl. **2**: 318 (1807). —Milne-Redhead in Bol. Soc. Brot., sér. 2, **28**: 103 (1954). —Mohlenbrock in Webbia **16**: 45, figs. 27, 28 (1961). —Milne-Redhead in F.T.E.A., Leguminosae, Pap.: 450 (1971). —Drummond in Kirkia **8**: 229 (1972). —Verdcourt in Kirkia **9**: 502 (1974). —Lock, Leg. Afr. Check-list: 121 (1989). Type from South Africa (Cape Province).

Tab. 3.6.43. ZORNIA SETOSA subsp. OBOVATA. 1, habit (× ½); 2, stipule (× 3); 3, bract (× 3); 4, calyx, spread out, internal face (× 5); 5, flower, longitudinal section (× 5); 6, standard, internal face (× 5); 7, wing, external face (× 5); 8, keel petal, internal face (× 5); 9, androecium, with style of gynoecium protruding (× 5); 10, one article of the fruit (× 10); 1–10 from *Bequaert* 3337. Drawn by J.M. Lerinckx. From Fl. Congo Belge. Reproduced with permission of Jardin Botanique National de Belgique.

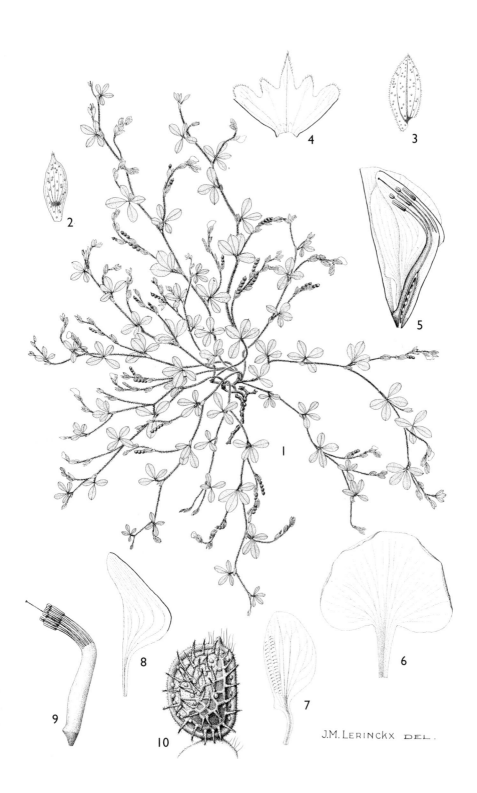

J.M. LERINCKX DEL.

Hedysarum tetraphyllum Thunb. in Nova Acta Regiae Soc. Sci. Upsal. **6**: 44 (1799). Type as above.

Perennial herb (rarely annual) with decumbent or prostrate stems 15–75 cm long. Stems mostly much branched at the base, usually glabrous but sometimes slightly pubescent (very rarely densely pubescent). Leaves 4-foliolate (rarely less); leaflets 5–30 × 2–8(9) mm, lanceolate, or narrowly elliptic, acute and apiculate at the apex, cuneate at the base, glabrous or slightly pubescent, densely glandular-punctate or sometimes mostly glandular-punctate along the margins; petiole 5–18 mm long; petiolules 0.2–1 mm long, glabrous or pubescent; stipules 5–15 mm long including the 1.5–5 mm long spur, 1.5–3 mm wide, narrowly to broadly lanceolate, glabrous, punctate or not. Inflorescences 1.5–20 cm long; peduncle 2–4 cm long; bracts 6–14 × 2.5–5.5 mm, elliptic to ovate, acute, glabrous or the margins distinctly ciliate, sparsely to densely pellucid-punctate. Calyx tube 1.5–2 mm long; lobes 2.5–3.5 mm long. Standard yellow, often lined with red or purple, 8.5–10 × 9–10 mm; wings yellow; keel yellow or greenish. Fruits 15–23 mm long, of 4–7 articles; each article 2–2.5(3.2) mm long and wide, glabrous and glandular or with few to many glabrous or retrorsely pubescent bristles. Seeds pale brown, 1.8–2.3 × 1.2–1.5 × 0.6 mm, ellipsoid-reniform, slightly beaked at one end beyond the small eccentric hilum.

Subsp. **capensis** —Verdcourt in Kirkia **9**: 503 (1974). —Lock, Leg. Afr. Check-list: 121 (1989).

Articles of fruit glabrous or with marginal cilia, reticulate and glandular but with no (or rarely very few) bristles. Leaflets conspicuously glandular-punctate.

Zimbabwe. W: Hwange Distr., Victoria Falls, fl. & fr. 9.ii.1912, *Rogers* 5707 (K). E: Mutare (Umtali) Commonage, fl. & fr. 18.ii.1956, *Chase* 5967 (BM; K; LISC; PRE; SRGH). S: Masvingo Distr., c. 4.8 km from Morgenster Mission, south of Mutirikwi (Kyle) Dam, fl. & fr. 18.xii.1970, *Müller & Pope* 1731 (K; LISC; SRGH). **Mozambique**. GI: Xai-Xai Distr., Praia Sepúlveda, fl. & fr. 14.viii.1957, *Barbosa & Lemos* 7844 (COI; LISC). M: Maputo (Lourenço Marques), road to Marracuene, fl. & fr. 5.xi.1964, *Marques* 36 (COI; LMU).

Also in South Africa (Cape Province and KwaZulu-Natal), and in Madagascar. Grassland and dambo edges, open mixed deciduous woodlands with *Acacia, Sclerocarya, Albizia*, etc., also on cultivated ground; 0–1080(1620) m.

Specimens with a few glabrous bristles on the fruit but having essentially the same kind of articles as subsp. *capensis* are not uncommon, e.g. Zimbabwe E: Chimanimani, fl. & fr. ix.1953, *Williams* 151 annotated by Mohlenbrock as *Z. milneana*. His key distinction using the size of the articles is mostly not usable. One very interesting specimen — Mozambique GI: Inharrime–Nhacoongo, Campo Exp. C.I.C.A., 9.x.1945, *Pedro* 277 (LMA; SRGH) which is cited by Mohlenbrock as *Z. milneana* has several shoots bearing fruits exactly like those of *Z. capensis* subsp. *capensis* and other shoots bearing fruits with pubescent bristles exactly as in *Z. milneana*. The PRE duplicate is normal *Z. capensis* subsp. *capensis*. All these shoots are undoubtedly from the same rootstock; it might be a curious hybrid but is probably a variant showing a tendency to *Z. milneana*. The record of *Z. capensis* subsp. *capensis* from Victoria Falls cited above is difficult to understand; *Rogers* 5606 is typical *Z. milneana*. *Z. capensis* subsp. *tropica* Milne-Redh. occurring in East Africa has the habit of *Z. capensis* but the fruits of *Z. milneana*. These three form a triangle of related taxa in which *Z. milneana* appears distinct enough to be treated at specific rank. Mohlenbrock treats *Z. tropica* as a distinct species.

5. **Zornia milneana** Mohlenbr. in Kew Bull. **15**: 325 (1961) (Oct.); in Webbia **16**: 26, figs. 12, 13 (1961) (July, anglice). —Torre in C.F.A. **3**: 216 (1966). —Schreiber in Merxmüller, Prodr. Fl. SW. Afrika, fam. 60: 125 (1970). —Drummond in Kirkia **8**: 229 (1972). —Verdcourt in Kirkia **9**: 504 (1974). —Lock, Leg. Afr. Check-list: 122 (1989). Type: Zimbabwe, Matopos, Mtshelele (Mtscheleli) Valley, *Plowes* 1410 (K, holotype; SRGH).

Perennial herb (rarely annual) with decumbent, prostrate or less often erect stems 8–75 cm long. Stems sparsely to densely pubescent rarely glabrous, usually much branched at the base. Leaves 4-foliolate; leaflets 7–46 × 1–9 mm, linear or narrowly lanceolate, less often narrowly elliptic, acute and apiculate at the apex, cuneate at the base, pubescent particularly beneath or glabrous, densely glandular-punctate; petiole 5–20 mm long; petiolules c. 1 mm long, mostly densely pubescent; stipules 5–19 mm long including the 1.5–7 mm long spur, 1.5–4 mm wide, narrowly to broadly lanceolate, glabrous or slightly hairy, punctate. Inflorescences 3.5–14 cm long; peduncle 1–6.5 cm long; bracts 6–16 × 2.5–8 mm, elliptic to ovate, acute, glabrous or glabrescent but the margins usually conspicuously ciliate, densely pellucid-punctate. Calyx tube 1.5–2 mm long; lobes 2.5–4 mm long. Standard yellow or orange, often lined with red or purple; wings yellow; keel greenish and yellow. Fruits 13–18 mm long of 4–7 articles; each article 3–3.5 × 2.5–3 mm, covered with retrorsely pubescent bristles up to 1 mm long. Seeds chestnut-brown, 1.7–2.1 × 1.5 × 0.8 mm, semicircular-reniform; hilum small, very eccentric.

Caprivi Strip. Mashi, fl. & fr. 4.xi.1962, *Fanshawe* 7121 (K; SRGH). **Botswana**. N: near Morombe–Kwando, lower road, 18°09'S 23°14', fr. 28.i.1978, *P.A. Smith* 2302 (K; SRGH). SW: Ghanzi, fr. 9.iv.1969, *de Hoogh* 240 (K). SE: 32 km WNW of Lobatse (Lobatsi) on Kanye road, fl. 18.i.1960, *Leach & Noel* 176 (K; LISC; SRGH). **Zambia**. B: Sesheke Distr., Mashi River area, fl. 1.xi.1964, *Verboom* 1080 (K; SRGH). C: Kabwe (Broken Hill), fl. xi.1928, *van Hoepen* 1261 (PRE) (atypical). S: Namwala Distr., 128 km north of Choma, Maala, fl. 6.i.1957, *E.A. Robinson* 2028 (K; SRGH). **Zimbabwe**. N: Gokwe South Distr., Sengwa Research Station, fl. & fr. 18.ii.1968, *Jackson* 145 (SRGH). W: Matopos, fl. & fr. xi.1921, *Eyles* 3751 (SRGH). C: Marondera Distr., Delta Farm, fl. & fr. 9.ii.1966, *Corby* 1438 (SRGH). E: Nyanga Distr., Inyanga Mountains Hotel, fl. & fr. 25.i.1966, *Corby* 1432 (SRGH). S: Masvingo Distr., *Monro* 908 (BM).

Also in Angola, Namibia and South Africa (KwaZulu-Natal, Transvaal). Grassland and floodplain grassland, *Brachystegia*, *Terminalia–Parinari* and other types of woodland, also in old cultivations, mostly on sandy soil; 900–1500 m.

I initially considered this should be maintained as a subspecies of *Z. capensis* and an intermediate has already been noted at the end of the account of that species. Moreover subsp. *tropica* Milne-Redhead in East Africa with pubescent bristles on the fruit articles could be considered to be a subsp. of either *Z. milneana* or *Z. capensis*; indeed the Kabwe specimen cited above is scarcely distinguishable from *Z. capensis* subsp. *tropica* and the single specimen seen from Angola is similar. On the other hand there is a great mass of material from Zimbabwe with uniformly condensed inflorescences, larger, broader bracts, narrower, longer leaflets and usually distinctly hairy stems and it seems practical to maintain *Z. milneana* as a species. Clearly, however, all three are derived from the same immediate ancestor.

81. STYLOSANTHES Sw.

Stylosanthes Sw., Nov. Gen. Sp. Pl. Prodr.: 108 (1788); Fl. Ind. Occid. **3**: 1280, t. 25 (1806). —Mohlenbrock in Ann. Missouri Bot. Gard. **44**: 299 (1957); in J. S. African Bot. **31**: 95 (1965). —Verdcourt in Kirkia **9**: 492 (1974).

Erect or spreading perennial herbs or subshrubs, often somewhat hispid with glandular hairs. Leaves pinnately 3-foliolate, or abnormally 1-foliolate; stipules adnate to the petiole for most of their length, persistent, biapiculate; stipels absent. Inflorescences usually dense, axillary or terminal, composed of 1-flowered groups (reduced inflorescence parts) in spikes or panicles; primary bracts 1(2–3)-foliolate, imbricate, persistent; secondary bracts lanceolate or irregularly 2–3-fid, hyaline, persistent, ciliate. Flowers subsessile, accompanied by 1–2 persistent linear hyaline ciliate bracteoles and sometimes by a plumose filiform axis representing a reduced part of an inflorescence now no longer present. Receptacle (hypanthium) long and filiform. Calyx 5-lobed; lobes joined at the base, membranous, the lowest lobe longer than the rest, the upper pair joined for about half their length. Corolla usually small, yellow; standard rounded or obovate, emarginate, narrowed into a basal claw; wings oblong or obovate, free, with a lateral basal spur and a small internal appendage and also with a series of small pockets on the blade; keel petals similarly spurred and appendaged. Stamens all joined; 5 anthers longer and sub-basifixed alternating with 5 shorter and versatile. Ovary linear, sessile, 2–3-ovuled; style long and filiform, breaking off towards the middle or nearer the base after flowering, the lower part persistent, recurved or revolute with the dilated apex simulating a stigma; true stigma terminal minute. Fruit oblong, compressed, beaked, 1–2-jointed, but usually either the upper or lower loment aborted, reticulate or muricate. Seeds approximately ovoid or irregularly oblong, compressed; hilum often eccentric; aril somewhat developed or absent.

A genus of about 25 species, mostly poorly defined (and sometimes estimated at 50) in the tropics and subtropics of both hemispheres.

Several species of this genus have been cultivated at agricultural research stations and have in other parts of Africa occurred as escapes. The most important of these are:

S. humilis Kunth (*S. sundaica* Taub.) characterised by the very long beak to the fruit and the prostrate habit (e.g. Marondera, Grasslands Research Station, 24.iii.1961, *Corby* 971 (LISC; SRGH)).

S. guianensis (Aubl.) Sw. with short broad not elongated inflorescences (this species occurs as an escape in the Flora Zambesiaca area and is treated more fully below).

S. grandifolia M.B. Ferr. & Sousa Costa (*S. guianensis* var. *robusta* L. 't. Mannetje), introduced from Brazil and grown in Zambia (Mt. Makulu Research Station, fl. 1.vi.1967, *van Rensburg* 3115 (K; SRGH)). It differs from *S. guianensis* in the more prominent leaflet venation and in the fruit articles which are papillate near the apex.

S. hippocampoides Mohlenbr. (*S. guianensis* var. *intermedia* (Vog.) Hassler), introduced from northern Argentine, Uruguay and Paraguay, has been grown in Zimbabwe (Matopos Research Station, fl. & fr. 11.iv.1967, *S.S.D.* 55 (K; SRGH)). It differs completely from *S. guianensis* in having reticulate articles, the four vertical curved ribs being particularly raised and also with apical papillae.

A number of other sheets are too young to determine. Collectors should not collect material without ripe articles.

1. Fruits with a long curved hairy beak, 5–7 mm long, clearly protruding from inflorescences; cultivated · *humilis*
- Fruits with short beaks not protruding · 2
2. Ribs of primary bracts clearly evident despite indumentum (very common native plant) · 1. *fruticosa*
- Ribs of primary bracts obscured by a dense subspreading indumentum (stipules often reddish); cultivated and also as an escape · 2. *guianensis*

1. **Stylosanthes fruticosa** (Retz.) Alston in Suppl. Trimen Handb. Fl. Ceylon: 77 (1931). — Mohlenbrock in Ann. Missouri Bot. Gard. **44**: 318 (1957). —Torre in C.F.A. **3**: 214 (1966). —Schreiber in Merxmüller, Prodr. Fl. SW. Afrika, fam. 60: 112 (1970). —Verdcourt in Kew Bull. **24**: 59 (1970); in F.T.E.A., Leguminosae, Pap.: 437, fig. 62 (1971). —Drummond in Kirkia **8**: 227 (1972). —Verdcourt in Kirkia **9**: 493 (1974). —Lock, Leg. Afr. Check-list: 121 (1989). TAB. 3.6.**44**. Types from India and Sri Lanka.
     Arachis fruticosa Retz., Observ. Bot. **5**: 26 (1789).
     Stylosanthes mucronata Willd., Sp. Pl. **3**: 1166 (1802). —J.G. Baker in F.T.A. **2**: 157 (1871). —E.G. Baker, Legum. Trop. Africa: 320 (1929). —Léonard in F.C.B. **5**: 348 (1954). — Hepper in F.W.T.A., ed. 2, **1**: 575 (1958). —Mohlenbrock in J. S. African Bot. **31**: 98 (1965). Type as for *A. fruticosa*.
     Stylosanthes bojeri Vogel in Linnaea **12**: 68 (1838). —J.G. Baker in F.T.A. **2**: 157 (1871). — E.G. Baker, Legum. Trop. Africa: 321 (1929). —Brenan, Check-list For. Trees Shrubs Tang. Terr.: 444 (1949). Type from Zanzibar.
     Stylosanthes flavicans Baker in F.T.A. **2**: 156 (1871). —E.G. Baker, Legum. Trop. Africa: 320 (1929). Type from Sudan, Kordofan.

Woody herb or subshrub, sometimes only a short-lived perennial, mostly erect but sometimes prostrate, 0.1–1 m long or tall; rootstock mostly thick and woody. Stems pubescent to densely spreading-hairy and often with sparse to dense bristles as well, frequently glandular. Leaflets 5–33 × 1–9 mm, elliptic or lanceolate, rounded to acute at both ends, usually pubescent and sparsely to rather densely bristly as well, rarely glabrescent; lateral nerves often conspicuously thickened towards the margins; free part of petiole and rhachis together 4–15 mm long; petiolules 0.5 mm long; stipules 8–16 mm long. Inflorescences dense; peduncle 0–5.5 cm long; rhachis 1–1.5(4) cm long; primary bracts 1.2–2 cm long, sparsely to densely pubescent and bristly; secondary bracts 4–6 × 1.5 mm; bracteoles 2, 3.5–5 mm long; plumose axis usually present, 3–10 mm long. Receptacle 5–7 mm long; calyx lobes 2.5–3 × 0.7–1 mm, oblong. Standard creamy-white to orange-coloured, with red veins and a red

Tab. 3.6.**44**. STYLOSANTHES FRUTICOSA. 1, flowering stem (× ²/₃); 2, leaf with stipules (× 2); 3, inflorescence (× 2); 4, flower with bracteoles, primary and secondary bracts and plumose axis (× 2); 5, section of flower (× 6); 6, upper part of calyx, opened out (× 6); 7, standard (× 6); 8, wings (× 6); 9, base of wing from inner side (× 6); 10, keel (× 6); 11, keel petal viewed from inner side (× 6); 12, androecium, spread out (× 6); 13, gynoecium (× 6), 1–13 from *Faulkner* 1998; 14, fruit with bracts and bracteoles (× 6), from *Peter* 49595. Drawn by Derek Erasmus. From F.T.E.A.

D.E.

mark near the base inside, 5–7 × 4–5 mm; wings and keel yellow or orange-yellow. Fruit 4–9 mm long, 1–2-jointed, the articles 3.5–4 × 2–2.5 mm, usually densely pubescent; beaks 1.5–2.5 mm long. Seeds chestnut-brown, shiny, 2.5–3 × 2–2.5 × 1–1.2 mm, compressed-ellipsoid, beaked or pointed near the hilum.

Botswana. N: Ngamiland Distr., Bushman Pits, 27.iii.1961, *Richards* 14885 (K). SE: Kweneng Distr., 8 km from turnoff towards Kanye from Gaborone–Molepolole road, fl. 31.iii.1977, *O.J. Hansen* 3105 (C; GAB; K; PRE; SRGH; UPS). **Zambia**. B: Sesheke, fl. & fr. i.1925, *Borle* s.n. (PRE). N: Kasama [cult. at Marondera (Marandellas), fl. 6.i.1965] *Corby* 1212 (SRGH). W: Luanshya, fl. & fr. 30.iii.1956, *Fanshawe* 2853 (K). C: 12.8 km east of Lusaka, fl. & fr. 4.iii.1956, *King* 336 (K). S: Mazabuka, 28.ii.1963, *van Rensburg* 1523 (K; SRGH). **Zimbabwe**. N: Binga Distr., Mwenda Research Station, fl. & fr. 7.vi.1966, *Grosvenor* 134 (SRGH). W: Matopos, fl. & fr. 21.iv.1952, *Plowes* 1426 (K; SRGH). C: Chegutu (Hartley), fl. & fr. 10.iv.1948, *Hornby* 2862 (K; PRE; SRGH). E: Mutare (Umtali), fl. 4.i.1956, *Chase* 5935 (BM; COI; K; LISC; PRE; SRGH). S: Chiredzi Distr., Gonakudzingwa, fl. & fr. 4.iv.1961, *Goodier* 1056 (K; SRGH). **Mozambique**. N: Mocímboa da Praia to Diaca, c. 53 km from latter, fl. & fr. 14.iv.1964, *Torre & Paiva* 11907 (LISC). Z: Quelimane, fl. & fr. 1908, *Sim* 20800b (PRE). MS: Manica Distr., between the R. Revué and R. Douro, fl. & fr. 18.iii.1948, *Barbosa* 1210 (LISC). GI: Bilene Distr., Planícies de Magul, fl. & fr. 6.ii.1948, *Torre* 7266 (LISC). M: Matutuíne Distr., Tinonganine, fl. & fr. 28.iii.1957, *Barbosa & Lemos* 7563 (COI; LISC; LMA).

Widespread in tropical Africa from West Africa to Sudan and Somalia southwards to South Africa and Namibia. Also in Madagascar, Arabia, India and Sri Lanka. Grassland and wooded grasslands, *Colophospermum* and *Acacia* woodlands and miombo, sometimes by river banks and mostly on sandy soil, also in abandoned cultivations; 0–1470 m.

This species would certainly be expected to occur in Malawi, but no specimens have been seen from there. It has also been cultivated. Over 90% of the numerous sheets seen from Mozambique are from M: where it must be very common.

Gonçalves in Garcia de Orta **5**: 114 (1982) has recorded this for Tete Province of Mozambique.

Léonard, in F.C.B. **5**: 348 (1954), considered that the "type material" of *Arachis fruticosa* differed from the African plant by its unilaterally pubescent stem and longer styles. As I have shown (Verdcourt in Kew Bull. **24**: 59–60 (1970)), the specimen Léonard saw from Retzius' Herbarium was in fact one of *S. hamata* (L.) Taub., but two other specimens he did not see because they were misplaced may be accepted as the types of *Arachis fruticosa* Retz. Retzius cited only a part of *Hedysarum hamatum* L. in synonymy and was hence dividing the species: he specifically excluded the Jamaican element now known as *S. hamata* (L.) Taub.

2. **Stylosanthes guianensis** (Aubl.) Sw. in Kongl. Vetensk. Acad. Nya Handl. **10**: 301 (1789). — Verdcourt in Man. New Guinea Leg.: 373 (1979). —Lock, Leg. Afr. Check-list: 121 (1989). Type from French Guiana.

Trifolium guianense Aubl. Hist. Pl. Guiane Française: 776, t. 309 (1775).

Erect perennial herb or subshrub 0.6–1.8 m tall, sometimes prostrate. Stems coarsely hairy to almost glabrous or with lines of short pubescence or scattered long bristles. Leaflets 5–30(45) × 2–10(20) mm, ovate, elliptic or lanceolate, glabrous to puberulous or bristly; stipules often reddish. Inflorescences dense, 1–40-flowered, mostly 1–1.5 cm long; axis rudiment absent. Receptacle 4–8 mm long. Standard yellow with red-brown lines; wings bright yellow; keel greenish; claws often red. Fruits 1-jointed; loment 2–3 × 1.5–2.5 mm, ovoid, glabrous or minutely pubescent at the apex but without papillae, with a minute beak 0.1–0.8 mm long, strongly bent inwards.

Malawi. N: Mzimba Distr., 4.8 km west of Mzuzu, at Katoto, fr. 24.x.1975, *Pawek* 10323 (K; MAL; MO; SRGH; UC). C: Lilongwe town, fl. 24.vi.1987, *Salubeni* 4949 (K; MAL).

Widespread in central and south America, now naturalised in parts of Africa and Asia. Rice paddy banks, *Brachystegia* woodland; 540–1350 m.

Extensively cultivated as a pasture legume under the name "stylo" and also Brazilian lucerne. The specific epithet is frequently spelt *guyanensis* and was in fact spelt *guyannense* on Aublet's plate.

Pawek 12872 (Malawi, Nkhata Bay Distr., 8 km south of Nkhata Bay junction, Limpasa Rice Scheme, fl. 13.viii.1977) has been determined by Sousa Costa as var. *pauciflora* Brandao, Sousa Costa & R. Schultze-Kraft whereas those cited, *Pawek* 10323 is var. *vulgaris* M.B. Ferr. & Sousa Costa and *Salubeni* 4949 is var. *pauciflora*.

Var. *vulgaris* has been grown in Zambia at Mt. Makulu Research Station, fl. & fr. 6.xi.1965, *van Rensburg* 3073 (BM; K; SRGH), fl. 3.xii.1965, *van Rensburg* 3085, 3087 (K), fl. 1.vi.1967, *van Rensburg* 3116 (K); in Zimbabwe at the Agriculture Experimental Station in Harare, fl. & fr.

14.i.1943, *Arnold* G512 (K; SRGH); and at the Matopos Research Station, fl. 11.iv.1967, *S.S.D.* 56 (K; SRGH).
Var. *pauciflora* has been grown at Marondera, Grasslands Research Station, 7.v.1961, *Corby* 1006 (K; SRGH).

82. ARACHIS L.

Arachis L., Sp. Pl.: 741 (1753); Gen. Pl., ed. 5: 329 (1754). —Burkart in Darwiniana 3: 261 (1939). —Hoehne in Fl. Brasilica 25 (II; 122): 3 (1940). —Hermann, U.S. Dept. Agric. Monographs: 19 (1954). —Verdcourt in Kirkia 9: 495 (1974).

Annual or perennial erect or prostrate herbs. Leaves paripinnately 4-foliolate, rarely 3-foliolate; stipules partly adnate to the petiole, membranous, apiculate, persistent and veined; stipels absent. Inflorescences axillary short dense sessile 2–7-flowered spikes; bracts membranous, the primary ones biapiculate; bracteoles absent. Flowers ± sessile, soon deciduous, small or medium-sized, yellow, sometimes striped with red. Receptacle long and filiform, pedicel-like; calyx membranous, filiform, 5-lobed, the 4 upper lobes joined, the lower ± free. Standard rounded, shortly narrowed at the base; wings free; keel beaked, incurved. Stamens 8–10, all joined; 4–5 anthers elongate and sub-basifixed, alternating with 4–5 short and versatile ones. Ovary subsessile, situated at the base of the receptacular tube, linear, (1)2–4(7)-ovuled; style filiform, very long, soon deciduous; stigma minute, terminal. Fruit oblong or sausage-shaped, 1–6-seeded, somewhat constricted between the seeds but not articulated, continuous inside, functionally indehiscent, the walls thick and reticulate, developing below the soil, having been pushed beneath by the considerable lengthening, reflexing and stiffening of the gynophore. Seeds irregularly ovoid or oblong cotyledons thick and fleshy, rich in oil.

A genus of c. 20 species in S America, one of which, the ground-nut, is widely cultivated throughout the warmer parts of the world.
What may be *Arachis diogoi* Hoehne has been grown at the Grasslands Research Station, Marondera.

Arachis hypogaea L., Sp. Pl.: 741 (1753). —J.G. Baker in F.T.A. 2: 158 (1871). —Harms in Engler, Pflanzenw. Afrikas [Veg. Erde 9] 3 (1): fig. 301 (1915). —E.G. Baker, Legum. Trop. Africa: 322 (1929). —Chevalier in Rev. Bot. Appl. Agric. Trop. 13: 689 (1933). —Hoehne in Fl. Brasilica 25 (II; 122): 18 (1940). —Léonard in F.C.B. 5: 351 (1954). —Hepper in F.W.T.A., ed. 2, 1: 576 (1958). —Meikle in Rhodesia Agric. J. 62: 109–113 (1965). —Torre in C.F.A. 3: 214 (1966). —Verdcourt in F.T.E.A., Leguminosae, Pap.: 442, fig. 63 (1971); in Kirkia 9: 496 (1974). —Gonçalves in Garcia de Orta 5: 64 (1982). —Lock, Leg. Afr. Checklist: 110 (1989). TAB. 3.6.**45**. Type a specimen cultivated in Sweden.

Annual, erect or straggling herb c. 30 cm long. Stems at first pilose, later glabrescent. Leaves 4-foliolate; leaflets 1–7 × 0.7–3.2 cm, obovate or elliptic, rounded or emarginate and mucronate at the apex, narrowly rounded to the base, glabrous, or sparsely pilose beneath, ciliate; free part of petiole 1.5–7 cm long; petiolules 1–2 mm long; stipules 1.5–4 cm long, free part 0.6–3.2 cm long, linear-lanceolate, very acute, ciliate. Flowers axillary, apparently solitary and stalked; primary bracts 10–14 × 4–5 mm, ovate-lanceolate, biapiculate; secondary bracts similar but 2-fid. Receptacle 0.2–4 cm long, pilose. Corolla yellow, usually with red nerves, 7–13 mm long. Stamens 8–9. Fruit 2–6 × 1–1.5 cm, the gynophore becoming 1–20 cm long. Seeds 1–2 cm long and c. two-thirds the length in diameter, irregularly ovoid.

Zambia. W: Ndola, fr. x.1949, *Trapnell* (K). E: Chipata Distr., Luangwa Valley, Munkanya (Mulila Munkanya), fl. & fr. 25.ii.1968, *R. Phiri* 23 (K). **Zimbabwe**. C: Marondera Distr., Grasslands Research Station, fl. 29.i.1964, *Corby* 1078 (SRGH). E: Chipinge Distr., Mt. Selinda (Silinda), Farm Gunguinyana, fl. xii.1937, *Obermeyer* in *Transvaal Mus.* 372676 (PRE). **Malawi**. S: Lower Shire Valley, 20.iv.1932, *Lawrence* 17 (K). **Mozambique**. N: Lago Distr., Maniamba, fl. 21.v.1948, *Pedro & Pedrógão* 3778 (EA; LMA). T: Boruma, *Menyharth* s.n. M: Inhaca Island, fl. 3.iv.1958, *Mogg* 31680 (K; PRE; SRGH).
Widely cultivated throughout the tropics and occasionally found as an escape. It has been included here with full treatment since it is so widespread.
Gonçalves in Garcia de Orta 5: 63 (1982) has recorded this for Tete Province of Mozambique.

Tab. 3.6.**45**. ARACHIS HYPOGAEA. 1, habit (× ²⁄₃), from *Dyson-Hudson* 51 & *Trapnell*; 2 & 3, stipules (× 1); 4, section through flower (× 1¹⁄₂); 5, calyx, opened out (× 1¹⁄₂); 6, standard (× 1¹⁄₂); 7, wings (× 1¹⁄₂); 8, keel (× 1¹⁄₂); 9, upper part of androecium, spread out (× 4); 10, upper part of style (× 4), 2–10 from *Greenway* 1105; 11, section through fruit (× ²⁄₃); 12, seeds (× ²⁄₃), 11 & 12 from *Trapnell*. Drawn by Derek Erasmus. From F.T.E.A.

INDEX TO BOTANICAL NAMES

AESCHYNOMENE L., 58,59
subgen. Aeschynomene, 59
subgen. Bakerophyton J. Léonard, 59
subgen. Ochopodium (Vogel) J. Léonard, 59
subgen. Rueppellia (A. Rich.) J. Léonard, 59,88
sect. Basiadhaerentes J. Léonard, 59
sect. Liberae J. Léonard, 59
sect. Marginulatae (Harms) J. Léonard, 59
sect. Rubrofarinaceae (P.A. Duvign.) Verdc.,59
sect. Samaroideae J. Léonard, 59
abyssinica (A. Rich.) Vatke, 65,104,tab.16 fig.6
afraspera J. Léonard, 61,72,73,tab.15 fig.5
americana L., 59,77
var. americana, 77
var. glandulosa (Poir.) Rudd, 77
aphylla Wild, 62,83
aspera L., 73
aspera sensu Baker, 72
baumii Harms, 62,82,83
var. baumii, 82
var. kassneri (Harms) Verdc., 82,83
bella Harms, 79
bracteosa sensu E.G. Baker, 111
bracteosa Welw. ex Baker, 64,86,87,97
var. bracteosa, 86,87
var. delicatula (Baker f.) Verdc., 86,87
var. major Verdc., 87
chimanimaniensis Verdc., 62,80,tab.20
cristata Vatke, 61,74,tab.15 fig.7
var. cristata, 74,tab.19
var. pubescens J. Léonard, 73,74
curtisiae Johnston, 114
delicatula Baker f., 87
dissitiflora Baker, 97
elaphroxylon (Guill. & Perr.) Taub., 59,60, 76,tab.16 fig.5
elisabethvilleana De Wild., 87
falcata (Poir.) DC., 59,60,117
fluitans Peter, 60,68
fulgida Welw. ex Baker, 65,100
gazensis Baker f., 62,84
glabrescens Welw. ex Baker, 64,114
var. glabrescens, 114
var. pubescens J. Léonard, 114
glauca R.E. Fr., 64,96,97
glutinosa Taub., 104,105
?goetzei Harms, 108,109
grandistipulata Harms, 62,80
heurckeana Baker, 64,96,97
hockii De Wild., 101
homblei De Wild., 101
humilis N.E. Br., 101
indica L., 61,67,70,tabs.15 fig.3; & 16 fig.3
inyangensis Wild, 62,83
kassneri Harms, 83
katangensis De Wild., 65,66,109,110
subsp. katangensis, 111

subsp. sublignosa (De Wild.) J. Léonard, 111
kilimandscharica Taub. ex Engl., 104
lateriticola Verdc., 60,86,tab.21
leptobotrya sensu E.G. Baker, 77
leptophylla Harms, 65,111,113,tab.28
var. crassituberculata Verdc., 113
var. leptophylla, 113
subsp. magnifoliolata J. Léonard, 113
mearnsii De Wild., 70
mediocris Verdc., 60,68,tab.17
megalophylla Harms, 62,79
micrantha (Poir.) DC., 60,116,117
mimosifolia Vatke, 65,101,tab.26
minutiflora Taub., 61,87
subsp. grandiflora Verdc., 61,88
subsp. minutiflora, 61,88
mossambicensis Verdc., 61,90,tab.23
var. longistipitata (Verdc.) Vollesen, 90
subsp. mossambicensis, 90
mossoensis J. Léonard, 105
var. pubescens J. Léonard, 65,106
multicaulis Harms, 65,105
nambalensis Harms, 86,87
nematopoda Harms, 61,90,92
newtonii Schinz, 109
nilotica Taub., 61,72,73,tab.15 fig.6
nodulosa (Baker) Baker f., 62,77
var. glabrescens J.B. Gillett, 78
var. nodulosa, 78
nyassana Taub., 65,113,114,tab.16 fig.8
nyikensis Baker, 103
var. gracilis Suess., 101
var. mossambicensis Baker f., 104
oligantha Welw. ex Baker, 67
oligophylla Harms, 64,94,tab.24
paludicola Harms, 70
pararubrofarinacea J. Léonard, 60,115,116
pawekiae Verdc., 64,88,tab.22
pfundii Taub., 59,60,76,tab.16 fig.4
pseudoglabrescens Verdc., 64,65,106
pygmaea Welw. ex Baker, 100
var. hebecarpa J. Léonard, 65,101
var. pygmaea, 101
racemosa De Wild., 111
recta N.E. Br., 101
rhodesiaca Harms, 61,64,92
rogersii N.E. Br., 111
rubrofarinacea (Taub.) F. White, 60,115, 116
ruppellii Baker, 104
schimperi Hochst. ex A. Rich., 61,70, tabs.15 fig.4; & 18
schlechteri Harms ex Baker f., 68
schliebenii Harms, 65,103
var. mossambicensis (Baker f.) Verdc., 104
var. schliebenii, 104
semilunaris Hutch., 62,78
sensitiva Sw., 60,66,tabs.15 fig.2; & 16 fig.2
shirensis Taub., 77
siifolia Baker, 114

solitariiflora J. Léonard, 60,**84**
sp. (Gardner 17) of Eyles, 92
sp. A, 61,**73**
sp. B, 61,**92**
sp. B of Verdcourt, 73
sp. C, 64,**93**
sp. C of Verdcourt, 92
sp. D, 64,**96**
sp. D of Verdcourt, 93
sp. E, 64,65,**109**
sp. E of Verdcourt, 88
sp. F of Verdcourt, 96,109
sp. G of Verdcourt, 109
sparsiflora Baker, 64,**93**
stipulosa Verdc., 61,**106**,tab.**27**
stolzii Harms, 64,**97**
subaphylla De Wild., 109
subaphylla sensu E.G. Baker, 111
sublignosa De Wild., 111
telekii Schweinf., 70
tenuirama sensu E.G. Baker, 111
tenuirama Welw. ex Baker, 66,**109**,110
  var. hebecarpa Verdc., 65,**110**
  var. huillensis Welw. ex Hiern, 110
  var. sculpta Welw. ex Hiern, 110
  var. tenuirama, **110**
trigonocarpa Taub., 108
trigonocarpa Taub. ex Baker f., 65,**108**,
  109,tab.**16** fig.**7**
uniflora E. Mey., 61,**66**,tabs.**15** fig.1; & **16**
  fig.1
  var. grandiflora Verdc., 66
  var. uniflora, **66**
venulosa Verdc., 64,**98**
  var. grandis Verdc., **98**,tab.**25** fig.B
  var. venulosa, **98**,tab.**25** fig.A
walteri Harms, 101
youngii Baker f., 101
zigzag De Wild., 87
AESCHYNOMENEAE (Benth.) Hutch., **50**,51
  subtribe Aeschynomeninae Benth., 50
  subtribe Bryinae B.G. Schub., 51
ALYSICARPUS Desv., 1,**36**
  glumaceus (Vahl) DC., 36,**42**,tab.**9** fig.5
    subsp. glumaceus, **42**
    var. glumaceus, **42**
    var. intermedius Verdc., **43**
  hochstetteri A. Rich., 42
  nummulariifolius (L.) DC., 38
  ovalifolius (Schumach.) J. Léonard, 36,**38**,
    tabs.**9** fig.2; & **10**
  rugosus (Willd.) DC., 36,**40**,42
    subsp. perennirufus J. Léonard, **41**,
    tab.**9** fig.4
    subsp. reticulatus Verdc., 36,41,**42**
    subsp. rugosus, **41**
    var. heyneanus (Wight & Arn.) Baker,
    41
  vaginalis (L.) DC., 36,tab.**9** fig.1
    var. *paniculatus* Baker f., 38
    var. parvifolius Verdc., **38**
    var. vaginalis, **37**
  vaginalis sensu auctt. mult., 38
  violaceus (Forssk.) Schindl., 40,42
    var. *pilosus* Schindl., 40
  zeyheri Harv., 36,**40**,tab.**9** fig.3
Anarthrosyne cordata Klotzsch, 8
ARACHIS L., 51,**169**
  diogoi Hoehne, 169

fruticosa Retz., 166,168
  hypogaea L., 166,**169**,tab.**45**

Brya P. Br., 51
  ebenus (L.) DC., 51

CULLEN Medik., **43**
  holubii (Burtt Davy) C.H. Stirt., 44
  obtusifolium (DC.) C.H. Stirt., 44
  tomentosum (Thunb.) J.W. Grimes, **44**,
    tab.**11**
CYCLOCARPA Afzel. ex Baker, 51,**157**
  stellaris Afzel. ex Baker, **158**,tab.**42**
Cytisus hispidus Willd., 53

Damapana aeschynomenoides (Welw. ex Baker)
  Kuntze, 128
  africana (Endl.) Kuntze, 119
  capitulifera (Welw. ex Baker) Kuntze, 137
  strigosa (Benth.) Kuntze, 133
  strobilantha (Welw. ex Baker) Kuntze, 130
  welwitschii (Taub.) Hiern, 144
DESMODIEAE (Benth.) Hutch., **1**
  subtribe Desmodiinae Benth., 1
DESMODIUM Desv., **1**,27
  abyssinicum (A. Rich.) Hutch. & Dalziel., 11
  adscendens (Sw.) DC., 4,**21**
    var. adscendens, 21
    var. robustum B.G. Schub., **21**,tab.**1**
    fig.10
  appressipilum B.G. Schub., 3,**14**
  barbatum (L.) Benth., 3,**23**,24
    var. argyreum (Welw. ex Baker) B.G.
    Schub., 3,**23**,24
    var. barbatum, 23,24
    var. dimorphum (Welw. ex Baker) B.G
    Schub., **23**,**24**
    var. procumbens B.G. Schub., 3,**23**,24,
    tab.**1** fig.12
    subsp. *dimorphum* (Welw. ex Baker)
    Laundon, 24
  barbatum sensu Taub., 22
  caffrum (E. Mey.) Druce, 21
    var. *schlechteri* Schindl., 22
  canum (J.F. Gmel.) Schinz & Thell., 2
  cordifolium (Harms) Schindl., 2,**16**
  delicatulum A. Rich., 13
  dichotomum (Willd.) DC., 3,**8**,tab.**1** fig.2
  diffusum (Willd.) DC., 8
  dimorphum Welw. ex Baker, 24
    var. *argyreum* Welw. ex Baker, 24
  discolor Vogel, **2**,4
  distortum (Aubl.) Macbr., **2**,4
  dregeanum Benth., 3,**21**,tab.**1** fig.11
  fulvescens B.G. Schub., 3,**22**
  gangeticum (L.) DC., 3,**9**,tab.**1** fig.4
  helenae Buscal. & Muschl., 3,**12**
  hirtum Guill. & Perr., 3,4,**13**
    var. *delicatulum* (A. Rich.) Harms ex
    Baker f., 3,**13**
    var. hirtum, **13**,tab.**1** fig.6
  hirtum sensu J.G. Baker, 15
  homblei De Wild., 16
  incanum DC., **2**,4
  intortum (Mill.) Urb., 4,**10**,11
    var. pilosiusculum (DC.) Fosberg, **2**,4
  lasiocarpum (P. Beauv.) DC., 8
  leiocarpum G. Don, 4
  mauritianum sensu Baker, 16

megalantha Taub., 28
nicaraguense Oersted, **2**,4
ospriostreblum Chiov., 3,**10**,11,tab.**1** fig.5
paleaceum Guill. & Perr., 17
pilosiusculum DC., 2
polygonoides Baker, 9
procumbens (Mill.) Hitchc., 11
psilocarpum Gray, 4,**12**
purpureum (Mill.) Fawc. & Rendle, 11
ramosissimum G. Don, 4,**16**,tab.**1** fig.8
repandum (Vahl) DC., 3,**6**,tabs.**1** fig.1; & **2**
salicifolium (Poir.) DC., 3,**17**,tabs.**1** fig.9;&**3**
    var. salicifolium, **17**
sandvicense E. Mey., 2
scalpe DC., 6
scorpiurus (Sw.) Desv., 3,**6**
sennaarense Schweinf., 8
setigerum (E. Mey.) Benth. ex Harv., 4,**15**,
    tab.**1** fig.7
spirale DC., 11
spirale sensu E.G. Baker, 11
spirale sensu J.G. Baker, 11
stolzii Schindl., 4,**15**
tanganyikense Baker, 3,**19**,tab.**4**
terminale sensu Guill. & Perr., 11
tortuosum (Sw.) DC., 4,**11**
tortuosum sensu Hepper, 11
triflorum (L.) DC., 3,4,**14**
uncinatum (Jacq.) DC., 4,**10**
velutinum (Willd.) DC., 3,**8**,tab.**1** fig.3
Diphaca Lour., 51
    kirkii (S. Moore) Taub., 55
    trichocarpum Taub., 54
Dolichos platypus Baker, 29
    pteropus Baker, 28,29
DROOGMANSIA De Wild., 1,**27**
    friesii Schindl., 30
    giorgii De Wild., 33
    hockii De Wild., 29
    longipes R.E. Fr., 29
    longirhachis Schubert, 29,34
    longistipitata De Wild., 30
    megalantha (Taub.) De Wild., **28**
        var. megalantha, **28**
    munamensis De Wild., 29
    platypus (Baker) Schindl., 29
    pteropus (Baker) De Wild., 27,**28**,33,34
        var. angustipetiolata Verdc., 28,30
        var. axillaris Verdc., 28,29,33
        var. giorgii (De Wild.) Verdc., 29,**33**
        var. platypus (Baker) Verdc., **29**
        var. pteropus, **29**,33
        var. quarrei (De Wild.) Verdc., 29,**30**,
            tab.**6**
        var. whytei (Schindl.) Verdc., 29,**30**,33,
            tab.**7**
    pteropus sensu De Wild., 30
    quarrei De Wild., 30
    stuhlmannii sensu R.E. Fr., 29
    tenuis Schubert, 29,34
    whytei Schindl., 30,33

Fabricia Scop., 36

Geissaspis auctt., 140
    apiculata De Wild., 149
    bakeriana De Wild., 149
    bequaertii De Wild., 143
    castroi Baker f., 145

chiruiensis R.E. Fr., 115
clevei De Wild., 115
descampsii De Wild. & T. Durand., 153
?descampsii sensu Brenan, 155
drepanocephala Baker, 146
elisabethvilleana De Wild., 146
homblei De Wild., 147
incognita De Wild., 151
    var. *latifoliolata* De Wild., 151
kassneri De Wild., 145
luentensis De Wild., 149
maclouniei De Wild., 115
megalophylla (Harms) Baker f., 145
minima Hutch., 155
robynsii De Wild., 151
rosea De Wild., 151
    var. *divergentiloba* De Wild., 151
rubrofarinacea (Taub.) Baker f., 115
scott-elliotii De Wild., 115,116
subscabra De Wild., 151
welwitschii (Taub.) Baker f., 144
    var. *kapiriensis* De Wild., 141
Glycine cordifolium Harms, 16

Hedysarum adscendens Sw., 21
    dichotomum Willd., 8
    diffusum Willd., 8
    gangeticum L., 9
    glumaceum Vahl, 42
    hamatum L., 168
    intortum Mill., 10
    lasiocarpum P. Beauv., 8
    micranthos Poir., 116
    ovalifolium Schumach., 38
    pictum Jacq., 34
    purpureum Mill., 11
    repandum Vahl., 6
    rugosum Willd., 40
    salicifolium Poir., 17
    scorpiurus Sw., 6
    sennoides Willd., 52,58
    tetraphyllum Thunb., 164
    tortuosum Sw., 11
    triflorum L., 14
    uncinatum Jacq., 10
    vaginale L., 36
    velutinum Willd., 8
    violaceum Forssk., 40,42
    violaceum L., 40
Herminiera Guill. & Perr., 58
    elaphroxylon Guill. & Perr., 76
HUMULARIA P.A. Duvign., 51,**140**
    sect. Rubrofarinaceae P.A. Duvign., **140**
    apiculata (De Wild.) P.A. Duvign., 141,**149**
    bakeriana (De Wild.) P.A. Duvign., 149
    bequaertii (De Wild.) P.A. Duvign., 141,
        143,144
        var. bequaertii, **144**
        var. purpureocoerulea (P.A. Duvign.)
            Verdc., **144**
    bianoensis P.A. Duvign., 115,116
    ciliato-denticulata (De Wild.) P.A. Duvign.,
        151
    descampsii (De Wild. & T. Durand.) P.A.
        Duvign., 141,**153**,tab.**40**
        var. abercornensis P.A. Duvign., **155**,
            tab.**35** fig.5
        var. acuta P.A. Duvign., 155
        var. descampsii, **153**,tab.**35** fig.4

forma pilosa P.A. Duvign., **153**
var. nyassica P.A. Duvign., 155
drepanocephala (Baker) P.A. Duvign., 141,
  146,tab.**35** fig.2
var. drepanocephala, 146,**147**,tab.**37**
var. forcipiformis P.A. Duvign., **147**
var. homblei (De Wild.) P.A. Duvign.,
  147,tab.**35** fig.3
forma denticulata P.A. Duvign., **147**
forma homblei P.A. Duvign., 147
elisabethvilleana (De Wild.) P.A. Duvign.,
  141,**146**,tab.**35** fig.1
flabelliformis P.A. Duvign., 157
kapiriensis (De Wild.) P.A. Duvign., 141,
  143,tab.**36**
var. kapiriensis, 141,**143**
var. repens Verdc., 140,**143**
kassneri (De Wild.) P.A. Duvign., 141,**145**
var. kassneri, **145**
var. kibaraensis P.A. Duvign., 146
var. perpilosa P.A. Duvign., 146
var. vanderystii (De Wild.) P.A. Duvign.,
  146
katangensis var. *glabrescens* P.A. Duvign., 149
landaensis P.A. Duvign., 145
luentensis (De Wild.) P.A. Duvign., 149
maclouniei (De Wild.) P.A. Duvign., 115,116
megalophylla (Harms) P.A. Duvign., 145
minima (Hutch.) P.A. Duvign., 141,**155**
subsp. flabelliformis (P.A. Duvign.)
  Verdc., **157**
subsp. minima, **157**
pseudaeschynomene Verdc., 140,**157**,tab.**41**
purpureocoerulea P.A. Duvign., 144
reptans Verdc., 153
rosea (De Wild.) P.A. Duvign., 140,**151**
var. *denticulata* P.A. Duvign., 151,153
var. reptans (Verdc.) Verdc., **153**,tab.**39**
var. rosea, **151**
rubrofarinacea (Taub.) P.A. Duvign., 115,
  116, 144
submarginalis Verdc., 140,141,**149**,tab.**38**
welwitschii (Taub.) P.A. Duvign., 141,**144**
var. landaensis (P.A. Duvign.) Verdc., **145**
var. welwitschii, **145**

KOTSCHYA Endl., 51, **117**
aeschynomenoides (Welw. ex Baker)
Dewit & P.A. Duvign., 118,**128**,130
africana Endl., 118,**119**,120,121
var. africana, **119**
var. bequaertii (De Wild.) Verdc., **119**,
  120, 121
var. latifoliola Verdc., 119,**120**,121
var. ringoetii (De Wild.) Dewit & P.A.
  Duvign., 119,**120**
africana sensu Dewit & P.A. Duvign., 120
bullockii Verdc., 117,118,**126**,tab.**30**
capitulifera (Welw. ex Baker) Dewit & P.A.
  Duvign., 117,**137**
var. capitulifera, **137**
var. grandiflora Verdc., 138
var. *robusta* Dewit & P.A. Duvign., 137
carsonii (Baker) Dewit & P.A. Duvign.,
  118,**125**
forma *multifoliolata* Dewit & P.A. Duvign.,
  125
subsp. carsonii, **126**
subsp. reflexa (Portères) Verdc., 126

coalescens Dewit & P.A. Duvign., 117,**136**
eurycalyx (Harms) Dewit & P.A. Duvign.,
  117,**136**,137
subsp. venulosa Verdc., **137**
imbricata Verdc., 118,**133**,tab.**33**
longiloba Verdc., 118,**125**
prittwitzii (Harms) Verdc., 118,**132**
var. parviflora Verdc., **132**
var. prittwitzii, **132**
recurvifolia (Taub.) F. White, 118, **122**
subsp. recurvifolia, **124**,tab.**29**
scaberrima (Taub.) Wild, 118,122,**124**,125
schweinfurthii (Taub.) Dewit & P.A.
  Duvign., **137**
sp. 1 of F. White, 128
sp. A, 118,**124**,125
speciosa (Hutch.) Hepper, 118,**135**
strigosa (Benth.) Dewit & P.A. Duvign.,
  118,**133**,135,136
var. grandiflora Dewit & P.A. Duvign.,
  118,**135**
var. strigosa, **135**
strobilantha (Welw. ex Baker) Dewit &
  P.A. Duvign., 118,**130**,tab.**32**
var. kundelunguensis Dewit & P.A.
  Duvign., 132
var. strobilantha, **132**
suberifera Verdc., 117,**128**,tab.**31**
thymodora (Baker f.) Wild, 118,**121**,122,
  124,125
subsp. septentrionalis Verdc., 118,**122**,
  124
subsp. thymodora, **122**
uguenensis (Taub.) F. White, 118,**120**,122,
  124
uguenensis sensu White, 120

Meibomia caffra (E. Mey.) Kuntze, 22
purpurea (Mill.) Vail, 11
tortuosa (Sw.) Kuntze, 11

Nicolsonia caffra E. Mey., 21
setigera E. Mey., 15

ORMOCARPUM P. Beauv., **51**
affine De Wild., 55
bibracteatum (A. Rich.) Baker, 55
bibracteatum sensu J. Léonard, 55
discolor Vatke, 55
guineense (Willd.) Hutch. & Dalziel ex
  Baker f., 53
kirkii S. Moore, 52,54,**55**,tab.**14**
mimosoidea S. Moore, 55
schliebenii Harms, 52,**53**,58
sennoides (Willd.) DC.,**52**
subsp. hispidum (Willd.) Brenan &
  J. Léonard, 53
subsp. sennoides, 53
subsp. zanzibarianus Brenan & J.B.
  Gillett, 53
setosum Burtt Davy, 54
sp. of Verdcourt, 52,**57**
trachycarpi (Taub.) Harms, 57
trachycarpum sensu Gonçalves, 57
trichocarpum (Taub.) Engl., 52,**54**,55
zambesianum Verdc., 52,**57**
OTHOLOBIUM C.H. Stirt., 43,**46**
foliosum (Oliv.) C.H. Stirt., **46**,tab.**12**
subsp. foliosum, **46**

subsp. gazense (Baker f.) Verdc., **48**
gazense (Baker f.) C.H. Stirt., 48
sericeum (Poir.) C.H. Stirt., 44

Patagonium racemosum E. Mey., 116
Phaseoleae (Bronn) DC., 1,33
PSEUDARTHRIA Wight & Arn., 1,**25**
confertiflora (A. Rich.) Baker, 27
cordata (Klotzsch) Walp., 8
hookeri Wight & Arn., **25**,27
var. argyrophylla Verdc., 27
var. hookeri, **25**,tab.**5**
PSORALEA L., 43,**48**
affinis sensu Hutchinson, 48
arborea Sims, **48**,tab.**13**
foliosa Oliv., 46
var. *gazensis* Baker f., 48
glabra E. Mey., 48,**50**
holubii Burtt Davy, 44
obtusifolia DC., 44
pinnata var. *glabra* (E. Mey.) Harv., 50
var. *latifolia* Harv., 48
tomentosa Thunb., 44
PSORALEEAE (Benth.) Rydb., **43**
subtribe Psoraleinea Benth., 43
Pterocarpus sp. of Pires de Lima, 53

Robinia guineensis Willd., 53
Rueppellia A. Rich., 58
abyssinica A. Rich., 104

Saldania Sim, 51
acanthocarpa Sim, 54
Sarcobotrya Viguier, 117
strigosa (Benth.) Viguier, 133
SMITHIA Aiton, 51,**138**
aeschynomenoides Welw. ex Baker, 128
africana (Endl.) Taub., 119
bequaertii De Wild., 120
burtii Baker f., 137
capitulifera Welw. ex Baker, 137
carsonii Baker, 125
chamaechrista Benth., 120
congesta Baker, 122
drepanophylla Baker, 122
elliotii Baker f., **138**
var. elliotii, **138**,tab.**34**
eurycalyx Harms, 136
eylesii, 136
goetzei sensu Hutchinson, 119
harmsiana De Wild., 125
kotschyi Benth., 119
kotschyi sensu Brenan, 120
megalophylla Harms, 145
mildbraedii Harms, 128,130
nodulosa Baker, 77
prittwitzii Harms, 132
recurvifolia Taub., 122
ringoetii De Wild., 120
riparia R.E. Fr., 119
rubrofarinacea Taub., 115

ruwenzoriensis Baker f., 128,130
scaberrima Taub., 124
setossima Harms, 125
speciosa Hutch., 135
sphaerocephala Baker, 128,130
strigosa Benth., 133
strobilantha sensu Hutchinson, 132,136
strobilantha Welw. ex Baker, 130
thymodora Baker f., 121
uguenensis Taub., 121
volkensii Taub., 128,130
welwitschii Taub., 144
Stylosantheae (Benth.) Hutch., 50
STYLOSANTHES Sw., 51,**165**
bojeri Vogel, 166
flavicans Baker, 166
fruticosa (Retz.) Alston, **166**,tab.**44**
grandifolia M.B. Ferr. & Sousa Costa, 166
guianensis (Aubl.) Sw., **166**,**168**
var. *intermedia* (Vog.) Hassler, 166
var. pauciflora Brandao, Sousa Costa &
R. Schultze-Kraft, 168,169
var. *robusta* L., 166
var. vulgaris M.B. Ferr. & Sousa Costa,
168
hamata (L.) Taub., 168
hippocampoides Mohlenbr., 166
humilis Kunth, 166
mucronata Willd., 166
sundaica Taub., 166

Trifolium guianense Aubl., 168
Trigonella tomentosa Thunb., 44

URARIA Desv., 1,**34**
picta (Jacq.) DC., **34**,tab.**8**

ZORNIA J.F. Gmel., 51,**159**
capensis Pers., 159,**162**,164,165
subsp. capensis, **164**
subsp. tropica Milne-Redh., 164,165
conjugata (Willd.) Sm., 160
diphylla (L.) Pers., 160
diphylla sensu auctt., 159
diphylla sensu Hutchinson, 161
diphylla sensu Robyns, 161
glochidiata C. Rchb. ex DC., **159**
milneana Mohlenbr., 159,**164**,165
obovata (Baker f.) Mohlenbr., 162
pratensis Milne-Redh., 159,**160**
subsp. barbata J. Léonard & Milne-Redh.,
161
subsp. pratensis, **161**
var. glabrior Milne-Redh., **161**
var. pratensis, **161**
setifera Mohlenbr., 161
setosa Baker f., 159, **161**
subsp. obovata (Baker f.) J. Léonard &
Milne-Redh., **162**,tab.**43**
subsp. setosa, **162**
tetraphylla sensu auctt., 162
var. *obovata* Baker f., 162